Springer Tracts on Transportation and Traffic

Volume 12

About this Series

The book series "Springer Tracts on Transportation and Traffic" (STTT) publishes current and historical insights and new developments in the fields of Transportation and Traffic research. The intent is to cover all the technical contents, applications, and multidisciplinary aspects of Transportation and Traffic, as well as the methodologies behind them. The objective of the book series is to publish monographs, handbooks, selected contributions from specialized conferences and workshops, and textbooks, rapidly and informally but with a high quality. The STTT book series is intended to cover both the state-of-the-art and recent developments, hence leading to deeper insight and understanding in Transportation and Traffic Engineering. The series provides valuable references for researchers, engineering practitioners, graduate students and communicates new findings to a large interdisciplinary audience.

**** Indexing: The books of this series are submitted to SCOPUS and Springerlink ****

More information about this series at http://www.springer.com/series/11059

Elena S. Prassas · Roger P. Roess

The Highway Capacity Manual: A Conceptual and Research History Volume 2

Signalized and Unsignalized Intersections

Springer

Elena S. Prassas
Tandon School of Engineering
New York University
Brooklyn, NY, USA

Roger P. Roess
Emeritus Professor
Tandon School of Engineering
New York University
Brooklyn, NY, USA

ISSN 2194-8119 ISSN 2194-8127 (electronic)
Springer Tracts on Transportation and Traffic
ISBN 978-3-030-34478-8 ISBN 978-3-030-34480-1 (eBook)
https://doi.org/10.1007/978-3-030-34480-1

This Springer imprint is published by the registered company Springer Nature Switzerland AG
The registered company address is: Gewerbestrasse 11, 6330 Cham, Switzerland

Like Volume I, this book is dedicated to the members, past and present, of the Committee on Highway Capacity and Quality of Service *of the Transportation Research Board. From the first Committee of eleven members, led by the legendary O. K. Normann, to the current Committee of 32—plus over 100 subcommittee members—led by Tom Creasy, the Committee is a family of dedicated and passionate professionals dedicated to advancing the state of the art of highway capacity analysis and it many critical applications in highway design, analysis, management, and operation. It has been our privilege to work among them and to call many of them friends.*

Preface

When Volume I was published in 2014, the *Highway Capacity and Quality of Service Committee* was beginning work on what would become the 6th edition of the *Highway Capacity Manual* (HCM). We originally anticipated that Volume II would be completed shortly thereafter. The 6th edition of the HCM appeared in 2016, while our work took considerably longer.

The documenting of material on interrupted flow was more difficult than anticipated, particularly when it became clear that we would have to include the 2016 HCM to be complete. Source material was quite extensive and was continually developing as this volume was being written. Some of the source material for the 2016 HCM was in the form of unpublished (at the time) documents and/or computer programs. The complexity of interrupted flow methodologies slowed things down as well.

Our intent, as with Volume I, is to document and explain as clearly as possible the development of core concepts and detailed methodologies of the HCM. We believe it is important that current and future researchers in the field of highway capacity analysis understand the history of the concepts and research that underpins the HCM. We hope we are providing a single source that can both provide an overview for their use and identify other documents that might be of interest.

Originally, our intent was that Volume II would complete the set. The amount of material, however, dictated that Volume II focuses solely on intersections—both unsignalized and signalized. Signalized intersections, of course, remain the most complex points in street and highway networks, and the methodology for their analysis is the most frequently used in the HCM. The increasingly complex analysis of facilities and networks must await a Volume III.

Collectively, our involvement with the *Highway Capacity and Quality of Service Committee* spans over 40 years as contributing researchers, committee, and sub-committee members. We hope that this volume highlights the accomplishments of this extraordinary group and provides a helpful source for those who join it in the future.

Brooklyn, USA Elena S. Prassas
Roger P. Roess

Contents

List of Figures

List of Tables

Chapter 1
The Highway Capacity Manual and the Committee on Highway Capacity and Quality of Service

The Highway Capacity Manual is one of the most fundamental documents used in the transportation and traffic engineering profession. It provides for the systematic design and evaluation of traffic facilities with regard to their capacity, and the quality of traffic service provided to facility users. This chapter provides an overview and history of the document itself, and its development over the years.

1.1 Introduction

The "Highway Capacity Manual" (HCM) is the principal publication of the Transportation Research Board (TRB) of the National Academies of Science and Engineering. It has become the single most important technical document used throughout the U.S. and other nations as a guide to the analysis of highway capacity and quality of service. Its use is mandated in the U.S. for all federally-aided highway projects. The manual was and is created and maintained by the Committee on Highway Capacity and Quality of Service (HCQSC) of TRB. A very detailed history of the manual is included in Volume 1 of this book [1]. A brief overview is provided here for readers of this volume.

1.2 The Development of a National Highway System

When the U.S. gained its independence from Great Britain, its transportation network consisted primarily of waterways, as all of the colonies were located along the east coast of North America. Inland transportation was limited to old Native American trails and a small number of trails built by the early settlers to provide some connectivity between communities and colonies.

E. S. Prassas and R. P. Roess, *The Highway Capacity Manual: A Conceptual and Research History Volume 2*, Springer Tracts on Transportation and Traffic 12,
https://doi.org/10.1007/978-3-030-34480-1_1

The founding fathers, however, were aware of the need for significantly improved inland trails. George Washington feared that without better inter-state transportation, the colonies would quickly drift apart and operate too independently to form a cohesive nation. In the early days of the nation, however, the financial burdens on the fledgling government did not allow expenditures on overland trails.

Originally, local systems of roadways were primarily built by local governments to serve the immediate needs of the local population. They focused on promoting local economies and agricultural trade.

Between 1790 and 1820, the need for improved inter-state trails was met by private developers of toll roads, who charged fees for use of their roadways. They were often built over public lands, and many obtained charters from the State governments to do so. Originally, such roads were little more than rough trails hewn through the wilderness, with poorly-compacted earth surfaces that made wagon travel perilous, particularly in bad weather. After 1820, a number of such private roadways used planks as a roadway surface [2–4].

The exact number of private toll roads built throughout the U.S. is not fully documented. Most historians believe that between 30,000 and 50,000 miles of such roadways were built, and that about 10,000 miles of it may have been plank-surfaced. However, by 1850, virtually all of the private toll roads had fallen into bankruptcy. It was up to State and local governments to take them over and maintain them. While this happened in some cases, in many, the roadways were simply abandoned.

With the development of Ohio and its achievement of statehood in 1803, the need for a road linking the east coast with the west became a national interest. The Potomac and Ohio Rivers were major conduits for commerce, but they were separated by the Allegheny Mountains. The idea for a "national road" connecting the two was developed over a period of years. In 1803, Congress proposed to allocate a portion of the funds raised through land sales in Ohio for the construction of a National Road that would travel from Cumberland, MD, to Wheeling, Virginia (now in West Virginia). After much discussion and controversy, President Jefferson authorized construction of the roadway on May 29, 1806, providing $30,000 in federal funds to do so. Construction began in 1811, and the road was completed to Wheeling in 1819. The road was the first in the U.S. to provide a surface of crushed stone, known as a "MacAdam" surface, which provided for greater durability than previous roadways. Congress authorized western extensions to the road in 1820 and 1825 [5–7].

In 1830, President Andrew Jackson vetoed the "Maysville Road Bill." The Maysville Road was to be part of the National Road, but was located entirely within Kentucky. Jackson indicated that the federal government could not fund any public projects that did not benefit "the entire nation," and particularly not a roadway that was located entirely within one state. If such funding were desired, he declared that a constitutional amendment would be required. This veto had a lasting effect, and has effectively kept the federal government out of the direct construction, maintenance, and administration of roadways in the U.S. After this veto, it was generally understood that road construction and maintenance were to be a function of the various states.

This view has prevailed, even though a subsequent Supreme Court decision (Wilson vs. Shaw, 1907) declared that the federal government could build and administer interstate highways under the provisions of the Commerce Clause of the U.S. Constitution [7].

Jackson did not oppose all funding of the National Road, and approved additional funds for interstate segments of the route. Construction of the National Road continued with federal support through 1838, when it reached its western terminus in Vandalia, Illinois. From the mid-1830s, various segments of the National Road were turned over to the states for maintenance and administration.

Over the course of its history, the National Road was often referred to as the Cumberland Road or the National Pike. As the road fell into poor repair and disuse in the late 1800s, its various parts were transferred to state and local jurisdictions. The majority of the original route is now part of U.S. Route 40.

Beginning in the 1880s, public interest and support for a more cohesive national highway system grew. Organized originally by bicycle enthusiasts through the League of American Wheelmen, the "Good Roads Movement" formally began. The League began publishing the "Good Roads" magazine, which continually decried the condition of town and inter-town roadways, and advocated for a national program to improve them and to add new roadways [8, 9]. The movement led to the Department of Agriculture conducting the first study of the nation's roadway systems in 1893.

By the early 1920s, the "Good Roads" movement had added another group of enthusiastic supporters—early motorists. With the introduction of the automobile came driving enthusiasts, and the inadequacy of most existing roadways became even clearer. Motor clubs began mapping out a series of named national highways, which consisted of preparing maps and directions for the heartiest hobbyists who would dare to drive across the nation. The famous Lincoln Highway [10] became the first among many such highways, which soon included the Dixie Highway and the Yellowstone Highway. With the publicity accorded these mapped routes, local and State governments found it useful to attract both tourists and businesses by upgrading these facilities.

The Office of Road Inquiry was established in the Department of Agriculture in 1893 [11]. In 1905, the Office was renamed the Office of Public Roads, and in 1915, it became the Bureau of Public Roads. In 1949, the Bureau was transferred from the Department of Agriculture to the Department of Commerce. In 1966, it became the Federal Highway Administration (FHWA), and was transferred to the newly-formed Department of Transportation.

The Federal-Aid Highway Act of 1916 marked the first official entry of the federal government in providing financing for highway construction. While it left the responsibility for planning, building, operating, and maintaining highway systems to the states, it provided significant federal-aid funds to spark expansion. Thus, by 1916, a federal infrastructure with responsibility for highways had been established, and federal funding was provided to the states to build new highways. From this point on, the participation and influence of the federal government on highway development only grew.

1.3 Developing the Interstate System

In 1922, the Bureau of Public Roads authorized General John. P. Pershing to construct a national highway map that would meet the nation's defense needs in time of war. The resulting Pershing Map became the starting point for defining the "Primary or Interstate" highway system. The Federal-Aid Highway Act of 1921 also required that the Secretary of Agriculture prepare a map of primary or interstate highways by 1923, and issue updates on an annual basis.

These initial studies were the historical antecedents to the Interstate System as we know it. In 1944, President Franklin Delano Roosevelt sent a historic report to Congress: "Interregional Highways: A Message from the President of the United States transmitting a report of the National Interregional Highway Committee outlining and recommending a National System of Interregional Highways." It recommended a national highway system of a little over 33,000 miles that would connect all cities of population 300,000 or more directly to each other [12]. While the Federal-Aid Highway Act of 1944 authorized construction of this system, no appropriations bill was enacted. The Federal-Aid Highway Act of 1956 finally provided both the authorization and appropriations needed to implement the system [13].

1.4 Formation of the Committee on Highway Capacity and Quality of Service

When President Roosevelt appointed the National Interregional Highway Committee in 1941, Thomas H. MacDonald, the Commissioner of Public Roads, was elected as its chairman. As head of the Bureau of Public Roads, MacDonald assigned several professionals from the Bureau to work full time with the Committee. Among those he selected was Olav K. Normann. Normann, then a young engineer, became responsible for much of the technical analysis that was the foundation of the Committee's historic report [14].

Normann later developed an innovative methodology for estimating the capacity of a multilane highway—a most difficult task given that virtually no such highways were ever congested at the time. Using innovative equipment he developed, he studied the relationship between the speeds of pairs of vehicles, and concluded that the headway between such pairs of vehicles reached a minimum of approximately 2s when both vehicles were traveling at 35 mi/h. At higher or lower speeds, he observed that the headway between vehicles increased. This led to the conclusion that the capacity of such highways was approximately 1,800 vehicles/hour/lane [15]. As the result of this and other work, Normann was a well-known expert in highway capacity by the time the National Committee on Interregional Highways completed its work in 1944.

With the emergence of interest in a national highway system, and the realization that new, better, and more uniform design and analysis methods were needed to estimate the capacity of various types of highways, the Highway Research Board established the Committee on Highway Capacity in 1944. Because of his prominence in the field, O.K. Normann was appointed as its first chairman. Powell Walker, also from the Bureau of Public Roads, became the Committee's first secretary.

The Committee had a very focused objective—to produce a document that would help practitioners to estimate the capacity of various types of highway facilities. The Bureau of Public Roads was very supportive, and allowed both Normann and Walker to spend significant amounts of their time working on the effort. They were also provided with a number of support personnel from the Bureau to assist them. Because of this, Normann and Walker wrote most of the first edition of the *Highway Capacity Manual*. The remaining members of the Committee reviewed and approved the work as it was completed.

1.5 The First Edition: The 1950 Highway Capacity Manual

The first edition of the *Highway Capacity Manual* [16] was published in 1950. It was actually first published in 1949 as a series of articles in *Public Roads* magazine [17, 18]. It was published in this form first at the insistence of Thomas H. MacDonald, head of the Bureau of Public Roads, due to the overwhelming contributions of the agency's personnel to the effort. It was then published as a single document jointly by the Bureau and the Highway Research Board.

Although only 147 pages long, the first-edition HCM provided a set of standard methodologies for the estimation of the capacity of a variety of common traffic facilities.

The 1950 HCM provided the first formal definition of capacity. Actually, it defined three levels of capacity: (a) basic capacity, (b) possible capacity, and (c) practical capacity. These three concepts are discussed in detail in Chap. 3.

While the 1950 HCM essentially allowed practitioners to estimate the capacity of various types of roadways, its application was primarily in design. It was intended to answer the fundamental question: how big should we build it? Boiled down further, it allowed practitioners to figure out how many lanes were needed to serve various levels of traffic demand. While some of the methodologies could have been applied in analysis, it was not the primary focus of the manual.

It is also important to note that the 1950 HCM dealt with long sections of facilities that were primarily operating under uninterrupted flow. Methodologies for signalized intersections, weaving sections, ramps, and ramp terminals dealt with points or very short segments of focused activity within an overall facility. Further, methodologies for weaving sections, ramps, and ramp terminals were not limited to such features on freeways: they could be applied to multilane and two-lane highways as well, and, in some cases, to arterials.

What the 1950 HCM provided, above all else, was uniformity. Now, when highway engineers across the country were considering how many lanes to provide on new and re-built facilities, they did so using a common set of methodologies and criteria. The

1950 HCM was a huge success, and over 26,000 copies were sold. It was eventually translated into nine other languages [19].

1.6 The Second Edition: The 1965 Highway Capacity Manual

When the 1950 HCM was released, it was thought that the Committee on Highway Capacity would be disbanded. This never happened, but the Committee was largely inactive between 1950 and 1952. With new questions on highway capacity emerging, and growing interest in highway capacity research and methodologies, the Committee was re-activated in 1953, with O.K. Normann remaining the chairman.

New members were added, and the Committee directed an active effort in fostering new and improved understanding of highway capacity issues. While Normann continued to devote a good deal of his time to the effort, he had also taken on additional responsibilities at the Bureau of Public Roads as its Deputy Director of Research (1957). Committee members devoted significant amounts of time to the effort, while maintaining their own professional positions. While it was becoming apparent that a second edition would be needed, the volunteer Committee faced increasing challenges of time and resources to devote to the task.

Two of the most significant efforts included:

- A detailed survey of traffic officials across the country was made in 1954. Detailed forms and instructions seeking data on the operation of signalized intersections were distributed. These resulted in obtaining data on 1600 signalized intersection approaches. The Bureau of Public Roads was able to use the results of this survey for detailed analysis.
- In 1957, the Committee sponsored the publication of *Highway Research Bulletin 167* [23]. It contained 6 papers on new analysis and suggested methodologies on highway capacity based on a wide range of research by interested professionals. Of the 6 papers, four involved O.K. Normann or Powell Walker as participating authors.

Highway Research Bulletin 167 was often referred to as the 1.5 edition of the HCM, and saw wide use over a ten-year period following its publication.

By the early 1960s, it was clear that the Committee would have a difficult time producing a new edition of the HCM based entirely on the volunteer efforts of its dedicated, but time-restricted, members. In 1963, the Bureau of Public Roads assigned five BPR employees to work full-time on the effort, directed by O.K. Normann.

As the development of a new HCM progressed, the Committee, after many long and heated discussions, resolved that the three capacity levels defined in the 1950 HCM—basic, possible, and practical capacity—had to be replaced with a single definition. This led to the concept of levels of service, which could be used to describe effective maximum volumes that could be accommodated while maintaining certain defined operational characteristics. While formally voting for this in 1963, the Committee remained divided over how to implement the concept in new methodologies. It was the task force of BPR employees who finally came up with the structure for implementing level of service.

It is important to point out that the general structure of the level of service concept was developed primarily for uninterrupted flow, and particularly with freeways in mind. The concept, of course, was then modified and applied to other facilities, such as signalized intersections.

O.K. Normann passed away in May of 1964, so he never saw the final results of the efforts of the Committee. The 1965 HCM [20], which was actually published in early 1966, was dedicated to Normann due to his massive efforts and commitment to the field of highway capacity research, and to the development of the *Highway Capacity Manual*.

The 1965 HCM provided a great deal of new material. The level of service concept was implemented. Freeways were treated as a separate category, based upon extensive new knowledge about freeway operations. New material was introduced concerning signalized intersections, arterials, and downtown streets. Material on weaving and ramps was extensively updated. A new chapter on bus transit was added.

Once again, methodologies such as weaving sections and ramp terminals could be applied to such configurations on freeways, and on other types of facilities as well. The treatment of at-grade intersections was expanded to include some material on unsignalized intersections as well as signalized intersections.

The 1965 HCM was every bit as successful as its predecessor. The length of the manual had increased to 411 pages. The 1965 HCM quickly became the most widely distributed publication of the then Highway Research Board. Like the 1950 manual, it was translated into numerous foreign languages.

1.7 The Third Edition: The 1985 Highway Capacity Manual

With the death of O.K. Normann, the Committee on Highway Capacity entered a new phase of its existence, and a new approach to developing material for future editions of the manual began to evolve. The Committee remained quite active as professional interest in the subject continued to grow. It did not, however, have access to full-time support of Bureau of Public Roads personnel, however.

The use of the 1965 HCM became virtually universal throughout the U.S., and was heavily relied on by federal, state, and local highway agencies. The level of service concept was extremely popular, and became a standard language of communication for quality of service provided by highway facilities.

With increased use, came increased scrutiny, and new issues began to emerge. While the level of service concept gained substantial acceptance, there was much discussion concerning the appropriateness of the operational measures used to define it. This was particularly true of signalized intersections, where the load factor parameter used in the 1965 HCM proved to be difficult to interpret and to measure. Interest in the use of a time-based parameter began to build.

The use of the manual in the analysis and design of interrupted flow facilities grew exponentially, exposing needs for significant improvements in methodologies for arterials and downtown streets. The inclusion of a chapter on bus transit in the 1965 HCM increased the interest in methodologies for other transportation modes, particularly pedestrian and bicycle facilities.

Significant organizational changes were also taking place. The Highway Research Board became the Transportation Research Board; the Bureau of Public Roads became the Federal Highway Administration, and the Committee on Highway Capacity became the Committee on Highway Capacity and Quality of Service (HCQSC).

Formal efforts towards publication of a third edition of the HCM began with the appointment of Jim Kell as Chairman in 1977. A new approach to the monumental task of researching new methodologies and preparing them for publication in the HCM developed: In cooperation with the Committee, research projects were sponsored by various government agencies, in particular, the National Cooperative Highway Research Program (NCHRP) and the Federal Highway Administration (FHWA).

By the late 1970's, a significant amount of new material had been developed and reviewed by the Committee on a variety of critical subjects. With a new manual still years in the future, the Committee published TRB Circular 212 [21], entitled "Interim Materials on Highway Capacity."

In 1983, Carlton C. Robinson, who had already served as a previous chairman of the HCQSC, again became chairman. His primary task was to lead the Committee to complete its work on a third edition, and insure its publication by 1985. Working with the contractor for NCHRP 3-28(B) (Polytechnic Institute of New York) and the members of the Committee, he managed the complex task of reviewing final materials, resolving critical issues, and, in general, pushing the Committee to never deviate from the intended publication date in 1985 (which had already been delayed from the original target of 1983).

The 1985 Highway Capacity Manual broke new ground in several significant ways:

1. It was the first manual developed with the assistance of a funded contractor using materials mostly developed by funded research agencies.
2. It was the first manual published in a loose-leaf format, with the initial expectation that updates would be frequent, and that changes might be made on a page-by-page basis.
3. It was the first manual containing methodologies that were somewhat too difficult and significantly time-consuming to implement manually. The *Highway Capacity Software* (HCS) package was subsequently developed with funding from FHWA to ease the computational burden of several of the procedures, most critically, the methodology for signalized intersections.

4. It was the *last* manual where all HCQSC members reviewed all chapters and all methodologies, and had intimate knowledge of all of its contents. At 506 pages, it was the last manual that had a length that could even be conceivably reviewed and thoroughly understood by each individual Committee member.

The last item was extremely significant. Shortly after publication of the 1985 HCM, the HCQSC created permanent subcommittees for the first time. Each HCM chapter had a subcommittee that was charged with reviewing comments from users, making recommendations to the full Committee on corrections and changes, and the responsibility for reviewing research results and proposed changes to the chapter methodology in the future. The creation of subcommittees greatly expanded the number of professionals who could participate in the Committee's work in a formal way. At the same time, it diminished the role of members of the parent Committee, which now had the principal role of reviewing recommendations from the various subcommittees. Over time, Committee members became focused on individual parts of the manual relevant to their interest and expertise. On any given chapter or at any given time, only a portion of the Committee was fully engaged in the review and discussion of specific source materials.

1.8 Updates to the 1985 Highway Capacity Manual

Highway Research Bulletin 167 (1957) was essentially a bridge between the 1950 and 1965 versions of the manual. Circular 212 (1980) performed the same bridging function between the 1965 and 1985 manuals.

By 1985, however, the pace of development of new materials was accelerating greatly. The enormous effort of producing a completely new edition of the HCM was a time-consuming process that included the physical details of publishing and distributing a document to the profession.

The Committee created the 1985 manual as a loose-leaf document, thinking that updates could then be issued almost immediately as they were approved, and inserted directly into the manual. This proved to be impractical, as questions of how to physically handle the updates posed significant barriers. How do purchasers of the manual get updated pages automatically? This was a necessary function if the Committee was to insure that every user had access to the *current* version, which could now change at frequent intervals. At the same time, when the 1985 manual was published, a new methodology for multilane highways was under funded development, and no one could conceive of waiting 15 or 20 years to see it released to the user community.

The leader charged with resolving these and other critical issues was Adolf D. May Jr., who was appointed as chairman of the HCQSC in 1989. With May leading the Committee, it quickly resolved two issues:

- A set of research priorities for a fourth edition HCM was developed and published as *Circular 371* in 1991. Working aggressively with Committee members and officials of funding agencies, particularly NCHRP and FHWA, funding for many

of the Committee's priorities was achieved, and a target date of 2000 was set for production of a fourth edition HCM.
- Working with Transportation Research Board officials, it was determined that the Committee could issue interim updates in the HCM directly, but that material had to be released as packages of new or updated chapters, not pages or sections within chapters.

The first major update to the 1985 HCM was issued in 1994. Virtually half of the chapters (7 of 14) of the 1985 HCM were replaced or significantly updated, including chapters on definitions and concepts, traffic characteristics, basic freeway segments, ramps and ramp junctions, multilane highways, signalized intersections, and unsignalized intersections. With the growing availability of personal computers, several of the chapters included in the 1994 update simply could not be efficiently implemented manually. Thus, it became critical that with each updated methodology, the *Highway Capacity Software* package had to be updated as well. This was enabled by having close collaboration between the Committee and the McTrans Center of the University of Florida, which had become the depository and maintenance facility for the HCS package.

John D. Zegeer became chairman of the HCQSC in 1995, almost immediately after release of the 1994 update. Discussion immediately ensued as to the need for a second updating of the third edition, which was scheduled for 1997. The 1997 update, which actually debuted in early 1998, contained new or updated chapters for basic freeway segments, signalized intersections, unsignalized intersections, and urban arterials.

The 1997 update ran into some difficulties that had been unanticipated. With a full fourth edition expected in 2000, many agencies were reluctant to change their practices twice within such a short period. Software development, which lagged about a year past the publication of the update, was so close to the publication of fourth edition, that it accentuated the problem of updating official practice so frequently. Indeed, the official process for mandating the use of new materials itself took some time, and varied from jurisdiction to jurisdiction. The Committee began to realize that the zeal to get the latest research results into practice had to be balanced with the practical realities of agencies and individual users updating their official and unofficial practices.

1.9 The Fourth Edition: The 2000 Highway Capacity Manual

By the time the fourth edition of the 2000 HCM [22] was published, the HCQSC was in an almost-continuous production mode, developing and reviewing new and updated material most of the time. The Signalized Intersection Subcommittee was particularly active, and produced a steady stream of proposed updates and addition to its methodology. Production of the fourth edition was led by John Zegeer as chairman of the HCQSC, and William Reilly of Catalina Engineering Inc., under NCHRP sponsorship.

The 2000 HCM was vastly different from any of its predecessors. It was written with a much broader view of potential users in mind. In addition to operational analysts and designers, planning use of the manual and its methodologies had increased greatly, and the tailoring of many procedures for planning use was built into the fourth edition. At the same time, the manual was expanded to provide at least overall coverage of the assessment of multiple facilities, corridors, and systems. These subjects were treated in more-or-less general terms, but some analytic procedures were included. The 2000 HCM also directly treated the subject of simulation and simulation packages, and how their use could supplement and/or complement HCM methodologies. The fourth edition of the HCM ballooned to over 1100 pages, and featured a new presentation format including marginal notes.

The manual also came in a CD-ROM that included not only the full text of the paper HCM, but multimedia components including hypertext links, video and audio clips, and animated tutorials and sample problems. Close coordination between the Committee and software developers insured that the *HCS* and other products became available in relatively close proximity to the release of the manual itself.

The organization of chapter materials also changed significantly. Introductory material was divided into six separate chapters, comprising Part I of the four-part document. The methodological chapters of the 1985 HCM (and its updates) were separated into to two parts: Concepts and Methodologies. Introductory sections of the various facility chapters were separated into a "concept chapter" in Part II of the manual, and a "methodology chapter" in Part III of the manual. Part IV of the 2000 HCM dealt with all new material on Corridor and Areawide Analysis, while Part V dealt with the issues involving simulation. In response to the then-current legislation requiring states to convert to metric units, the fourth edition was also produced in two versions: standard English units, and metric units. Of course, Congress eventually backed off this requirement, and although several states had already changed to metric systems, use of English units continued in most places (in the U.S.).

New methodologies introduced in 1994 and 1997 transferred with only minor and editorial changes into the 2000 HCM. Earlier methodologies were significantly re-written and updated based upon new research, again, primarily funded by NCHRP and FHWA.

Of particular note, the 2000 HCM introduced a major new methodology for the analysis of freeway facilities containing basic freeway and weaving segments, as well as ramp junctions. The new methodology was quite complex, and virtually impossible to implement using manual computations. Unfortunately, no user-friendly software was developed for this methodology, which was mostly ignored by the user community. Once again, the importance of having usable software to implement complex computational methods was hammered home.

1.10 The Fifth Edition: The 2010 Highway Capacity Manual

John Zegeer was succeeded as chairman of the HCQSC by Richard G. Dowling in 2001. After the frenzy of activity that had dominated the time between the 1985 and 2000 HCMs, the Committee decided to forego interim updates, and targeted the production of the fifth edition manual in 2010. This represented the shortest gap between full editions of the HCM in its history.

The pace of research continued to accelerate, with 9 NCHRP and 2 FHWA major studies completed after 2000. One of these broke new ground, producing the first principal document other than the HCM with the support and supervision of the HCQSC. The *Highway Capacity Manual Applications Guidebook* provided comprehensive case studies showing how the HCM and its methodologies should be implemented and interpreted in complex cases involving multiple facilities. It also highlighted the use of the HCM in conjunction with other models that might be applied to some types of situations. While representing a significant achievement, the distribution of the Guidebook created another problem for the Committee: now, every time the manual was updated, the Guidebook would also have to be updated.

The development of the fifth edition was guided by Richard Dowling, as chairman of the HCQSC and Mark Vandehey of Kittelson and Associates, the prime contractor for its production. If the 2000 HCM represented a significant expansion of material and situations covered by the HCM, the 2010 version was a veritable explosion. New material included, but was not limited to:

- A methodology for analysis of interchange ramp terminals over a wide range of interchange types,
- A new methodology for analysis of roundabouts,
- Inclusion of a multimodal approach to analysis of urban streets, including intersections; this included a methodology for specifically predicting user perceptions of quality of service,
- A new methodology for analysis of freeway weaving segments,
- Development of comprehensive sets of default values for use in planning applications, and in cases where some aspects of required input information were not available,
- Development of service volumes based upon AADT for planning use,
- Additional guidance on use of alternative tools and models in capacity and quality of service analysis,
- A new methodology for shared-use (pedestrian and bicycle) paths, and
- New material on the impacts of Active Traffic Management approaches to increase capacity and improve performance.

In addition, the Committee determined to *include* the applications guidebook as part of the HCM itself.

The 2010 HCM [23] was released in a radically different format from any of its predecessors. The manual includes four separate volumes, only three of which were available in print form. The first three volumes are in separate loose-leaf binders, and the fourth is available only electronically at www.HCM2010.org. Only purchasers of the print volumes have access to the electronic part of the manual.

Part 1, Concepts, is greatly expanded from the 2010 HCM. It moves material on alternative models and their use from Part IV of the 2000 HCM, and expands it with new research in Part 1. Also, some of the conceptual material in the 2000 HCM had originally been located with the methodological chapters. It was felt that the separation of chapter-specific concepts from the methodologies themselves did not produce a more effective flow of information. Thus, any conceptual material specific to a methodology was moved back into the methodology chapters of Part 2.

Parts 2 and 3, Uninterrupted Flow and Interrupted Flow respectively, contain all of the traditional methodological chapters of the HCM. Greatly expanded, and with more emphasis on multimodal analysis, the methodologies are now organized by type of facility.

There are a number of significant changes in the presentation format of Part 3, Interrupted Flow. Arterials and downtown streets are now simply referred to as "urban streets" without differentiation. Analysis methodologies are now divided into separate chapters, one treating "urban street facilities" and the other "urban street segments." Segments refer to one signalized intersection and the adjacent approach segment(s), while facilities refer to a sequence of several segments with multiple signalized intersections. In the signalized intersection chapter, the focus was now on actuated signal systems. In fact, analysis of pre-timed signals is now possible only by entering data that forces an actuated signal to perform on a fixed timing.

Part 4, Applications Guide, has provided a means to greatly expand the availability of information to manual users. For every methodology, there is a chapter in Part 4 containing supplemental material. The most critical supplemental material relates to computational details needed by software developers to properly implement the methodologies of Parts 2 and 3. In some complex cases, this material includes executable spreadsheets or code, where the details needed by software developers could not be efficiently communicated in English. Supplemental materials also include research details not needed to describe the methodology, but which might be of interest to some users.

Part 4 also contains other valuable information. The case studies that form the primary basis of the *Highway Capacity Manual Applications Guidebook* are included here, in updated form. A current summary of official HCQSC interpretations is also included. These result from the many user comments and questions submitted to the Committee for its formal interpretation,

Perhaps the most significant aspect of Part 4 is the inclusion of a technical reference library. Here, source materials, including final research reports from the many NCHRP- and FHWA—sponsored projects are included for the benefit of both HCM users and future researchers. Other important source documents are included here as well.

1.11 The 2016 HCM

In early 2013, the HCQSC was already planning work on a major interim update to the 2010 HCM. The Committee had already approved new work on travel time reliability, and NCHRP Project 3-115 has provided significant funding aimed at producing an interim update by the end of 2016 (originally 2015).

By 2015, it had become clear that the "update" was becoming so massive that it was virtually completely new. The Committee decided to turn the update into a full 6th Edition, and retitled it: *Highway Capacity Manual, 6th Edition: A Guide for Multimodal Mobility Analysis*. The 6th Edition was released in late 2016.

There was some controversy over making the "update" a new edition. Because it was developed as an update, there was no systematic consideration of its overall organization, or of overall concepts, such as level of service, despite considerable discussion and argument within the Committee over both. Even the name change was controversial: although there are methodologies for assessment of non-auto/highway modes (pedestrians, bicycles, transit), these analyses yield only single-mode evaluations. While the concept is discussed, there are no procedures that actually produce a complete and integrated multimodal analysis.

The major additions to the 6th edition include methodologies to assess travel time reliability of facilities, and the use of these methodologies to assess advanced traffic demand management strategies and control systems. Virtually every methodological chapter, however, was substantially updated from the 5th edition.

The HCM has now grown from a 147-page document to one that now comprises over 3000 pages of material. Its use and importance to the profession has grown with its size. The sheer volume of material now included, however, presents a critical problem: there is no way that any one individual, including members of the HCQSC, can have intimate knowledge of *every* bit of that information. It is no longer even reasonable to assume that each committee member has even read all of it. Yet, the members of the Committee, make the final decisions on what is in it, and with those decisions, have enormous influence on the profession, and on the nation as its decisions are implemented in concrete and control systems throughout the U.S. and elsewhere.

The initial Committee, with its eight members, had intimate knowledge of every word that was included in the 1950 HCM. This was substantially true through the publication of the 503-page 1985 HCM. It is no longer the case, nor could anyone expect it to be. When the 2016 HCM was published, the Committee had 34 regular and emeritus members. Through its subcommittees, over 120 additional members are involved in the development and approval of materials. Project panels of the NCHRP and FHWA provide additional oversight of the research and development of materials, and are substantially integrated with the Committee. Yet, the fact remains, that on any single methodology, section, or chapter of the manual, of the 34 voting members of the Committee, many simply have to rely on the advice of the subcommittees and researchers who were more intimately involved in its development. It explains why conceptual issues, such as level of service criteria, are hotly and extensively debated, while major changes in a specific methodology receive little discussion at the full committee level before approval.

It is simply a fact of life. As the scope and range of the manual has expanded, more experts willing to devote their time and effort to the HCM were and are needed. Yet, it is very uncomfortable to have to vote on a particular methodology when the voter has only a cursory knowledge of the details of its content and development. The Committee will have to deal with this reality as it moves forward into an ever-more complex field.

References

1. Roess RP, Prassas ES (2014) The highway capacity manual: a conceptual and research history. In: Uninterrupted flow, vol 1. Springer, Heidelberg, Germany
2. Beito DT, Beito LR (1998) Rival road builders: private toll roads in Nevada, 1852–1880. Nevada Hist Soc Q 41(Summer)
3. FHWA (2011) Toll Roads in the United States: history and current policy. Federal Highway Administration Publication FHWA-PL-11-032. Washington D.C.
4. Klein DB, Majewski J (2010) Turnpikes and toll roads in nineteenth—century America. EH.net Encyclopedia, Economic History Association, 2 Feb 2010
5. Raitz KB et al (1996) The national road. Johns Hopkins University Press, Baltimore MD
6. Sky T (2012) The National road and the difficult path to sustainable national investment. Delaware University Press
7. Colby, Steve, The Cumberland Road.com, online
8. Wilson v. Shaw—204 U.S. 24 (1906) On-line at Justia, U.S. Supreme Court Center
9. Hilles WC (1958) The good roads movement in the U.S. Thesis. Duke University, Durham NC
10. Wallis M (2007) The Lincoln highway: coast-to-coast from times square to the golden gate. W. Norton and Company, New York, NY
11. Statute of the 52nd Congress (1893) 3 Mar 1893
12. United States Government Printing Office (1944) Interregional highways: a message from the president of the united states transmitting a report of the national interregional highway committee outlining and recommending a national system of interregional highways. United States Government Printing Office, Washington D.C.
13. Weingroff R (Jan–Feb 2006) The year of the interstate Public roads 69(4). Federal Highway Administration, Washington D.C.
14. Kittelson WK (2001) Historical overview of the committee on highway capacity and quality of service. In: Transportation research circular E-CO18: proceedings of the fourth international conference on highway capacity, Transportation Research Board, Washington D.C.
15. Normann OK (June 1942) Results of highway capacity studies, Public Roads 23(4). Bureau of Public Roads, Washington D.C.
16. Transportation Research Board (1950) Highway capacity manual, 1st edn. Transportation Research Board, Washington D.C.
17. Bureau of Public Roads (Oct 1949) Public roads, Bureau of public roads, 25(10). Washington D.C.
18. Bureau of Public Roads (Nov 1949) Public roads, 25(11), Bureau of public roads. Washington D.C.
19. Normann OK (1965) A tribute. Highway capacity manual, special report 87. Transportation Research Board, Washington D.C., p vii
20. Transportation Research Board (1965) Highway capacity manual, 2nd edn. Special report 87, Transportation Research Board, Washington D.C.

21. Transportation Research Board (1980) Interim materials on highway capacity. Transportation research circular 212. Transportation Research Board, Washington D.C.
22. Transportation Research Board (2000) Highway Capacity Manual. Transportation Research Board, Washington D.C.
23. Transportation Research Board (2010) Highway Capacity Manual. Transportation Research Board, Washington D.C.

Chapter 2
Fundamental Concepts for Interrupted Flow

Any discussion of capacity analysis for signalized intersections must begin with a common vocabulary concerning key factors and measures describing the operation of traffic signals. Unfortunately, this vocabulary has changed over the years. It is useful, however, to use a common nomenclature in discussing any of the historic or current analysis methodologies. The textbook "Traffic Engineering" [1] *provides a good source for basic concepts and definitions.*

2.1 Discharge Headways, Saturation Flow Rates, Lost Times, and Capacity at Signalized Intersections

The most fundamental factor governing interrupted flow is the process by which vehicles depart a standing queue that forms whenever a traffic signal turns RED. The traffic signal is the principal cause of "interruptions" on surface streets and arterials, although there are others, such as STOP signs. Because at every signalized intersection, traffic is systematically stopped and restarted, platoons of vehicles are created and discharged. While platoons tend to disperse over time and distance, flow on arterials and streets is dominated by platoons of vehicles moving together, with gaps between platoons that accommodate generally lower flows made up of vehicles who have entered the traffic stream between platoons from driveways, parking places, or other unsignalized entry points.

Figure 2.1 illustrates the basic mechanism of queue discharge at a signalized intersection. No matter how good signal coordination is, there are always some vehicles that will queue on an intersection approach while the signal is RED. When the light turns GREEN, these vehicles will discharge from the intersection as illustrated in the figure.

The mechanism used to quantify discharge characteristics is the headway of vehicles as they cross over the intersection line (in some cases, the STOP line may be used as an alternative) to enter the intersection. For consistency, common practice is

E. S. Prassas and R. P. Roess, *The Highway Capacity Manual: A Conceptual and Research History Volume 2*, Springer Tracts on Transportation and Traffic 12,
https://doi.org/10.1007/978-3-030-34480-1_2

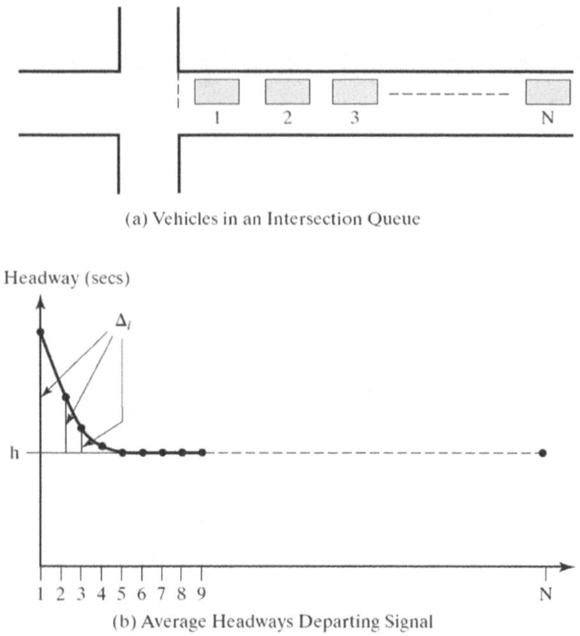

Fig. 2.1 Discharge of vehicles at a signalized intersection. *Source* Roess, R.P., and Prassas, E. S., *Traffic Engineering*, 4th Edition, Pearson-Prentice Hall, Upper Saddle River, NJ, 2011 [1]. Figure 20.1, Pg. 465. Reprinted with permission of Pearson Education Inc

to measure headways as the *front wheels* of each vehicle cross the intersection line. Time measurements begin when the signal turns GREEN. Then:

- The first headway is the time between the initiation of the GREEN signal and the time the front wheels of the first vehicle in queue cross the intersection line.
- The second headway is the time between the front wheels of the first vehicle crossing the intersection line and the front wheels of the second vehicle doing so.
- Subsequent headways are similarly measured.

Figure 2.1b plots the observed average headway vs. the discharge position in the queue. Obviously, these values vary from signal cycle to signal cycle with individual driver characteristics and other factors. The trend of average values is, however, quite typical.

The first headway is generally the largest, as the first driver must go through the perception-reaction sequence and accelerate the vehicle from a standing stop. The second headway is lower, as the second driver can overlap the reactions of the first driver somewhat. Each headway (on average) gets smaller until equilibrium sets in. This occurs at a point when vehicles have fully accelerated to their desired speeds *before* crossing the intersection line. While this characteristic varies considerably, most studies show that equilibrium is established after the 3rd or 4th headway.

A stable constant headway between vehicles occurs after equilibrium, and is referred to as the *saturation headway* (h) in units of seconds/vehicle. If every vehicle consumed "h" seconds of green time, and if the signal were GREEN at all times, the number of vehicles (per lane) that could be discharged during an hour would be:

$$s = \frac{3600}{h} \tag{2.1}$$

where

s saturation flow rate (veh/hg/ln),
h saturation headway (s/veh), and
3600 number of seconds per hour.

The units for saturation flow rate are "vehicles per hour of green time per lane". Of course, the signal (for any given approach) is *not* green all of the time. Further, the saturation headway of "h" s/veh is valid only if the signal is *never* RED and no one ever stops.

Vehicles must stop, however, once during each signal cycle when the light turns RED. They must then start up again when the light turns GREEN. Thus, flow through a signalized intersection approach must deal with the periodic stopping and starting of the traffic stream. There is time consumed when the traffic stream is stopped, and time consumed when it starts up again.

As indicated in Fig. 2.1b, the first few vehicle in the queue experience headways that are *more* than "h" seconds, due to the perception-reaction time and the need to accelerate to a desired speed. The headway for each of these first few vehicles is expressed as "$h + \Delta_i$" where Δ_i is the increment over h seconds consumed by vehicle "i". The total additional time consumed by the first three or four vehicles in the queue (the only ones that experience a Δ) is:

$$\ell_1 = \sum_{i=1-3,4} \Delta_i \tag{2.2}$$

where

ℓ_1 start-up lost time (s/phase), and
Δ_i additional headway consumed by vehicle "i" in queue (s).

A second lost time occurs when the flow of vehicles stops. This value is referred to as the *clearance lost time* (ℓ_2). It is more difficult to measure, as it requires that the queue of vehicles discharges continuously throughout the GREEN signal. When this condition exists, it is measured as the time between the front wheels of the *last* vehicle in queue crossing the intersection line and the initiation of GREEN on the next signal phase.

In effect, this simplified model of vehicle discharge into a signalized intersection presumes that each vehicle consumes "h" seconds of green time, not including the start up or clearance lost times. This leads to the concept of "effective green time," described in a later section. The "capacity" of an intersection approach lane depends upon the effective green time and the saturation headway (or saturation flow rate).

2.2 Fundamental Signal Definitions

A traffic signal essentially allocates time to competing movements that would, if not controlled, attempt to occupy the same space at the same time. A rotational set of signal lights allows each movement or set of movements to move in turn. After every movement has had at least one chance to move, the rotation begins again.

The most fundamental unit describing signal operation is the *cycle*. A signal cycle is *one* pass through the rotational set of lights. A critical parameter describing the signal is the amount of time it takes to go through one cycle, referred to as the *cycle length* (C), in seconds.

For each legal intersection movement, there are two basic signal indications:

G_i green time for movement or approach "i" (secs), and
R_i red time for movement or approach "i" (secs).

As a practical matter, however, signals cannot change from GREEN to RED instantaneously. Drivers cannot bring vehicles to a stop in an instant. It takes time for the driver to perceive that the signal is now RED, and time to decelerate the vehicle to a stop. Thus, there are transition signals between the GREEN and the RED to allow this to occur safely. Between the GREEN and the RED for every movement or approach, there are two additional indications:

y_i yellow time for movement or approach "i" (secs), referred to as the *change interval*, and
ar_i all red time for movement or approach "i" (secs), referred to as the *clearance interval*.

While one movement or approach "i" is GREEN or YELLOW, all other movements or approaches are RED. During the all-red interval (ar), *all* movements or approaches are RED.

In terms of function, the YELLOW interval is timed to allow a vehicle which cannot safely stop when the YELLOW begins to legally enter the intersection. The all-red interval allows a vehicle that has just entered the intersection on YELLOW to cross the intersection plus one car length before vehicles are released from a conflicting movement or approach.

A signal "phase" is a period of time during which a GREEN plus the yellow and all-red transition intervals are assigned to one set of movements or approaches. The simplest signals have two phases: one for each or two intersecting streets. First, all movements from one street are permitted to move, then all movements on the other are permitted to move. Figure 2.2 shows how the various signal intervals in a two-phase signal overlap each other.

Figure 2.2 makes one key relationship clear: the cycle length is the sum of the GREEN, YELLOW, and ALL-RED intervals:

$$C = \sum_{i=1-N} (G + y + ar)_i \tag{2.3}$$

where "N" is the number of phases, and all other terms are as previously defined.

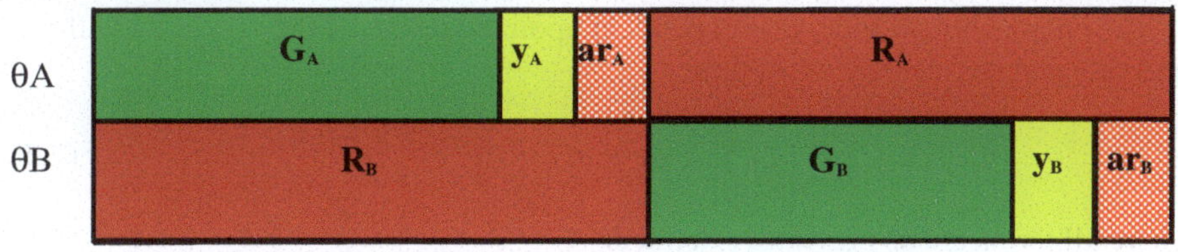

Fig. 2.2 Relationship between signal intervals illustrated: two-phase signal

2.3 The Concept of Effective Green Time

Drivers do not enter the intersection only on the GREEN interval. It is not reasonable to expect that, as the YELLOW interval is specifically provided to allow drivers and vehicles to enter the intersection if they cannot safely stop when the YELLOW is initiated. In some cases, they will even encroach on the all-red time, when they are not legally supposed to enter the intersection.

In terms of driver behavior, there are really only two intervals: effective green time and effective red time. During effective green time, they will enter the intersection; during effective red time, they will not. Further, effective green time, for convenience, is *defined* as the signal time during which they will enter the intersection at a rate of one vehicle every "h" seconds per lane, or at the saturation flow rate of "s" vehicles per hour of green.

The question is, then, how does the effective green time (g) compare to the actual green time (G)? There are two adjustments that have to be made:

- At the beginning of the green interval (G), the start-up lost time (ℓ_1) must be *subtracted* from G, to account for the additional time beyond "h" seconds per vehicle used by the first 3 or 4 vehicles in queue.
- At the end of the green interval (G), drivers and vehicles will use some of the yellow and all-red time to enter the intersection. This must be *added* to G.

Then

$$g_i = G_i - \ell_1 + e \tag{2.4}$$

where

g_i effective green time for movement or approach "i" (secs),
G_i actual green time for movement or approach "i" (secs),
ℓ_1 start-up lost time (secs), and
e encroachment time into yellow and all-red (secs).

The encroachment time, e, is dependent upon the length of the yellow and all-red intervals and the clearance lost time:

$$e = y_i + ar_i - \ell_2 \tag{2.5}$$

where all terms are as previously defined.

The saturation flow rate, s, was previously defined as the rate at which vehicles may enter the intersection during effective green time. This simple definition allows us to estimate the *capacity* of an approach or lane in terms of the saturation flow rate and the signal timing:

$$c_i = s_i\left(\frac{g_i}{C}\right) \tag{2.6}$$

where all terms are as previously defined.

2.4 Options for Signal Phasing

While it is not the purpose of this book to go into the details of signal phasing, it is important to note that the principal difficulty at many intersections is the handling of left turns. At a two-phase signal, left turns are made through a conflicting opposing through flow. When a left turn cannot be immediately made due to opposing traffic, the left-turning vehicle waits, blocking or partially blocking a lane. Thus, where the number of left turns is high and opposing flows are also high, left turns can create considerable congestion and delay at a signalized intersection.

In designing a signal phase plan, there are two fundamental options for each legal left-turning movement:

- The left turn may be *permitted* to be made against an opposing through flow.
- The phasing may be arranged to allow the left turn to be *protected* against the opposing flow, by stopping the opposing flow when left turns are being made.

Of course, there are signal timings that allow left turns to be made on a *permitted* basis during part of the cycle and on a *protected* basis during another part of the cycle. Protected turns are made far more efficiently than permitted turns. They require, however, additional signal phases, which introduce additional lost time into the cycle, and may increase delays.

Protected left turns are almost always accompanied by exclusive left-turn lanes. Exclusive left-turn lanes *may* be provided even where left turns are permitted, as this allows waiting left-turners to do so out of the way of through vehicles.

There are many texts [1] and manuals [2] that describe the complexities of signal timing. Suffice it to say, however, that any capacity and level of service analysis methodology must be capable of looking at any of the many options which exist in the field.

2.5 Types of Signal Control

Capacity and level of service analysis of signalized intersections depends upon a critical feature: the timing of the signal. There are two main types of signal control (with many sub-categories as well):

> *Pretimed Signals*: For any given time period, each signal cycle is exactly the same. All GREEN, YELLOW, ALL-RED, and RED intervals are constant, repeating themselves in each cycle.
>
> *Actuated Signals:* Using information from detectors in, over, or at the side of the street, the signal timing can change on a cycle-by-cycle basis. While YELLOW and ALL-RED intervals, which are tied to safety, are constant, GREEN and RED intervals can change based upon information provided by detectors. Thus, cycle lengths and green times are not fixed, but may be different during each cycle.

This was not a problem for the early highway capacity manuals (1950 and 1965), as virtually all signals were of the pretimed variety, and all signal parameters were known.

As time progressed, however, actuated signals became more popular, and their capabilities expanded. Today, many, if not most, urban and suburban traffic signals are of the actuated variety. In order to conduct a capacity or level of service analysis of an actuated signal, the *average* signal timing must be estimated and used in most models.

In the 1985 and 2000 HCMs, this was an explicit step in the analysis methodology. By 2010 (and 2016), however, the assumption of actuated control was used, and the estimation of average timings is never explicitly done. Rather, the steps in the process are imbedded within the methodology. In addition, applying the 2016 HCM to a pretimed signal analysis is not straightforward, and involves entering information that causes the assumed actuated controller to behave as a pretimed signal.

In essence, it is important to note that the methodology for analysis of signalized intersections in the HCM is based upon a model of *pretimed* signal operation. It essentially uses *average* signal timing for actuated signals. The model *does not* consider individual cycles that may differ considerably over any given period of analysis on a cycle-by-cycle basis. A simulation would be required to do so. More will be said about this when the 2010 and 2016 HCM methodologies are discussed.

2.6 The Concept of Critical Lanes and/or Movements

All signalized intersection capacity methodologies have been based upon the concept of "critical" lanes or movements since 1985. The concept is a very simple one: each defined signal phase handles several lanes and movements at the same time. *One* of these lanes or movements requires the largest amount of green time. That lane or movement is called "critical," and it determines the green time requirements of that phase.

Consider the intersection shown in Fig. 2.3. It shows the intersection of two arterials, and has *three* signal phases: (A) a left-turn phase for the EB and WB left turns, (B) a through phase for the EB and WB through and right-turn movements, and (C) a single phase for all movements NB and SB.

In the first phase (Phase A), EB and WB left-turning vehicles are moving on a protected basis, each occupying one lane. *One* of these left turns will require the largest green time for the phase. In this case, it is the WB LT, which is shown with a bolder arrow. In the second phase (Phase B), six lanes of traffic serving the EB and WB through vehicles and right turning vehicles are permitted to move. Only *one* of these six lanes requires the longest amount of green time, depicting here as the WB middle lane (bolder arrow). In the third phase (Phase C), all NB and SB vehicles are moving in four lanes. *One* of these lanes will require more green time than the others. In this case, it is the left hand SB lane, depicted with a bolder arrow.

Signal timing will be based upon the flows in the three identified critical lanes, one for each signal phase (V_A, V_B, and V_C). If sufficient time is provided to handle these three critical lane volumes, then every other lane volume will be adequately served—because they require *less* green time.

Because of this, all signal analysis methodologies since 1985 focus on the identification of critical movements and critical lane volumes. The identification of critical movements and critical lane volumes can sometimes be complex, but the following rules generally apply:

1. For each discrete signal phase, there is one and only one critical lane volume.
2. During all effective green time in the cycle, one critical movement must always be moving; at no time can there be two critical movements moving.

It is *not* always the highest lane volume (in vehicles per hour) that is critical for a given signal phase. All vehicles are not created equal: trucks take more time

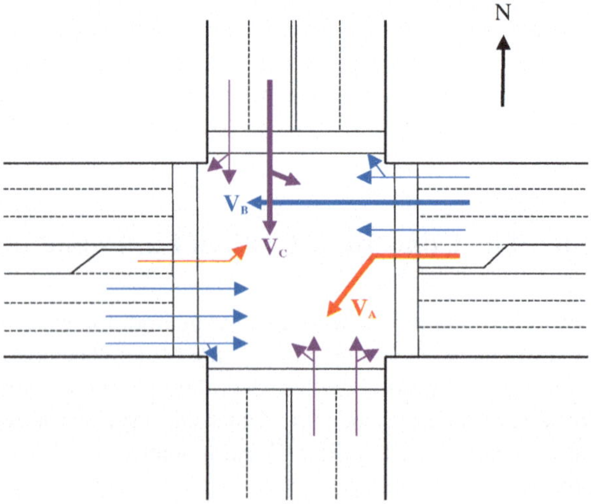

Fig. 2.3 Critical lanes or movements illustrated

than cars, right-turning vehicles take more time than through vehicles, left-turning vehicles take more time than through vehicles, vehicles in narrow lanes take more time than vehicles in wider lanes, etc. Because of this, the lane requiring the most green time (in any phase) may or may not be the lane with the highest demand volume.

The most important characteristic of a signal timing is the cycle length (C). It must be sufficient to accommodate the sum of the critical lane volumes in the signal phase plan. Because each phase must accommodate its critical lane volume, than all of the phases comprising the cycle must accommodate the sum of these volumes. The sum of the critical lane volumes (V_c) for Fig. 2.3 is $V_A + V_B + V_C$. In more general terms:

$$V_c = \sum_N V_i \tag{2.7}$$

where

V_c sum of the critical lane volumes (veh/h),
V_i critical lane volume for Phase i (veh/h), and
N number of discrete signal phases.

A reasonable cycle length is directly related to this sum:

$$C = \frac{L}{1 - \left[\frac{V_c}{s^* PHF^* (v/c)}\right]} \tag{2.8}$$

where

C cycle length (s),
L sum of all lost times in the cycle (s); $L = \sum_N (\ell_{1i} + \ell_{2i})$
ℓ_{1i} start-up lost time in Phase i (s),
ℓ_{2i} clearance lost time in Phase i (s),
N number of discrete phases,
PHF peak hour factor,
v/c desired ratio of demand to capacity, and
s average saturation flow rate for all lanes (veh/hg/ln).

Dividing the sum of critical lane volumes by the peak hour factor (PHF) converts V_c from a full-hour volume to a peak flow rate within the hour. Dividing by a desired v/c ratio recognizes that using 100% of available green time is not a desirable situation, as any small variation in cycle-to-cycle flows could lead to the failure of one or more phases on one or more cycles. The usual range for desired v/c ratios is between 0.80 and 0.95. Note that in the equation, both critical lane volumes and the saturation flow rate are described *per lane*.

There are many issues involved in implementing these principles in the analysis of signalized intersections. There are many complicating factors, including the existence of overlapping phases, the differences between pretimed and actuated signals, and many local prevailing conditions, such as pedestrian flows, lane widths, grades, bicycles, local buses, and others. These and other factors, including more detail on the identification of critical movements and lanes, are discussed further in Chaps. 7–10.

2.7 Delay Concepts

Since 1985, the service measure used to define signalized intersection level of service has been delay. Delay, however, is a relatively complex concept, and there are numerous ways in which it can be defined, estimated, and measured.

There are five different types of "delay" that are commonly used:

- *Stopped Time Delay (s/veh)*: average time a vehicle spends stopped at a signalized intersection waiting to pass through it; stopped time does not include the time spend inching forward in a queue of vehicles waiting to proceed through the intersection.
- *Approach Delay (s/veh):* average time a vehicle spends stopped at a signalized intersection plus the time loss involved in decelerating from the approach speed to a stop and accelerating back to the approach speed.
- *Time-in-Queue Delay (s/veh)*: average time a vehicle spends in a queue of vehicles before entering the intersection on the green signal; includes both stopped time, and the time spent inching forward in the queue – essentially the total time between joining the back end of the queue and departing it by entering the intersection.
- *Travel Time Delay (s/veh)*: the average difference between the time drivers expect to spend traversing the intersection and the actual time spent traversing the intersection.
- *Control Delay (s/veh)*: this is the sum of time-in-queue delay plus the time lost in deceleration and acceleration; called control delay because it represents the total delay caused by the signal.

All delay measures apply to a specific approach lane group. Delay measures can be averaged over several lane groups on a single approach, and averaged over all approaches for the intersection as a whole.

When delay was introduced as the determinant of level of service in 1985, stopped time delay was the specific measure used. In the 1994 update to the 1985 HCM, the measure was changed to control delay, which remains the service measure in the 2016 HCM.

Because delay is the service measure that determines level of service for signalized intersections, all analysis methodologies since 1985 have included a model for the estimation of delay. What follows is a very brief history of early delay prediction models.

2.7.1 Webster's Uniform Delay Model

Delay estimation begins with Webster's Uniform Delay Model [3]. Webster's model makes several key simplifying assumptions:

- Arrivals are uniformly distributed at a rate of v (veh/s).
- During the green signal, vehicles depart from the intersection at a rate equal to the saturation flow rate, s (veh/s) until the queue of waiting vehicles is dissipated.
- There is no overflow, i.e., all vehicles arriving during the cycle depart during the same signal cycle.

Figure 2.4 illustrates the model. Arrivals are shown as a constant slope on the cumulative vehicle vs. time (in seconds) plot. The arrival and departure curves are the same until the signal turns red. During the red interval, arrivals continue at a rate of v (veh/s), but there are no departures. When the light turns green again, vehicles queuing during the red dissipate at the saturation flow rate, (veh/s) until the departure curve "catches up" with the arrival curve. Thereafter, the departure and arrival curves are the same—at least until the next red signal.

The area between the arrival and departure curves is the aggregate delay (in vehicle-seconds) to all vehicles arriving during the cycle and queue clearance time (t_c). This aggregate delay is essentially *arrival delay*, which includes both stopped time and acceleration-deceleration delay components. The area of the triangle is:

$$UD_a = \frac{1}{2} RV \tag{2.9}$$

where:

UD_a aggregate uniform delay (veh-s),

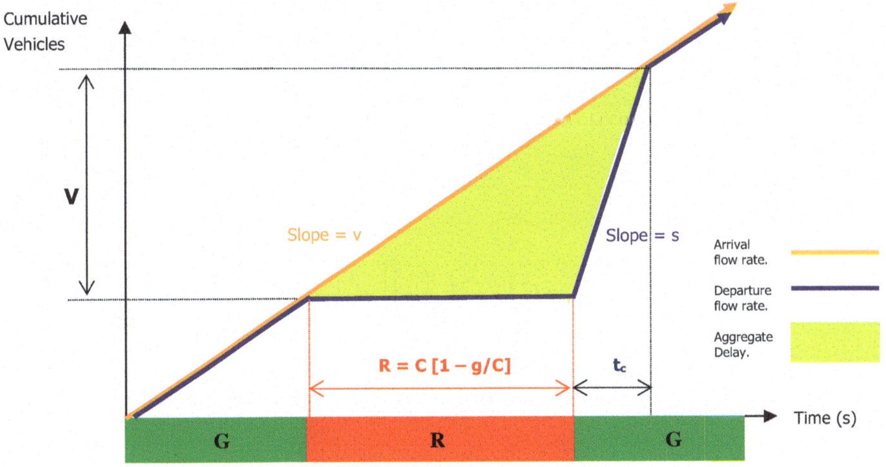

Fig. 2.4 Derivation of Webster's Uniform Delay Model

R length of the red interval (s), and
V number of vehicles arriving during the red interval (vehs).

Because most models do not explicitly work with the red interval, it is convenient to represent the red interval as *the portion of the cycle length which is not effectively green*, or:

$$R = C\left[1 - \left(\frac{g}{C}\right)\right] \quad (2.10)$$

where

R red interval (s),
C cycle length (s), and
G effective green time (s).

To determine the key factor, V—the number of vehicles in the queue—the time interval t_c must be determined. This is done by setting the number of vehicles arriving during the interval $R + t_c$ to the number of vehicles departing during the interval t_c, or:

$$v\,(R + t_c) = st_c$$
$$R + t_c = \left(\frac{s}{v}\right)t_c$$
$$R = \left(\frac{s}{v}\right)t_c - t_c = t_c\left[\left(\frac{s}{v}\right) - 1\right]$$
$$t_c = \frac{R}{\left(\frac{s}{v} - 1\right)}$$

Substituting for t_c:

$$V = v\,(R + t_c) = v\left[R + \frac{R}{\left(\frac{s}{v} - 1\right)}\right] = vR + \frac{vR}{\left(\frac{s}{v} - 1\right)}$$

$$V = R\left[v + \frac{v}{\left(\frac{s}{v} - 1\right)}\right] = R\left[\frac{v\left(\frac{s}{v} - 1\right) + v}{\left(\frac{s}{v} - 1\right)}\right] = R\left[\frac{s - v + v}{\left(\frac{s}{v} - 1\right)}\right] = R\left[\frac{s}{\left(\frac{s}{v} - 1\right)}\right]$$

$$V = R\left[\frac{sv}{s - v}\right]$$

Substituting for R:

$$V = C\left[1 - \left(\frac{g}{C}\right)\right]\left[\frac{vs}{s - v}\right]$$

The aggregate delay can now be estimated as:

$$UD_a = \frac{1}{2}RV = \frac{1}{2}C^2\left[1 - \left(\frac{g}{C}\right)\right]^2\left[\frac{vs}{s - v}\right] \quad (2.11)$$

The average uniform delay *per vehicle* is found by dividing the aggregate delay by the number of vehicles arriving during the signal cycle, vC:

$$UD = \frac{1}{2}C\left[1 - \left(\frac{g}{C}\right)\right]^2\left[\frac{s}{s - v}\right] = \frac{1}{2}C\frac{\left[1 - \left(\frac{g}{C}\right)\right]^2}{1 - \left(\frac{v}{s}\right)} \tag{2.12}$$

This equation most often appears in a slightly different form. Note that the saturation flow rate, s, is equal to the capacity divided by the g/C ratio, or s = c/(g/C). Substituting this into Eq. (2.12) yields the more familiar form of the equation:

$$UD = \frac{1}{2}\frac{C\left[1 - \left(\frac{g}{C}\right)\right]^2}{1 - \left(\frac{g}{C}\right)\left(\frac{v}{c}\right)} \tag{2.13}$$

where

UD average uniform delay, s/veh,
C cycle length, s,
g effective green time, s,
v demand flow rate (assumed to be uniform), veh/s or veh/h, and
c capacity, veh/s or veh/h.

The units for v and c must be the same. In some texts, the v/c ratio is given the symbol "X" for simplicity.

2.7.2 Introducing Randomness into the Equation

Webster's formula assumes that the arrival rate, v, is *uniform*. This means that the inter-vehicle arrival times are constant. Thus, if the demand flow rate is 1800 veh/h, it is assumed that the vehicles arrive at exactly 3600/1800 = 2.0 s intervals. This is rarely true.

Even for isolated intersections, where no platooning from adjacent signals is expected, arrivals are never uniform. They are more likely to be randomly distributed around an average. Thus, a flow rate of 1800 veh/h would not produce uniform arrivals every 2.0 s, but a random distribution of arrivals with an *average* of 2.0 s. Such an arrival pattern would result in *higher delay* than that caused by uniform arrivals. A number of early researchers developed stochastic models to address this situation, including Newall [4] and Miller [5, 6]. Webster himself developed such a model [3]:

$$RD = \frac{X^2}{2v(1 - X)}$$
$$D = 0.90(UD + RD) \tag{2.14}$$

where

RD random delay, s/veh,
v flow rate, veh/s
X v/c ratio, and
D sum of random plus uniform delay, s/veh.

Equation (2.14) applies a 0.90 "adjustment" to reflect a systematic overestimation of delay using the equation.

There is an obvious flaw in the equation: As the v/c ratio (X) approaches 1.00, the random delay term increases rapidly to infinity.

2.7.3 Overflow Delay Concepts

"Overflow" occurs when the demand flow (v) arriving at the intersection exceeds the capacity (c) to discharge them. Under this condition, the vehicles arriving in a cycle are *not* fully discharged in the cycle, and a residual of vehicles is carried over to be served on the *next* green phase. When the condition exists over a number of cycles, a queue continues to grow.

Webster's equation applies to cases where v/c $\leq$ 1.00. Overflow delay occurs when v/c > 1.00, even for one or two cycles. If one considers a series of signal cycles over an extended time period, there are three possible scenarios that may arise:

- All cycles, as well as the time period as a whole, operates with v/c $\leq$ 1.00.
- The time period as a whole operates with v/c $\leq$ 1.00, but some cycles within the time period have v/c > 1.00.
- Every cycle within the time period, and the time period as a whole, operates with v/c > 1.00.

Figure 2.5 illustrates these three scenarios. In each case, all areas between the arrival (blue) and departure (violet) curves illustrate aggregate delay to all vehicles involved. The portion of those areas that lie *below* the capacity (black) line constitute *uniform delay*, while the portion of those areas *above* the capacity line constitute *overflow delay*.

To obtain a complete description of delay, a methodology to estimate the *overflow* delay component is needed.

2.7.4 Estimating Overflow Delay

Overflow delay occurs when some vehicles cannot be processed on the green phase immediately during or after their arrival. This delay affects vehicles and their occupants that must wait for *more than one* green phase to be discharged. Such overflow exists when the v/c ratio for *any* single cycle is greater than 1.00. Overflow delay

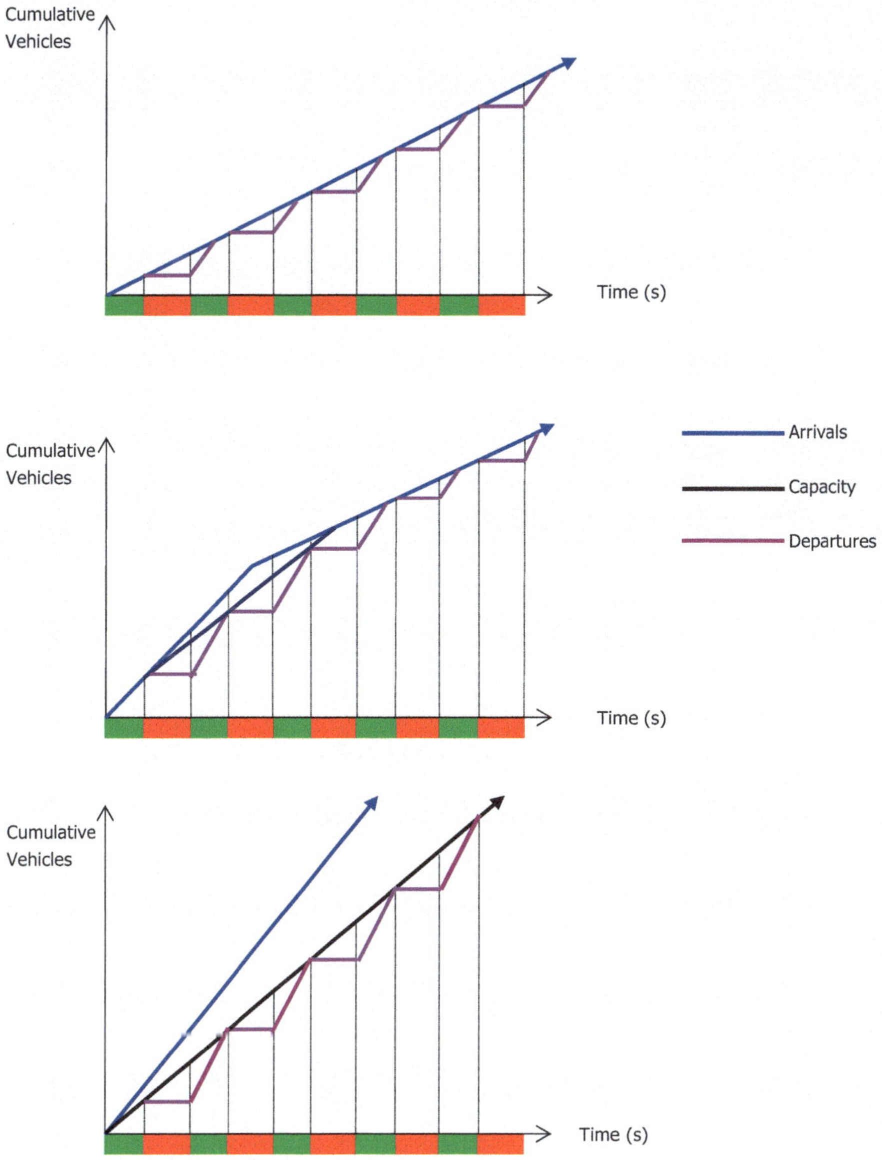

Fig. 2.5 Delay scenarios illustrated

becomes significant when arrival flows exceed the capacity of the signalized approach for a number of consecutive signal cycles.

Theoretically, estimating overflow delay is relatively straightforward. Given that it is the area between the capacity arrival curve and the capacity curve on a plot of cumulative vehicles vs. time (Fig. 2.5), two scenarios can be modeled. The first

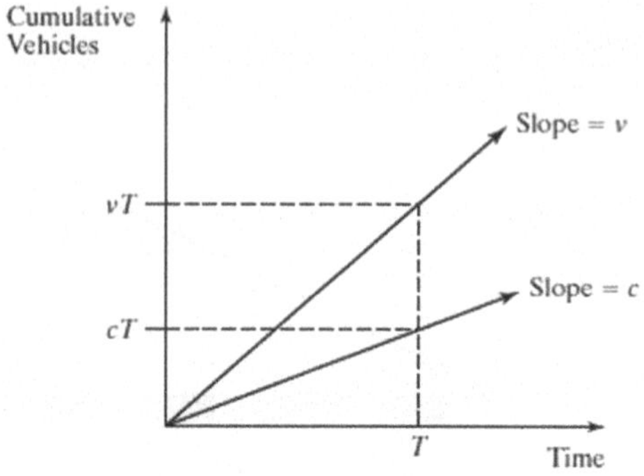

Fig. 2.6 Overflow delay in time "T". *Source* Roess, Prassas, and McShane, *Traffic Engineering*, 4th Edition, Pearson-Prentice Hall, Upper Saddle River, NJ, 2011 [1], Fig. 20.15, Pg. 482, Used with permission of Pearson Education Inc

involves overflow delay for some time period "T", where the overflow begins at T = 0. This is illustrated in Fig. 2.6.

Aggregate overflow delay in time "T" is simply the area in the triangle bounded by the arrival and capacity curves to time T:

$$OD_a = \left(\frac{1}{2}\right)T(vT - cT) = \frac{T^2}{2}(v - c) \tag{2.15}$$

where

OD_a aggregate delay (s),
T time period over which overflow delay exists(s),
v vehicle arrival rate (veh/s), and
c capacity of the signalized intersection approach (veh/s).

To convert this to average delay per vehicle, this is divided by the number of vehicles discharged within time T, or cT, yielding:

$$OD = \frac{T}{2}(X - 1) \tag{2.16}$$

where

OD average overflow delay (s/veh), and
X v/c ratio.

It should be noted that this formulation includes delay to vehicles that have *arrived* within time "T", but have not been *discharged* within time "T."

The second simple view of overflow delay produces an estimate of the overflow delay occurring *between* times T_1 and T_2. In this case, the aggregate delay area between the arrival and capacity curves is in the form of a trapezoid, as illustrated in Fig. 2.7.

The resulting model for average individual delay between times T_1 and T_2 becomes:

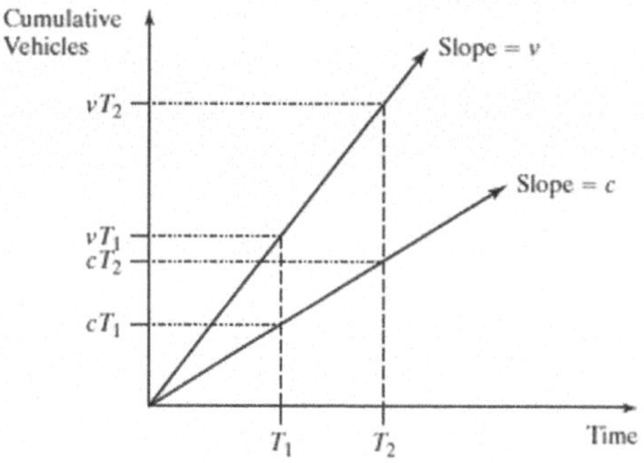

Fig. 2.7 Overflow delay between times T_1 and T_2. *Source* Roess, Prassas, and McShane, *Traffic Engineering*, 4th Edition, Pearson-Prentice Hall, Upper Saddle River, NJ, 2011, Fig. 20.16, Pg. 483, Used with permission of Pearson Education Inc

$$OD = \frac{T_1 + T_2}{2}(X - 1) \tag{2.17}$$

where all terms are as previously defined. The latter equation is important: Vehicles arriving early during time "T" experience far less delay than those arriving late in time period "T". Using the second equation allows the estimation of overflow delay during any slice of an overflow period that is T seconds long. Like the first equation, note that the delay includes vehicles that may have arrived *before* T_1 that are discharged *before* T_2, as well as vehicles that may have arrived *before* T_2 but are discharged *after* T_2.

2.7.5 Problems with Simple Delay Estimations

All of the delay formulas presented so far rely on a number of simplifying assumptions. All treat arrival and discharge rates as constants, and that such arrivals and departures are completely *uniform*. This is, of course, rarely true. In most cases, arrival rates vary over time, perhaps even cycle-to-cycle.

Further, the models deal with uniform arrivals over time. Even when the underlying arrival rate (v) is constant, individual arrivals are randomly distributed around a mean. This can be addressed by adding a component estimating the additional delay of random arrivals compared to uniform arrivals, such as in Webster's Random Delay formula, which was presented previously. Unfortunately, most arrival patterns are neither uniform nor random, but the result of a platoon pattern established by adjacent nearby signals.

While total delay has three distinct components—uniform delay, random delay, and overflow delay—most modeling approaches have historically combined random delay and overflow delay into a single component, called *overflow delay*. This, unfortunately, leads to some confusion, as the same term sometimes refers to pure overflow delay, and sometimes to the sum of random and overflow delay. When combined into a single term, however, the sum of Webster's random delay term and the theoretical overflow term presents a significant problem, as illustrated in Fig. 2.8.

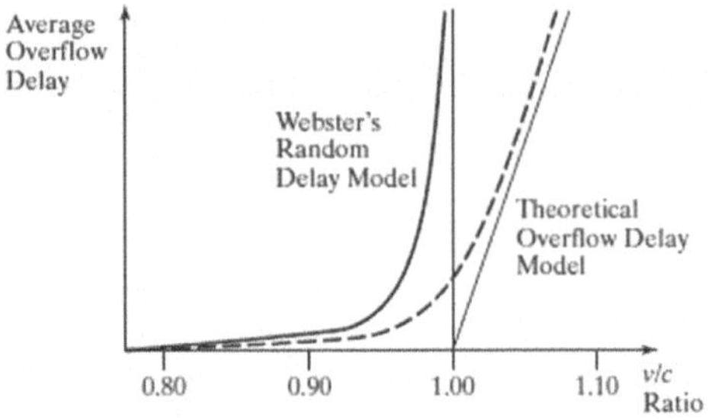

Fig. 2.8 Problems with the sum of random and overflow delay. *Source* Adapted from Hurdle, V.F., "Signalized Intersection Delay Model: A Primer for the Uninitiated, *Transportation Research Record 971*, Transportation Research Board, Washington D.C., 1984, p. 101, Copyright National Academy of Science, used with permission of the National Academy of Science, Courtesy of the National Academies Press

As illustrated in Fig. 2.8, Webster's random delay model blows up to infiniti (∞) as it approaches a v/c ratio of 1.00. The theoretical delay model predicts an overflow delay of zero (0) at a v/c ratio of 1.00. When the sum is considered, results are not reasonable or accurate in the vicinity of v/c = 1.00. This, however, is a very important "vicinity" in terms of operations. Most studies show that Webster's simple uniform delay model is a sufficiently accurate predictor when the v/c ratio is less than approximately 0.85. Similarly, the simple theoretical overflow delay model (when added to uniform delay) is a sufficient estimator when the v/c ratio exceeds 1.15. The problem is in the middle—a range in which many well-timed signals in urban areas operate.

Indeed, most modern efforts to model signalized intersection delay focus on appropriately dealing with this mid-range of v/c ratios between approximately 0.85 and 1.15. For many years, an Australian model developed by Akcelick [7] was used in the HCM, although it was modified several times. This and subsequent delay models are discussed in the context of specific HCM methodologies presented in other chapters. The fundamental concepts have been presented here to give the reader a background in the critical problems that must be addressed to accurately predict average individual delays to motorists at signalized intersections.

2.8 Queues and Stops

There are two parameters that describe operations on a signalized intersection approach that are closely related to delay: queues and stops. During a typical red phase, a queue of vehicles forms on the approach. Under good operations, all would be serviced (i.e. discharged from the intersection) on the following green phase. When the green phase cannot discharge all vehicles queued on the preceding red phase, the queue begins to grow, and will continue to grow until arrivals on subsequent phases are fewer than the departures. Obviously, as queues get larger, the average delay per vehicle also grows. The size of the queue, however, is of significant importance. If the queue gets too close to the adjacent upstream intersection, it will affect discharge from the upstream intersection. If it grows through the upstream intersection, effectively blocking discharges, a gridlock situation may emerge. Queues in a left-turn lane can spill back into through/RT lanes if the left-turn lane is not long enough. Similarly, queues in through/RT lanes can block access to a left-turn lane as well. Clearly, any intersection analysis must deal with queues and their impact on operations.

The number of stops made at a signalized intersection is also of interest. In a well-timed progressive signal network, many vehicles will arrive on green and pass through the intersection without stopping. The number of vehicles that actually stop on a signalized intersection approach is a relative measure of how good the progression is, the capacity sufficiency of the approach, and the strength of platooning in the traffic stream.

2.9 Intersections, Segments, and Facilities

The 2010 HCM [8] defines points (intersections), segments, and facilities for interrupted flow as follows:

> For the purpose of analysis, the urban street is separated into individual elements that are physically adjacent and operate as a single entity for the purpose of serving travelers. Two elements are commonly found on an urban street system: points and links. A point represents the boundary between links and is usually represented by an intersection or ramp terminal. A link represents a length of roadway between two points. A link and its boundary intersections are referred to as a segment. An urban street facility is a length of roadway that is composed of contiguous urban street segments and is typically functionally classified as an urban arterial or collector street. [9, p. 16-5]

The 2016 HCM [9] definition is somewhat different: The 2016 HCM, unlike the 2010 edition, only allows the analysis of urban street segments in both directions (on a two-way street). For these cases, the segment is defined as two adjacent signalized intersections and the links that connect them. For one way streets, the definition is conceptually the same as the 2010 definition above.

This generic, all-encompassing, definition is far less complicated than it appears. The 2010 and 2016 HCMs provide analysis methodologies for signalized intersections, unsignalized intersections (2-way STOP control, multiway STOP control, roundabouts), urban street segments, and urban street facilities.

Signalized intersections are treated as a point, although the analysis looks at each approach, and indeed lane groups within each approach, on an individual basis. All of the unsignalized analysis methodologies also do so.

An urban street *segment* is defined as a signalized intersection approach plus the street link to the next upstream signalized intersection (but not including it). An urban street segment, therefore, includes only *one* signalized intersection. It may, however, include several unsignalized intersections. An urban street *facility* involves the combined analysis of several adjacent urban street segments.

The 2010 and 2016 HCMs also look at interrupted flow in a multimodal way: analyses consider not only vehicular traffic, but pedestrians, bicycles, and transit—within each methodology. This was done to encourage the view of urban streets as multimodal or "complete" entities serving a variety of traffic modes.

The chapters of this volume detail the development history of current practice involving capacity and level of service analysis of interrupted flow points. Segment and facility analysis will be treated in Volume III.

References

1. Roess R, Prassas E, McShane W (2019) Traffic engineering, 5th Edn. Pearson—Prentice Hall, Upper Saddle River, NJ
2. Koonce P et al (2008) Manual of traffic signal timing. Final report, FHWA Contract No. DTFH61-98-C-00075, Kittelson and Associates, Portland OR, June 2008 (available online)
3. Webster F (1958) Traffic signal setting. Road research paper no. 39, Road Research Laboratory, Her Majesty's Stationery Office, London UK
4. Newall G (1965) Approximation methods for queues with application to fixed time traffic light. SIAM Rev vol 7
5. Miller A (1968) Settings for fixed-cycle traffic signals. ARRB Bulletin 3, Australian Road Research Board, Victoria, Australia
6. MillerA (1968) The capacity of signalized intersections in Australia. ARRB Bulletin 3. Australian Road Research Board, Victoria, Australia
7. Akcelik R (1980) Time dependent expressions for delay, stop rate, and queue lengths at traffic signals. Report no. AIR 367-1, Australian Road Research Board, Victoria, Australia
8. Transportation Research Board (2010) Highway capacity manual, 5th edn. Transportation Research Board, National Academy of Sciences, Washington DC
9. Highway Capacity Manual (2016) A guide for multimodal mobility analysis, 6th edn. Transportation Research Board, National Academy of Sciences, Washington DC

Chapter 3
Concepts of Capacity and Level of Service for Interrupted Flow

The general concepts of capacity and level of service are discussed in great detail in Chaps. 2 and 3 of Volume 1 of this manuscript. This chapter reviews some of the development history of these critical concepts, and specifically discusses their application to interrupted flow facilities.

The critical concepts of capacity and level of service have evolved over the years. The current use and understanding of these concepts is not the same as when they were first introduced. Capacity was introduced as a formal measure in the *1950 Highway Capacity Manual* [1]. The formal use of level of service to characterize the quality of traffic operations began with the *1965 Highway Capacity Manual* [2]. Both concepts had been under discussion and study for some time, as the profession struggled with the need to develop systematic methodologies for use in the planning, design, and analysis of the nation's growing highway systems.

It is interesting to note that both concepts were primarily developed with uninterrupted flow facilities in mind. They were later adapted for use in describing interrupted flow facilities as well. This fact of history helps explain why the application of these critical concepts has changed more radically over time for interrupted flow facilities than for uninterrupted flow facilities, and why their application to interrupted flow sometimes feels like pounding a square peg into a round hole. Yet, in practical terms, there are far more miles of interrupted flow facilities than uninterrupted flow facilities in a typical highway network, especially in urban areas where the problems of congestion are of paramount importance.

3.1 The Capacity Concept

One of the earliest researchers of traffic characteristics was A. N. Johnson, Dean of Engineering at the University of Maryland. As early as 1930, Johnson framed the discussion of highway capacity:

E. S. Prassas and R. P. Roess, *The Highway Capacity Manual: A Conceptual and Research History Volume 2*, Springer Tracts on Transportation and Traffic 12,
https://doi.org/10.1007/978-3-030-34480-1_3

> We can visualize a road carrying but a few vehicles and agree that there is no congestion. But as the number of vehicles increases, there will be a point reached at which some vehicles will be delayed because they are immediately unable to pass slower-moving vehicles. Such a point indicates the beginning of congestion or what may be called 'working capacity' or 'free-moving capacity' of the highway. [1, p. 218]

Johnson's view of capacity was the maximum traffic volume that could be accommodated by a roadway without the interaction of individual vehicles causing delay to others. His view, formulated at a time when there were relatively few vehicles, reflected the expectation that highways should provide for free flow at all times. In the modern view, this is hardly a reasonable expectation. In fact, such operation does not even describe the "onset of congestion" in modern terms.

His work, however, began to set the framework for discussion and early development of the capacity concept. There was relative agreement among professionals that capacity should signify a maximum hourly volume that a roadway could accommodate under some defined operational condition. The discussion focused on whether capacity should be defined as the maximum hourly volume under *any* operating conditions, or the maximum hourly volume for some defined set of *acceptable* operating conditions.

The legendary O. K. Normann, commonly known as the father of highway capacity analysis and the driving force behind the first two editions of the *Highway Capacity Manual*, was one of the first to discuss the dichotomy between an absolute version of capacity and one that was tied to reasonable operating conditions [3]. He also developed one of the earliest methods for measuring speeds and headways in a traffic stream—an approach that was used by the earliest researchers to establish capacity values. Bruce Greenshields further developed and applied these measurement methodologies, and was the first to suggest that capacity be defined as the peak of a speed-volume or volume-density curve [4, 5].

Over time, two defining characteristics for capacity were developed:

- Capacity represented a maximum hourly volume in veh/h or veh/h/lane. The concept of subhourly flow rates did not emerge until much later.
- Capacity should be stated for prevailing conditions. It should be stated in mixed veh/h or veh/h/lane, and would be dependent on the percentage of trucks in the traffic stream. Other "prevailing conditions" included lane widths, lateral clearances, and terrain.

The issue of an absolute capacity regardless of operating conditions, and one tied to reasonable operating conditions remained unresolved for some time.

3.1.1 The 1950 Highway Capacity Manual

The *1950 Highway Capacity Manual* [1] formally introduced the concept of capacity to the profession. Because the issue of absolute capacity vs. reasonable operating conditions was not yet settled, the manual actually defined *three* levels of capacity.

1. Basic Capacity: Basic capacity was defined as the maximum number of passenger cars that can pass a given point on a lane or roadway during one hour under the most nearly ideal roadway and traffic conditions which can possibly be attained.
2. Possible Capacity: Possible capacity was defined as the maximum number of vehicles that can pass a given point on a lane or roadway during one hour under the prevailing roadway and traffic conditions.
3. Practical Capacity: Practical capacity was defined as the maximum number of vehicles that can pass a given point on a roadway or lane during one hour without the traffic density being so great as to cause unreasonable delay, hazard, or restriction to the drivers' freedom to maneuver under prevailing roadway and traffic conditions.

All three definitions relate to "prevailing roadway and traffic conditions." Roadway conditions referred to those conditions determined by the physical features of the roadway, which could not be changed without reconstruction work of some kind. Traffic conditions referred to those dependent on the characteristics of the traffic using the roadway, which could change over time.

Roadway conditions included such features as horizontal and vertical alignment, often represented by the design speed or average highway speed of the roadway segment in question. Average highway speed is the average, weighted by length of sub-segment, of the design speeds of individual sub-segments included in the segment of roadway under consideration. It also included cross-sectional elements of the roadway such as lane widths and lateral clearances.

Traffic conditions included primarily truck presence in the traffic stream, but later came to include differences in driver populations, particularly as it reflected differences between weekday commuters and weekend drivers.

Basic capacity was defined relative to "nearly ideal" roadway and traffic conditions. While not explicitly stated, the 1950 HCM implied that this included 12 ft lane widths, 6 foot lateral clearances at the roadside, and level terrain (at least for 2-lane highways). It also meant that there were no trucks in the traffic stream, which is why basic capacity is the only level to be defined in terms of "passenger cars per hour" rather than "vehicles per hour." For any given type of roadway, basic capacity was a single fixed value. Possible capacity recognized the impacts of prevailing roadway and traffic conditions on maximum volumes. Possible capacity, therefore, changed any time that any one of the prevailing roadway and traffic conditions changed. It reflected, however, the maximum volume that could be achieved, regardless of the operating conditions that occurred when that happened.

Practical capacity was the most important of the three. The 1950 states:

> It is the practical capacity which is of primary interest to those striving to provide adequate highway facilities. The design engineer will plan his improvements with an adequate practical capacity to meet anticipated volumes of traffic on the facility. [1, p. 8]

The manual further recognizes that the definition of practical capacity is somewhat subjective, being dependent on individual judgment on when impingements to safety, delay, or freedom to maneuver become unreasonable. To the extent possible, the

1950 HCM based values of practical capacity on studies that considered the extent to which speed and other operating parameters were impacted by other vehicles on the roadway.

The commonality in all three capacity definitions, however, is that they all relate to *uninterrupted flow*. From these definitions, the application to *interrupted flow* is hardly clear. The *1950 HCM*, however, introduced some critical concepts, and framed the way in which professionals have thought about interrupted flow capacity ever since:

1. The critical component to the capacity of *any* interrupted flow facility is the signalized intersection. Signalized intersections are the points at which discharge of vehicles to the next street link are limited, because the signal limits the amount of time available for vehicles to proceed.
2. Signalized intersections *do not* have a capacity. Rather, capacity is most appropriately applied to each of the approaches to the signalized intersection. It is unlikely that all approaches to a signalized intersection will peak at the same time, thus the capacity of the intersection as a whole—which might be thought of as the sum of the capacities for the individual approaches—has little practical meaning or significance.
3. As vehicles may discharge from an intersection approach only during the *green* interval, capacities are stated in terms of *vehicles per hour of green time*—either per lane, or per foot of approach width. This is, of course, what we now refer to as *saturation flow rate*, as defined in Chap. 2. The capacity of the signalized intersection approach is the product of the saturation flow rate and the proportion of time that is *green*—the G/C ratio. In the *1950 HCM*, there is no discussion of actual green time (G) and effective green time (g), although the capacity values prescribed clearly take the latter into account.

The *1950 HCM* also stated that the unit of roadway width that most accurately determined capacity of an approach was its width in feet, measured from the curb line. For one-way streets, approach width was the curb-to-curb width; for two-way streets, approach width was the curb-to-centerline (or median) width, regardless of lane width, parking conditions or other features.

The *1950 HCM* provided values for basic, possible, and practical capacities of intersection approaches. Using a representative average discharge headway of 2.4 s/veh (a value which included consideration of start-up lost times), it set a value of 1250 pc/hg/10 ft of approach width as the capacity (actually saturation flow rate). Possible capacity had to account for the impact of parked vehicles, turning movements, commercial vehicles, pedestrians, and inclement weather. Practical capacity was defined as the capacity (over a full hour) that could be achieved without *any* vehicle having to wait for more than one *red* phase to proceed through the intersection. It stated that this could generally be achieved at approximately 80% of the possible capacity. Chap. 5 discusses how these were specifically estimated in the 1950 and subsequent editions of the *Highway Capacity Manual*.

3.1.2 *The 1965 Highway Capacity Manual*

The 1965 HCM [2] introduced a single definition of capacity:

> Capacity is the maximum number of vehicles which has a reasonable expectation of passing over a given section of a lane or a roadway in one direction (or in both directions for a two-lane or three-lane highway) during a given time period under prevailing roadway and traffic conditions. [2, p. 5]

It abandoned the three types of capacity defined in the 1950 HCM, and replaced them with a single value. For comparative purposes, the new definition approximately coincided with the definition of *possible capacity* in the 1950 HCM. The idea of *practical capacity* with reasonable operational quality was replaced by the concept of level of service, which is discussed later in this chapter.

The revised definition added the concept of "reasonable expectancy" for the first time—a concept that is still in use. This meant, in general terms, that the capacities predicted by the manual would apply to similar segments anywhere in the U.S. and at any time (exclusive of bad weather, accidents, or incidents). It recognized that with such a definition, it was expected that observed volumes would sometimes be higher than predicted capacities. Without specifically discussing the issue, it implicitly recognized that capacity had a stochastic variation component, and that standard values should reflect capacities that could be achieved most of the time in most places.

Like the 1950 HCM, however, the definition was oriented towards uninterrupted flow. The treatment of interrupted flow capacity in the 1965 HCM was not entirely straightforward. It cites a general value for saturation flow rate of 1500 pc/hg/ln, without specifying the lane width. This saturation flow rate, like that of the 1950 HCM, took into account lost times, and was converted to volume using the actual G/C ratio, not the g/C ratio based upon *effective* green time. In its methodology for signalized intersections, however, *lanes* are not used—approach width in feet is the key variable, as was the case in the 1950 HCM.

Both uninterrupted flow and interrupted flow capacity in the 1965 HCM are based upon full-hour volumes. However, the peak hour factor (PHF) is used within most methodologies (including signalized intersections), such that the resulting analysis *does in fact* consider peak flow rates for some cases. For signalized intersections, computational methodologies (see Chap. 5) actually focus on the peak 15 min within the hour; for uninterrupted flow, the peak 5 min within the hour was used.

3.1.3 *The 1985 Highway Capacity Manual*

In many ways, the 1985 HCM [6] made significant changes to the concept of capacity. While the general definition did not change much from previous editions, in 1985, capacity was clearly stated in terms of a maximum *flow rate* during the peak 15 min period of the peak hour (or other hour of interest). This applied equally to uninterrupted and interrupted flow facilities.

The methodology for signalized intersection analysis involved a complete change in modeling approach. Critical movement analysis (CMA) was used for capacity analysis. While the technique, in one form or another, had been in use for some time in signal timing, it had not been applied to capacity analysis. The capacity was also tied to *lanes* as opposed to approach width, although the average lane width was used to implement an adjustment where necessary. The 12 ft lane was used as the standard, for which no adjustment was necessary.

The capacity of an intersection approach was clearly stated in terms of its *saturation flow rate*, as defined and discussed in Chap. 2:

$$c = s * N * (g/C) \tag{3.1}$$

where

c capacity of the intersection approach, veh/h,
s saturation flow rate for the intersection approach under prevailing roadway and traffic conditions, veh/hg/ln,
N number of lanes on the approach (or lane group),
g effective green time for the approach, s/cycle, and
C signal cycle length, s.

In terms of the methodology, models including adjustment factors for a variety of prevailing conditions, were used to estimate the saturation flow rate under prevailing conditions from an *ideal* or *base* value that was taken to be 1800 pc/hg/ln. This value applied where the following conditions existed:

- 12 ft lanes,
- Level approach grade,
- No commercial vehicles,
- No turning vehicles,
- No local transit buses stopping within the intersection, and
- No curb parking.

The standard saturation flow rate adopted cannot be compared to the values stated in the 1950 and 1965 HCMs, as both were based upon actual green times, NOT effective green times. The former make allowances for lost times; the latter use green time that must be adjusted to reflect lost times.

The move from actual green time to effective green time is of great importance. When actual green time is used, capacity does not vary with cycle length. If the G/C ratio is 0.50, for example, the actual green time per hour does not vary with cycle length. However, lost times occur each cycle; the shorter the cycle length, the more cycles, and (therefore) the more lost time in the hour. This critical change allows consideration of additional capacity associated with longer cycle lengths as part of the analysis methodology.

3.1.4 The 1994 and 1997 Updates to the HCM

The 1994 and 1997 updates to the 1985 HCM did not produce any profound changes in the view of capacity for either uninterrupted or interrupted flow. Some values of uninterrupted flow capacity were increased to reflect more current data. In terms of interrupted flow, the base saturation flow rate was increased from 1800 pc/hg/ln to 1900 pc/hg/ln, reflecting additional data suggesting that drivers were becoming more aggressive in leaving standing queues at signalized intersections.

3.1.5 The 2000, 2010, and 2016 Highway Capacity Manuals

The 2000 [7], 2010 [8], and 2016 [9] HCM produced significant changes in the fundamental concept of capacity and its treatment in uninterrupted and interrupted flow analysis. For interrupted flow, saturation flow rate remained the key capacity variable. The base value remained at 1900 pc/hg/ln for the 2000 and 2010 HCMs. In the 2016 HCM, 1900 pc/hg/ln was applied to intersections in areas with a Metropolitan Area Population of 250,000 or more. In smaller areas, a default value of 1750 pc/hg/ln is specified, but locally-calibrated values are preferred.

3.2 The Level of Service Concept

The first edition of the *Highway Capacity Manual*, published in 1950 [1] did not mention the words "level of service." It focused on defining the capacity of various types of roadways with the limited information and research available at the time. Nevertheless, in making the case for improved knowledge of roadway capacity, it acknowledged the need to understand operating quality better:

> To be of value in the sound economic and functional design of new roadways, or in adapting to present or future demands many of the existing roadways which must continue with us for long periods into the future, the capacity criteria must include measures of such factors as speed and the relative interference between vehicles in addition to the number of vehicles that can pass a point on a given roadway in a specified period of time. It is of little value to know the quantifying measure without knowing the quality of service provided. [1, pp.1 and 2]

In a real sense, this statement clearly lays out the roadmap behind over 60 years of research and development in the highway capacity field.

3.2.1 The 1965 Highway Capacity Manual

The formal history of the level of service concept begins with the second edition of the *Highway Capacity Manual* in 1965. The single concept of "practical capacity" was replaced with the now-familiar 6-letter scale referred to as "level of service." The 1965 HCM defined the concept as follows:

> Level of service is a term which, broadly interpreted, denotes any one of an infinite number of different combinations of operating conditions that may occur on a given lane or roadway when it is accommodating various traffic volumes. Level of service is a qualitative measure of the effect of a number of factors, which include speed and travel time, traffic interruptions, freedom to maneuver, safety, driving comfort and convenience, and operating cost. [2, p. 7]

While the definition refers to a variety of qualitative measures, limitations in the state-of-the-art resulted in only a partial implementation, at best. Speed and travel time measures were indeed used to define levels of service for freeways, multilane highways, and two-lane highways—even though the analysis methodologies for these facilities relied even more heavily on the v/c ratio. The v/c ratio (or volume or flow rate) is not mentioned in the definition because drivers (and passengers) cannot discern these from within the traffic stream—they cannot be *perceived* by users.

Other measures noted in the definition were not used in 1965. Freedom to maneuver was later related to *density*, which was introduced in subsequent editions of the manual. Traffic interruptions, which would have been most relevant to interrupted flow, is accounted for only as part of "overall travel speed" for arterials and streets. Driving comfort and convenience is a very subjective term, and is only considered in terms of other variables that are directly measurable. Safety and operating cost have never been used in level of service criteria, and have subsequently been eliminated from the definition.

In many ways, the 1965 HCM methodology for analysis of signalized intersections was an aberration and a radical departure from previous and subsequent approaches.

Conceptually, it is important to note that there is a clear relationship between signal timing and design and capacity analysis of signalized intersections. In signal timing, demand volumes (or flows) are known, and the signal must be timed to provide sufficient capacity to each movement. In capacity analysis, the signal timing is known, and the capacity provided to each movement is estimated. Signal timing and capacity analysis of a signalized intersections should be inverse applications of the same model.

While capacity analysis was not a well-developed application for signalized intersections prior to 1965, signal timing methodologies, from the very beginning, had been done using a range of techniques generally classified as *critical movement approaches* (CMA). Virtually all were based upon the early work of Greenshields [10]. In these approaches, the specific movements that controlled the timing of individual signal phases were identified, and the timing endeavored to provide sufficient capacity to these. Conceptually, these approaches recognized that while multiple movements moved in each signal phase, only one required the most green time, and that that phase had to be timed to accommodate it. In the 1985 and subsequent HCMs, CMA was used in capacity analysis as well. It 1965, it was not.

The 1965 methodology for signalized intersection analysis was based upon a consultant report [11] using an extensive data set compiled by the Highway Capacity Committee. Level of service was defined for each signalized intersection approach on the basis of the "load factor." The load factor for a given approach (or lane group) was defined as follows:

> It is the ratio of the number of green phases that are loaded, or fully utilized by traffic (usually during the peak hour) to the total number of green phases available for that approach during the same period. [2, p. 116]

The load factor was viewed as a measure of degree of utilization (a fancy way of describing a v/c ratio for a signalized intersection approach). The 1965 HCM gave a good deal of advice on how to recognize when a phase was "fully loaded." In general, a green phase on an approach was to be considered fully loaded if:

- Vehicles are waiting to enter the intersection in all lanes when the light turns green.
- Vehicles are continuously available in all lanes during the entire phase, with no unusually long spacings between vehicles at any time due to lack of demand.

The judgment as to whether there was unused green time at the end of the phase was particularly difficult. The manual gave the following advice:

> Usually, the ending of a loaded cycle will force some vehicles to stop, but in a perfect progressive signal system, the last of a platoon of vehicles may pass through just as the signal changes. [2, p. 116]

The manual further noted that it was not necessary that vehicles be *moving* continuously for a phase to be considered fully loaded. A left-turning vehicle may be stopped waiting for a gap in opposing traffic for an entire phase, which would then be considered fully loaded (at least for the left lane). On the other hand, traffic in some or all lanes could be prevented from moving due to queues on the downstream street link. The data from such a case is termed "meaningless."

The value of load factor could vary from a value of 0.0 to 1.0. Table 3.1 shows the relationship between level of service and load factor for the 1965 HCM.

Over time, difficulties with the interpretation of these criteria arose. The classification of a phase as "fully loaded" was very subjective, particularly regarding the end of the phase. Trained observers frequently made different judgments when identifying such phases. This led to many inconsistencies in the way load factor was measured in the field—resulting in inconsistencies in level of service designations.

Conceptually, load factor was difficult. If every phase during a peak hour were 95% loaded, the load factor was 0.0, as no one phase was 100% loaded. Even as a surrogate for v/c ratio, the measure was not effective. Essentially, these definitions guaranteed that even at LOS B, *some* vehicles would have to wait for a second green phase to get through the intersection. At best, the measure was counterintuitive. At worst, it was severely misleading as a measure of operational quality.

Table 3.1 Level of service for signalized intersection approaches—1965 highway capacity manual

Level of service	Description	Load factor
A	Free flow	0.0
B	Stable flow	≤0.1
C	Stable flow	≤0.3
D	Approaching unstable flow	≤0.7
E[a]	Unstable flow	≤1.00
F	Forced flow	–[b]

[a]Capacity, [b]Not applicable

Source "Highway Capacity Manual," *Special Report 87*, Transportation Research Board, Washington D.C., 1965 [2], Table 6.3, p. 131; used with permission of the National Academy of Science, Courtesy of National Academies Press

3.2.2 The 1985 Highway Capacity Manual

By 1970, considerable debate and study had been devoted to the load factor and its relationship to delay. It was argued by May [12, 13] and others [13] that delay was a much better descriptor of intersection operational quality, and that it was not very well correlated to load factor. The dichotomy between signal timing practice and the signalized intersection method of the 1965 HCM was also noted. Berry [14], Messer and Fambro [15] and others had renewed the study and development of critical movement analysis as a modeling approach for signalized intersections.

While there was general agreement that "delay" should be the measure of effectiveness for signalized intersection level of service, there were several definitions of delay that were in use, including:

- Stopped-Time Delay: the time spent by vehicles fully stopped while waiting to proceed through a signal, expressed either as an aggregate, or an average per vehicle.
- Approach Delay: Stopped time delay plus the time spent by vehicles decelerating to a stop and re-accelerating to ambient speed, expressed either as an aggregate, or an average per vehicle.
- Time-in-Queue Delay: Total time spent by vehicles in a queue waiting to proceed through a signal (both moving and stopped), expressed either as an aggregate, or an average per vehicle.
- Travel Time Delay: The difference between drivers' expected travel time through an intersection and the actual time, expressed either as an aggregate, or an average per vehicle.

Travel time delay was quickly rejected, as it was largely conceptual and very difficult to measure. The average stopped-time delay per vehicle was selected, as it was the easiest to measure (at the time), and the easiest to explain and understand. Table 3.2 shows the relationship between stopped-time delay per vehicle and level of service established in the 1985 HCM.

Table 3.2 Levels of service for signalized intersection approaches—1985 HCM

Level of service	Stopped delay per vehicle (s/veh)
A	≤5.0
B	5.1–15.0
C	15.1–25.0
D	25.1–40.0
E	40.1–60.0
F	>60

Source "Highway Capacity Manual," *Special Report 209*, Transportation Research Board, Washington D.C., 1985 [6], Table 9.1, pp. 9–4; used with permission of the National Academy of Science, Courtesy of National Academies Press

The introduction of delay as a level of service measure, however, led to some complications. Through the 1965 HCM, and indeed for uninterrupted flow facilities through all editions of the HCM, the boundary between levels of service E and F has been uniformly defined as "capacity." This was possible, because the LOS E/F boundary always occurred when the v/c ratio was 1.00—even for signalized intersection approaches through the load factor. Because delay varies widely with a number of variables, this was no longer true.

Thus, it was possible to label an intersection approach as LOS F even if the v/c ratio was below—even considerably below—1.00. The reverse was also true: for short analysis periods, a v/c ratio marginally higher than 1.00 could produce delays that were less than 60 s/veh. Technically, a "failure" could occur (i.e. v/c > 1.00) with a level of service of E or better.

3.2.3 The 1994 Update Through the 2016 HCM

In terms of interrupted flow, the 1994 update to the HCM made a change in the type of delay used to define the level of service of a signalized intersection approach. The change affected arterials and streets, as intersection delay continued to be a component of overall travel time and overall travel speed.

Recognizing that stopped time delay was only one component of the delay experienced by users when traversing a signalized intersection, the 1994 update introduced a new term: control delay. Control delay was intended to define the total delay caused by the control device—in this case a traffic signal. None of the existing delay measures encompassed all aspects of delay caused by the signal. In simple terms, control delay was time-in-queue delay plus the additional delay due to deceleration and acceleration. It, therefore, included the total time that the vehicle spent in the queue (including time spent inching forward in the queue) as well as the deceleration time loss when joining the back of the queue, and the acceleration time loss when departing it. This new definition was applied in the 1997 update to the HCM and the 2010

Table 3.3 Levels of service for signalized intersection approaches—1994 through 2016 HCMs

Level of service	Control delay (s/veh)
A	≤10
B	>10–20
C	>20–35
D	>35–55
E	>55–80
F	>80

Source "Highway Capacity Manual," Transportation Research Board, Washington D.C., 2010 [8], Exhibit 16–2, pp. 16–2; used with permission of the National Academy of Science, Courtesy of National Academies Press

HCM, and was continued in the 2016 HCM. The criteria, which were adjusted to reflect the new definition, used are shown in Table 3.3.

One additional change was introduced in the 2010 HCM: After debating the issue for many years, the HCQSC decided that *all* situations in which the v/c ratio exceeded 1.00 must be labeled as LOS F. Thus, the criteria for LOS F was changed to ">80 or v/c > 1.00." No longer could demand exceed capacity and still produce a level of service other than F. This still permits, however, cases in which the control exceeds 80 s/veh being labeled as LOS F, even if the v/c ratio is <1.00.

3.3 Closing Thoughts

Chapter 3 of Volume 1 of this book contains a more detailed examination of the overall concept of level of service for all types of facilities. This chapter focuses primarily on level of service concepts as they affect interrupted flow facilities, particularly signalized intersections.

The 2010 and 2016 HCMs have a new and increased focus on facility-and system-level performance measures that do not necessarily relate to the segment measures that have been traditionally used. Further, the relationship between facility and system measures and user perceptions is largely uncharted territory at this point. The 2010 HCM includes, for example, a performance measure for autos at signalized intersections that essentially predicts user perceptions. It was *not* adopted to define level of service however. On the other hand, similar measures for pedestrians and bicyclists at signalized intersections *were* adopted for use to define level of service for these user groups. The ability of the state-of-the-art to predict *many* performance measures for many types of segments and facilities also forces reconsideration of the concept that level of service should be defined by a single performance parameter.

When it was initially adopted, the level of service concept helped provide an overview of operational quality across a wide range of points and segments of the highway system. In many cases, levels were described only verbally, as models for the prediction of specific performance measures did not exist. Today, prediction of many performance measures is possible, perhaps even practical. Selecting one, or some subset of these measures to define level of service elevates the importance of those selected over all others. This may or may not be appropriate going forward. The interesting element of the discussion is that as agencies have more and more options at their disposal, they are looking more and more to the HCQSC to make decisions on relative importance for them, or to codify decisions made locally for the entire country.

For better or worse, level of service is part of the language used for almost 60 years to communicate operational quality, and to help set priorities. The problem is that it has changed significantly over the years, with little or no change in its use. Misunderstandings among non-transportation-professionals, and even some professionals, abound. Whatever the path going forward turns out to be, it is imperative that all interested parties understand the language, and that its use and application are relatively uniform throughout the country.

References

1. Bureau of Public Roads (1950) Highway capacity manual. Bureau of Public Roads, U.S. Government Printing Office, Washington, D.C.
2. Transportation Research Board (1965) Highway capacity manual. Transportation Research Board, Special Report 87, Washington, D.C.
3. Johnson AJ (1930) Traffic capacity. In: Proceedings of the Highway Research Board, vol 10. Transportation Research Board, Washington, D.C.
4. Normann OK (1942) Results of highway capacity studies. In: Public roads, vol 23(4). Bureau of Public Roads, U.S. Department of Commerce, Washington, D.C.
5. Greenshields BD (1935) A study of highway capacity. In: Proceedings of the Highway Research Board, vol 14. Transportation Research Board, Washington, D.C.
6. Transportation Research Board (1985) Highway capacity manual. Transportation Research Board, Special Report 209. Washington, D.C.
7. Transportation Research Board (2000) Highway Capacity Manual. Transportation Research Board, Washington, D.C.
8. Transportation Research Board (2010) Highway capacity manual. Transportation Research Board, Washington, D.C.
9. Transportation Research Board (2016) Highway capacity manual. 6th edn: A Guide for Multimodal Mobility Analysis. Transportation Research Board, Washington, D.C.
10. Greenshields B (1947) Traffic performance at intersections. Yale Bureau Technical Report #1, Yale Bureau of Highway Traffic, New Haven, CT
11. Intersection Traffic Flow (1960) Final report to the Bureau of Public Roads. C.E.I.R Inc., Arlington, VA
12. May AD Jr, Gyamfi P (1969) Extension and preliminary validation of load factor at signalized intersections. In: Traffic engineering, Institute of Transportation Engineers, Washington, D.C.
13. Sagi GS, Campbell LR (1969) Vehicle delay at signalized intersections—theory and practice. In: Traffic engineering, Institute of Transportation Engineers, Washington, D.C.
14. Berry D, Gandhi P (1973) Headway approach to intersection capacity. Highway Research Record 453, Transportation Research Board, Washington, D.C.
15. Messer C, Fambro D (1977) Critical lane analysis for intersection design. Transportation Research Record 644, Transportation Research Board, Washington, D.C.

Chapter 4
Unsignalized Intersections: Two-Way STOP Control (TWSC)

By definition, an "unsignalized" intersection is any at-grade intersection that is not controlled by a traffic signal. The term covers a wide variety of situations, including (a) uncontrolled intersections, (b) two-way STOP-controlled intersections (TWSC), c) two-way YIELD-controlled intersections, (c) all-way STOP-controlled intersections (AWSC), and roundabouts. The 1950 HCM [1] *did not even mention any of these. Some material was included in the 1965 HCM* [2], *but much of it was simply in the form of general advice, and no specific analysis methodologies were introduced. Circular 212* [3], *published in 1980, contained a detailed methodology for analysis of TWSC intersections (and two-way YIELD-controlled intersections). The 1985 HCM* [4] *included updated methodologies for TWSC intersections, and provided a methodology for analysis of AWSC intersections. These were modified and improved in the 1994 and 1997 updates, and were included in the 2000 HCM* [5]. *The 2000 HCM also introduced procedures for consideration of roundabouts, while the 2010 HCM* [6] *contains an updated methodology for these. The 2016 HCM* [7] *includes these methodologies with some revisions and updates.*

This chapter details the development of procedures for analysis of TWSC intersections. Chapter 5 will cover AWSC intersections, and Chap. 6 will cover roundabouts.

It is indeed difficult to compare methodologies for unsignalized intersections across the years and editions of the HCM. As previously noted, until Circular 212 in 1980, there were no methodologies that actually took into account specific characteristics of an "unsignalized" intersection. Since then, methodologies have developed, but did not always share common modeling approaches, although most have relied on gap acceptance modeling in one form or another. Even in the 2016 HCM, not *all* types and configurations of "unsignalized" intersections are treated.

The early lack of attention to unsignalized intersections was born from the logic that any such intersections experiencing breakdowns or severe delays would be converted to a signalized intersection. Most such intersections serve low-volume situations that would not be expected to experience capacity deficiencies:

E. S. Prassas and R. P. Roess, *The Highway Capacity Manual: A Conceptual and Research History Volume 2*, Springer Tracts on Transportation and Traffic 12,
https://doi.org/10.1007/978-3-030-34480-1_4

> In a sense, then, capacities and the larger service volumes of unsignalized intersections may be considered of academic interest only: in practice, by the time such levels are reached at important intersections, signals ordinarily will be installed. [Ref. 2, p. 155]

In this chapter, the development of analysis methodologies for TWSC intersections is explored.

4.1 In the Beginning—the 1965 HCM

As noted, the 1950 HCM does not mention any type of unsignalized intersection. Thus, the history of methodologies to address these begins with the 1965 HCM.

The 1965 HCM recognized that signalization in many suburban and rural situations was relatively rare, and that some guidance needed to be given concerning the operation of unsignalized intersections. The manual did, however, recognize a key operational element that was not taken into account until 1985 (although it did not address it in any meaningful way). While unsignalized intersections in rural areas were often far enough away from nearby signals to operate independently of them (e.g. random arrivals):

> …. signal coordination causes traffic to pass an unsignalized location in a regular pattern. In one case, this pattern may produce simultaneously in both directions regular gaps during which the cross street can clear, but in another, may create a condition in which simultaneous gaps seldom, if ever, occur. [Ref. 2, p. 155]

In essence, where unsignalized intersections are located on arterials with a coordinated signal system, the passing of platoons in two directions affects the operation of the unsignalized location.

4.1.1 Uncontrolled Intersections

The 1965 HCM is the only edition that provided any guidance on the analysis of completely uncontrolled intersections. In such instances, the intersection operates under "basic rules of the road," in which the driver on the left must give way to the driver on the right if they are close enough to each other to present an impending hazard.

The manual suggests treating such intersections using the signalized intersection methodology (of the 1965 HCM), and employing a split of green time equal to:

$$\frac{G_1}{G_2} = \left(\frac{V_1}{V_2}\right) * \left(\frac{W_2}{W_1}\right) \tag{4.1}$$

where:

G_1 green time, street 1 (s),

G_2 green time, street 2 (s),
V_1 volume, street 1 (veh/h),
V_2 volume, street 2 (veh/h),
W_1 width, street 1 (ft), and
W_2 width, street 2 (ft).

This equation, however, ignores the issues of cycle length and lost times, which are critical to the signalized intersection methodology of the 1965 HCM. Implementation of the signalized intersection methodology requires that a G/C ratio for each signal cycle be known. The cycle length for a signal may be computed as:

$$C = \sum_{i=1-N} (G_i + y_i + ar_i) \tag{4.2}$$

where:

C cycle length (s),
G_i green time for phase i (s),
y_i yellow time for phase i (s),
ar_i all red time for phase i (s), and
N number of phases (for the application to uncontrolled intersections, this value is always assumed to be 2).

The ratio of G_1/G_2 in Eq. 4.1 describes the division of green time. However, the cycle includes green time, yellow time, and all-red time. The division of the green time can only be implemented if the yellow and all-red times are known. Then, the amount of green time to be divided becomes:

$$G_{TOT} = C - \sum_{i=1-N} (y_i + ar_i) \tag{4.3}$$

where:

G_{TOT} total green time in the cycle (s), all other variables as previously defined.

Of course, selecting values of y_i and ar_i for an uncontrolled intersection is also a difficulty. Standard means for estimating these from location-specific criteria could be used, but it is unlikely that such information would be available for an uncontrolled location. A conservative approach would be to use a default of $y_i = 2.0$ s and $ar_i = 2.0$ s. Actually, for this approach, the split between the two is irrelevant. It merely assumes that each phase adds 4.0 s of yellow and all-red time to the imagined signal.

Consider an uncontrolled intersection with characteristics as shown in Table 4.1.

The sum of the yellow and all-red times is 8.0 s. Thus, the total green time to be split is C – 8.0, where the cycle length, C, is obviously a variable. Using Eq. 4.1, the split ratio is found as:

$$\frac{G_1}{G_2} = \left(\frac{850}{600}\right) * \left(\frac{36}{50}\right) = 1.02$$

Table 4.1 Uncontrolled Intersection—an example—1965 HCM

Characteristic	Street 1	Street 2
Yellow time (s)	2.0	2.0
All-red time (s)	2.0	2.0
Volume (veh/h)	850	600
Street width (ft)	50	36

Table 4.2 Green splits for trial cycle lengths—sample problem

C	G_{TOT} (s)	G_1 (s)	G_2 (s)	G_1/C	G_2/C
30	22	11.1	10.9	0.36	0.36
60	52	26.3	25.7	0.44	0.43
90	82	41.4	40.6	0.46	0.45
120	112	56.6	55.4	0.47	0.46

The cycle length, is obviously variable, meaning that the values of G_1 and G_2 are also variable. Table 4.2 shows the values of G_1 and G_2 for cycle lengths of 30, 60, 90, and 120 s. Note that the arithmetic involved in obtaining these values involves several steps, which are not shown here.

With the two relevant G/C ratios determined, the problem can now be treated as a signalized intersection, using the signalized intersection methodology of the 1965 HCM (see Chap. 7 of this book). Of course, a range of results would be obtained, depending upon the cycle length (C). This is, at best, a very approximate analysis approach, but it remains the only suggested methodology for analysis of a completely uncontrolled intersection ever given in the HCM.

4.1.2 YIELD-Controlled Intersections

The 1965 HCM does not provide a separate methodology for considering YIELD-controlled intersections. Rather, it states:

> Where, at a YIELD location, volumes have become sufficiently heavy that capacity must be considered, practically every vehicle on the secondary YIELD legs will be making a full stop, just as under STOP control. [Ref. 2, p. 156]

In other words, at YIELD locations, the user is told to analyze them as if they were two-way STOP-controlled locations.

Obviously, this ignores the key difference between YIELD signs and STOP signs. YIELD signs do not require a full stop. In most states, however, they do require that drivers slow to 10 mi/h at the sign.

4.1.3 *Two-Way STOP-Controlled Intersections (TWSC)*

The 1965 HCM directs YIELD-controlled intersections to be treated as if they were two-way STOP-controlled intersections. Unfortunately, beyond a general discussion, the 1965 HCM does not really provide a methodology for TWSC intersections either.

The manual once again suggests that the intersection be treated as a signalized intersection, with the balance of green times as indicated in Eq. 4.1. Essentially, the methodology for uncontrolled intersections is suggested, with a caveat that it would not be valid in cases where substantial delay is experienced by drivers on STOP-controlled approaches. No guidance is given for what to do in this case.

4.2 The Road to New Approaches Based on Gap Acceptance

The 1965 HCM put consideration of unsignalized intersections on the map. The treatment was general and introductory, but provided an important platform for further work and development that would be reflected in subsequent editions of the manual.

A specific methodology for analysis of TWSC intersections first appeared in 1980 as part of the "Interim Materials on Highway Capacity," released in that year. The methodology relies on gap acceptance modeling, and was adapted from a European publication [8], which was itself a translation of an earlier German publication [9]. The methodology was modified and simplified for U.S. use by Jim Kell as part of the initial effort to produce a 1985 HCM.

While the generic name for intersections at which one of the intersecting streets is controlled by a STOP sign is "two-way STOP-controlled" (TWSC), this is not the only configuration covered by this classification. Also included are T-intersections where the stem of the T is controlled, and intersections of two one-way streets where one approach is controlled. Further, some of the analysis methodologies also cover such configurations when the control is provided by a YIELD sign or signs.

The inclusion of YIELD-controlled intersections in this category has changed over the years. The first appearance of a methodology in 1980 included YIELD-controlled intersections. However, these were subsequently deleted, and did not return to the HCM until 2010.

The methodology of Circular 212 had been updated with additional U.S. and European research along with the efforts of the Unsignalized Intersection Subcommittee of the Highway Capacity and Quality of Service Committee (HCQSC) of the Transportation Research Board. After its 1980 appearance, the methodology was updated for the 1985 HCM, the 1994 update (of the 85 HCM), the 1997 update (of the 85 HCM), the 2000 HCM, 2010 HCM, and the 2016 HCM. Because the 1997 update and the 2000 HCM were virtually identical (except for editorial changes), they are not treated separately herein.

Because the fundamental structure of the methodology has remained constant throughout the development of the methodology, this chapter presents the structure of the methodology as a series of modules, then discusses the development of each module through the 2016 HCM.

4.3 Basic Structure of the Methodology

The TWSC methodology is based on the fact that minor street movements must seek gaps in the major street traffic stream through which to execute their desired maneuvers. Some minor street movements must also seek gaps through other minor street movements as well. As several potential minor street movements may seek to use the same gap in major street traffic, a clear priority scheme for the use of these gaps must be established. Minor movements use available gaps in the major traffic stream strictly in the following priority order:

1. Right turns onto the major street.
2. Left turns from the major street.
3. Traffic crossing the major street.
4. Left turns into the major street.

Use of the methodology requires that data be collected to identify all demand flows at the intersection, and the major street approach speed. The computational procedure has the following key steps:

1. *Conversion of Demands into Flow Rates in Passenger-Car Equivalents*
   The methodology differs from other capacity analysis approaches in that some computations are made with demand in units of veh/h, while others are made in terms of equivalent pc/h. Early versions of the method allow for analysis using either full-hour volumes or flow rates for the worst 15 min of the analysis hour, while later versions require that flow rates be used.

2. *Determining Conflicting Volumes (Flow Rates)*
   There are several movements at a TWSC intersection that seek gaps through one or more conflicting movements. These generally include:

   - Left turns from the major street—seeking gaps through the opposing major street through and right-turn movements.
   - Right turns from the minor street—seeking gaps in the right-most lane of the major street.
   - Through movements from the minor street—seeking gaps through all major street movements.
   - Left turns from the minor street—seeking gaps through all major street movements and the opposing minor street through movement.

In each case, the conflicting volume (or flow rate) is different. Thus, a significant computational step involves defining the conflicting volume that each movement seeking gaps faces.

3. *Determining Critical Gaps and Follow-Up Times*
The fundamental modeling approach used for TWSC intersections is gap acceptance. Thus, a critical determination is the size of the critical gap—the average gap that a vehicle seeking gaps will accept (or choose to pass through). Initially called a "critical gap," by 2010, this was being referred to as a "critical headway" to avoid confusion with other uses of the term. Specifically, a "gap" is traditionally taken to mean the time difference between the rear of a lead vehicle and the front of a following vehicle, while a headway deals with the time difference between the front of both vehicles. While the initial versions of the methodology rely completely on the critical gap, later versions also use the follow-up time, the average time before a subsequent queued vehicle can use the same gap, when the gap is large enough to accommodate multiple vehicles.

4. *Determining Potential Capacities*
The approach to estimating capacities for each lane serving vehicles that seek gaps involves three separate steps. The first involves determining potential capacities, which assume that each movement has an exclusive lane and access to all available gaps in the appropriate conflicting volume. This is the key computation based upon gap acceptance modeling.

5. *Determining Movement Capacities: Impedance Issues*
Movement capacities are different from potential capacities. Movement capacities recognize that some gaps in the appropriate conflicting volume may be used by competing volumes of higher priority. For example, a left turn from the minor street seeks a gap in the major street flow, but an opposing minor street through vehicle may be seeking to use the same gap. Since the minor street through vehicle has a higher priority, it uses the first gap that appears, which is now unavailable to the minor street left-turning vehicle. This issue is generally called impedance, and potential capacities must be reduced to account for this effect. In more recent versions of the methodology, the impact of platooned flow on the major street is also treated as an impedance effect.

6. *Determining Shared-Lane Capacities*
In almost all cases, vehicles on a STOP- or YIELD-controlled approach share lanes. Rarely are separate lanes provided for right, through, and left movements on such an approach. Movement capacities, which still assume an exclusive lane for each movement, must be reduced to account for the shared-lane effect.

7. *Delay and Level of Service*
The final result of an analysis of a TWSC intersection is the assignment of a LOS to each controlled lane. In 1980, this was based simply upon unused capacity available. In subsequent versions, a delay estimation model was added,

and LOS was based on control delay. Once delay estimations were available, individual lane values could be averaged (weighted by demand flow rates) to produce estimates for each controlled approach, and for the intersection as a whole. The latter, while possible, has generally been discouraged, as most major street vehicles (which are still the dominant flow) will have no delay, and will create an intersection average that may look very good, despite the controlled approaches operating very poorly.

Because of the very modular structure of the TWSC model, it is more instructive to describe the development of the methodology over the years *by module*, rather than to treat the entire procedure as a whole by version.

4.4 Movement Designations

Because the methodology is highly structured around a very strict set of priorities in the use of available gaps, it is important to provide a consistent notation for referring to the various flows at a TWSC intersection. This notation has survived all versions of the methodology, and is illustrated in Fig. 4.1.

The analysis intersection is positioned to reflect the controlled approach(es) on the vertical, regardless of the cardinal directions of the intersection. Thus, for a 4-leg intersection, movements 7, 8, 9, 10, 11, and 12 are controlled movements, while movements 1, 1U, 2, 3, 4, 4U, 5, and 6 are on the major street. This allows for a consistent notation throughout the methodology.

Note that U-turns from the major street are separately designated. This only occurs in the 2010 and 2016 HCMs, where U-turns are handled as a separate movement. In previous versions, movement 1U is included in movement 1, and movement 4U is included in movement 4.

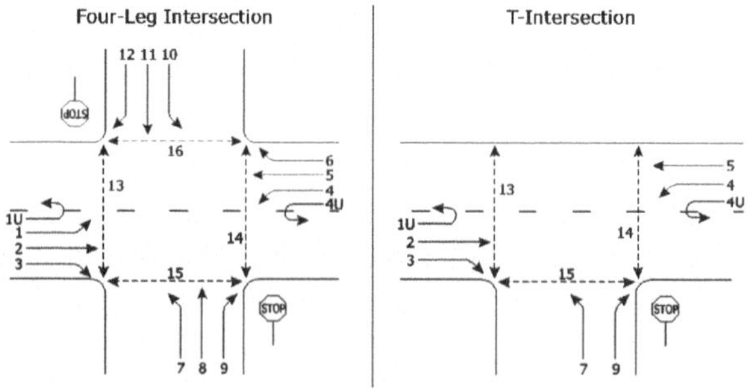

Fig. 4.1 Revised movement numbers for TWSC intersections. *Source* "Highway Capacity Manual, 6th Edition: A Guide for Multimodal Mobility Analysis", Transportation Research Board, Washington D.C., 2016 [7], Exhibit 20-1, pp. 20–5, used with permission of the National Academy of Science, Courtesy of National Academies Press

Also, pedestrian movements are ignored in the initial version of the methodology (1980), but are accounted for in later versions, beginning in 1997. In 2016, however, their inclusion is optional, and may be ignored where the analyst believes their impact is negligible.

4.5 Conversion to Passenger Car Equivalents and Flow Rates

Through the 2000 HCM, every version of the TWSC methodology required that movement demands be converted to passenger car equivalents. In the 2010 and 2016 HCMs, no such conversions are made, and all computations are done using mixed veh/h.

While the 1980 version of the methodology allowed for the use of full-hour demand volumes, use of flow rates was permitted. Later versions required that flow rates be used.

Demand volumes can be converted to flow rates using the familiar equation shown:

$$v_i = \frac{V_i}{PHF} \tag{4.4}$$

where:

v_i demand flow rate for movement i, veh/h,
V_i hourly demand volume for movement i, veh/h, and
PHF peak hour factor.

Conversion to passenger car equivalents, through the 2000 HCM, is done using the equivalents shown in Table 4.3.

These equivalents, which first appeared in Circular 212 (1980), are based primarily upon European research, and were used through the 2000 HCM. Unlike the usual

Table 4.3 Passenger car equivalents for unsignalized intersections: 1980–2000

Type of vehicle	Approach grade (%)				
	−4%	−2%	0%	+2%	+4%
Motorcycles	0.3	0.4	0.5	0.6	0.7
Pass. Cars	0.8	0.9	1.0	1.2	1.4
Trucks/RVs	1.0	1.2	1.5	2.0	3.0
Tractor-trailers	1.2	1.5	2.0	3.0	6.0
Motor Veh.*	0.9	1.0	1.1	1.4	1.7

*Approximate values used when vehicle distribution is not known

Source "Interim Materials on Highway Capacity—Unsignalized Intersections," *Transportation Research Circular 212*, Transportation Research Board, Table 1, p. 37, used with permission of the National Academy of Science, Courtesy of National Academies Press

U.S. practice, motorcycles are treated as a separate group, and have pce values of less than 1.00. Moreover, there is an adjustment for passenger cars when a grade exists, and single-unit trucks and RVs are considered to be in the same category. Tractor-trailer combinations are treated separately. If the exact vehicle composition is not known, the average equivalents shown for "Motor Vehicles" were used as a rough estimate.

Note that the only vehicles that were converted to pce's were those that *used* gaps in the major traffic stream. The major traffic stream was never converted, as the gap distribution in this traffic stream, through which all other streams must pass, is more dependent upon vehicles than on equivalent passenger cars.

These equivalents were never revised in subsequent versions of the methodology through the 2000 HCM. In 2010 and 2016 HCMs, the table was eliminated, as all computations were converted to a veh/h basis.

4.6 Defining Conflicting Traffic Streams

Because the basic methodology is one based gap acceptance, it is critical that the conflicting traffic stream for each movement using gaps be defined. This varies for each subject movement. Right turns from the minor street, for example, merge into through traffic in the right-most lane of the major street, while left turns from the minor street conflict with *all* major street movements, and the opposing minor street through and right turn movements.

The computation of conflicting volumes or flow rates has changed over the years. In 1980, pedestrian flows were not considered. These were added in 1985 and subsequent versions. In 2010, U-turn movements were added as a separate movement to be considered. In 2016, the consideration of pedestrian flows was made optional.

4.6.1 Circular 212—1980

The computation of conflicting volumes in Circular 212 is illustrated in Table 4.4. Note that conflicting volumes are computed in *vehicles per hour*, with no conversions to pc/h included.

There are some interesting concepts included in these equations. The right turn movement approaching from the left on the major roadway, for example, is discounted by ½. One might argue that it should be left out altogether, as the movement does not truly *conflict* with any minor street movement. The argument, however, is a practical one: many approaching right-turning vehicles will not engage their directional signals, and a waiting minor street driver might perceive some of these as potential conflicts. If a separate right-turn lane is provided for these vehicles, the term can be eliminated entirely from the equation.

Table 4.4 Conflicting volumes defined for unsignalized intersections: Circular 212. *Source* Modified from "Interim Materials on Highway Capacity—Unsignalized Intersections," *Transportation Research Circular 212*, Transportation Research Board, Washington D.C., 1980, Fig. 1, p. 37, used with permission of the National Academy of Science, Courtesy of National Academies Press

Subject movement	Conflicting volume (veh/h)	Illustration
Right turns onto the major street	$v_{c9/12} = v_{2/5} + \frac{1}{2} v_{3/6}$	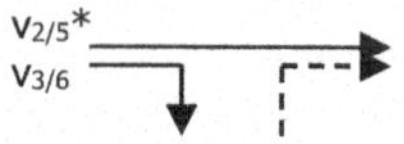
Left turns from the major street	$v_{c1/4} = v_{5/2} + v_{6/3}$	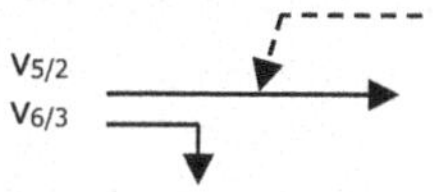
Crossing the major street	$v_{c8/11} = \frac{1}{2} v_{3/6} + v_{11/8} + v_{10/7} + v_{6/3} + v_{2/5} + v_{5/2} + v_{4/1}$	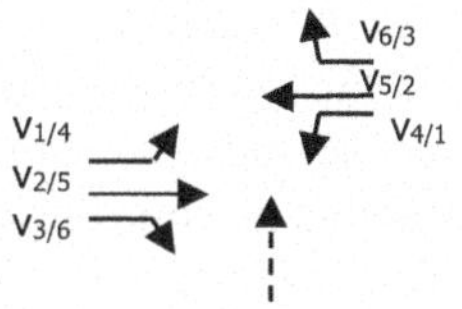
Left turns onto the major street	$v_{c7/10} = \frac{1}{2} v_{3/6} + v_{2/5} + v_{1/4} + v_{6/3} + v_{5/2} + + v_{4/1} + v_{11/8} + v_{12/9}$	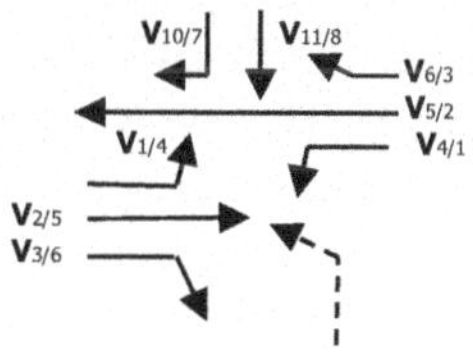

*Include only the volume in the right-hand lane.

Note Subscripts refer to the movements defined in Fig. 4.1. Mvt 1U is included in mvt 1; mvt 4U is included in mvt 4.

Most of the other conflicting volume components have a direct crossing or merging conflict with the subject movement.

The *conflicting volume* defines the total flow of vehicles through which each subject vehicle must find appropriate gaps to execute the desired movement. Obviously, as the total conflicting volume increases, there will be fewer gaps for STOP- or YIELD-controlled vehicles pass through. Further, as the complexity of the conflicting traffic streams increases, drivers will tend to seek larger gaps through which they can safely pass.

4.6.2 The 1985 Highway Capacity Manual and the 1994 Update

There were no changes in the computation of conflicting volumes introduced in the 1985 HCM, nor in the 1994 update to the 85HCM.

4.6.3 The 1997 Update to the 85 HCM and the 2000 HCM

The 1997 update is the first version of the TWSC methodology to consider the impact of pedestrian flows. Pedestrian flows are, for the first time, included in the computation of conflicting volumes, as shown in Fig. 4.2. The footnotes to this illustration are important, and must be carefully implemented based upon prevailing conditions.

For the minor street through and left-turn movements, conflicting traffic is shown in two stages. This reflects another change introduced into the 1997/2000 version of the methodology—the possibility of two-stage minor street movements. This occurs in cases where the major street has a median wide enough to store one or more vehicles. In such cases, minor street vehicles cross the first major street roadway, then wait in the median storage area to seek gaps in the second major street movement. When this occurs, minor street movements, specifically through and left-turn movements, actually perform two separate maneuvers, each of which involves a different conflicting volume or flow rate.

Thus, conflicting volumes are shown in two stages for these movements. Where a median with storage exists, the entire analysis is treated as two stages; where no such median exists, the conflicting volume is the sum of the values shown for Stages I and II. The two-stage movement case is discussed in greater detail when capacity computations are presented.

Because the conflicting volumes now include both vehicles and pedestrians, the units of conflicting volume (or flow rate) are now stated as conflicts/h.

4.6.4 The 2010 and 2016 Highway Capacity Manuals

The 2010 HCM introduced a number of changes to the algorithms to compute conflicting volumes (or flow rates). U-turns from the major street are now treated as a separate movement. Further, the 2010 manual introduced some subtle modifications based upon whether the major street has 2 lanes, 4 lanes, or 6 lanes. The revised algorithms for conflicting volume are shown in Table 4.5. The 2016 HCM made no changes to these computations.

Once again, two-stage values are shown for minor street through and left-turn movements. Where a one-stage process (no major street median with storage) exists, the total conflicting volume is the sum of the values shown for Stages I and II.

4.7 Critical Gaps (Headways) and Follow-up Times

For any modeling approach using gap acceptance, the gaps selected by drivers for use, and the manner in which they use them, is of paramount importance.

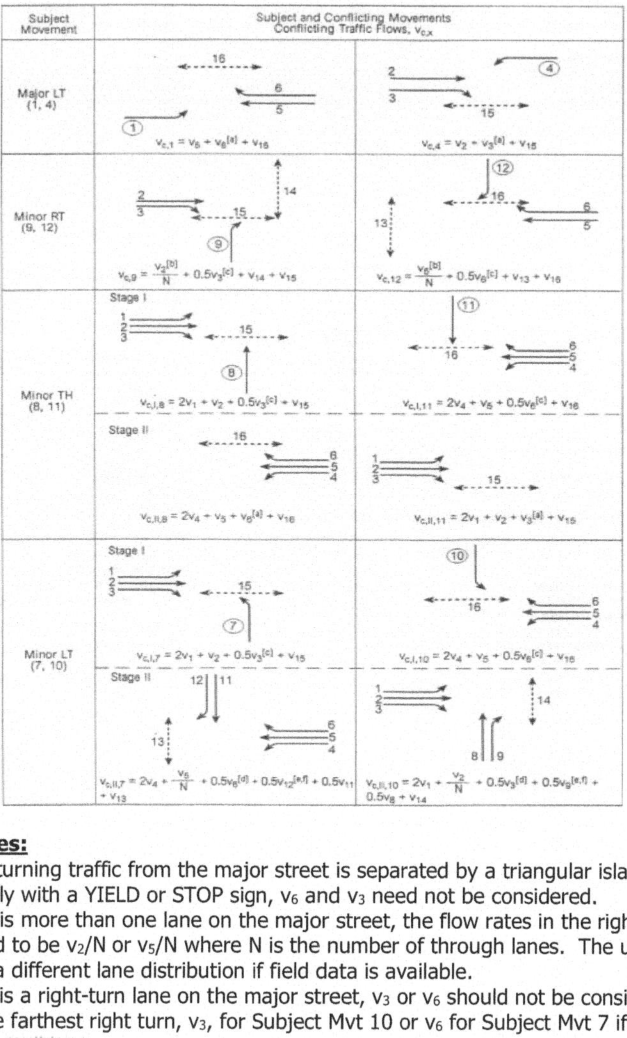

Footnotes:

(a) If right-turning traffic from the major street is separated by a triangular island and has to comply with a YIELD or STOP sign, v_6 and v_3 need not be considered.

(b) If there is more than one lane on the major street, the flow rates in the right lane are assumed to be v_2/N or v_5/N where N is the number of through lanes. The user can specify a different lane distribution if field data is available.

(c) If there is a right-turn lane on the major street, v_3 or v_6 should not be considered.

(d) Omit the farthest right turn, v_3, for Subject Mvt 10 or v_6 for Subject Mvt 7 if the major street is multilane.

(e) If right-turning traffic from the minor street is separated by a triangular island and has to comply with a YIELD or STOP sign, v_9 and v_{12} need not be considered.

(f) Omit v_9 and v_{12} for multilane sites, or use ½ their values if the minor approach is flared.

Fig. 4.2 Computing Conflicting Volume for TWSC Intersections 1997 Update and 2000 HCM. *Source* "Highway Capacity Manual," *Special Report 209*, Transportation Research Board, Washington D.C., 1997, Fig. 10-2, pp. 10–8 and 10–9, used with permission of the National Academy of Science, Courtesy of National Academies Press

Table 4.5 Conflicting flow rates for TWSC Intersections: 2010 and 2016 HCMs

Rank 2—LTs from the major street

$$v_{c1} = v_5 + v_6 + v_{16}$$

$$v_{c4} = v_2 + v_3 + v_{15}$$

Note Where major street RT is separated by a triangular angle, and controlled by a YIELD or STOP sign, v_6 and v_3 may be taken as 0

Rank 2—RTs from the minor street

$$v_{c9} = v_2 + 0.5v_3 + v_{14} + v_{15} \quad (2 - lane\, major\, streets)$$

$$v_{c9} = 0.5v_2 + 0.5v_3 + v_{14} + v_{15} \quad (4, 6 - lane\, major\, streets)$$

$$v_{c12} = v_5 + 0.5v_6 + v_{13} + v_{16} \quad (2 - lane\, major\, streets)$$

$$v_{c12} = 0.5v_5 + 0.5v_6 + v_{13} + v_{16} \quad (4, 6 - lane\, major\, streets)$$

Rank 2—U-Turns from the major street

$$v_{c1U} = v_5 + v_6 \quad (4 - lane\, major\, streets)$$

$$v_{c1U} = 0.73v_5 + 0.73v_6 \quad (6 - lane\, major\, streets)$$

$$v_{c4U} = v_2 + v_3 \quad (4 - lane\, major\, streets)$$

$$v_{c4U} = 0.73v_2 + 0.73v_3 \quad (6 - lane\, major\, streets)$$

Notes No data available for U-turns on 2-lane major streets; if the major street has an exclusive lane for RTs, v_3 and/or v_6 should be assumed as 0

Rank 3—TH movements from the minor street

$$v_{cI8} = 2(v_1 + v_{1U}) + v_2 + 0.5v_3 + v_{15}$$

$$v_{cII8} = 2(v_4 + v_{4u}) + v_5 + v_6 + v_{16}$$

$$v_{cI11} = 2(v_4 + v_{4u}) + v_5 + 0.5v_6 + v_{16}$$

$$v_{cII11} = 2(v_1 + v_{1U}) + v_2 + v_3 + v_{15}$$

Note Where no storage median on the major street exists, there is only one stage. Conflicting flow rates are the sum of the Stage I and Stage II values shown here

Rank 4—LTs from the minor street

2-Lane major streets

$$v_{cI7} = 2v_1 + v_2 + 0.5v_3 + v_{15}$$

$$v_{cII7} = 2v_4 + v_5 + 0.5v_6 + 0.5v_{12} + 0.5v_{11} + v_{13}$$

$$v_{cI10} = 2v_4 + v_5 + 0.5v_6 + v_{16}$$

$$v_{cII10} = 2_{v1} + v_2 + 0.5v_3 + 0.5v_9 + 0.5v_8 + v_{14}$$

4, 6-Lane major streets

$$v_{cI7} = 2(v_1 + v_{1U}) + v_2 + 0.5v_3 + v_{15}$$

$$v_{cII7} = 2(v_4 + v_{4U}) + 0.5v_5 + 0.5v_{11} + v_{13}$$

$$v_{cI10} = 2(v_4 + v_{4U}) + v_5 + 0.5v_6 + v_{16}$$

$$v_{cII10} = 2_1(v_1 + v_{1U}) + 0.5v_2 + 0.5v_3 + 0.5v_8 + v_{14}$$

Note Where no storage median exists on major street, only one stage is present. Total conflicting flow rate is the sum of those shown here for Stage I and Stage II

The critical gap (t_c), called the critical headway in the 2010 and 2016 HCMs, is the average minimum gap that a driver accepts (chooses to use) in the conflicting traffic stream. Obviously, driver behavior varies widely, and a gap that one driver accepts might be rejected by another. Thus, the calibration of the critical gap is based upon observed averages of gaps that are rejected and accepted by drivers. In terms of modeling, it is assumed that drivers will *always* reject a gap smaller than the defined critical gap, and *always* accept one that is equal to or larger than the critical gap.

The follow-up time (t_f) is equally important. There are gaps in conflicting flows that are large enough for more than one vehicle to pass through. The follow-up time assumes that such a gap is available, and that there are vehicles queued on the controlled approach that can use it. It is defined as the average minimum time for a second vehicle to use a gap, given no impedance from higher priority movements. The follow-up time is analogous to the saturation headway for a signalized intersection. Given that there is a continuous gap in conflicting traffic, and a queue of vehicles waiting to use it, it defines the amount of time each vehicle will consume passing through the conflicting flow.

Values of critical gap and follow-up time have been continuously adjusted over the development of TWSC methodologies.

4.7.1 Circular 212

The first methodology used only critical gaps, and did not refer to or use follow-up times in computational procedures. Circular 212 defined the critical gap as "the minimum gap (in seconds) needed by drivers to negotiate the conflicting flows," which is consistent with the general description given previously. Other approaches have defined the critical gap as the median or average gap size accepted by drivers of a given subject movement. Table 4.6 relates critical gap size to the subject movement, the type of control present, and the width of the major street (in lanes). YIELD-controlled approaches were specifically addressed. In some later versions, the ability to analyze YIELD-controlled approaches was removed.

The size of the critical gap increases with the complexity of the subject movement and its conflicting flows, the prevailing speeds on the major street, and the width of the major street. Note that no gaps are given for 6-lane major streets in Circular 212.

While some modifications were done, these critical gaps are primarily based upon the European values. Subsequent revisions would be based upon more extensive U.S. data.

Table 4.6 Critical gaps for unsignalized intersections: circular 212

Vehicle maneuver and type of control	Prevailing speed on major road (mi/h)			
	30 mi/h		55 mi/h	
	Number of lanes on major roadway			
	2	4	2	4
RT from minor road:				
STOP control	5.0 s	5.0 s	6.0 s	6.0 s
YIELD control	6.0 s	6.0 s	7.0 s	7.0 s
LT from major road:				
No control	5.0 s	5.5 s	5.5 s	6.0 s
Crossing major road:				
STOP control	6.0 s	6.5 s	7.0 s	8.0 s
YIELD control	7.0 s	7.5 s	8.0 s	9.0 s
LT from minor road:				
STOP control	6.5 s	7.0 s	8.0 s	9.0 s
YIELD control	7.5 s	8.0 s	9.0 s	10.0 s

Source "Interim Materials on Highway Capacity—Unsignalized Intersections," *Transportation Research Circular 212*, Transportation Research Board, Washington D.C., 1980 [3], Table 2, p. 38, used with permission of the National Academy of Science, Courtesy National Academies Press

4.7.2 The 1985 Highway Capacity Manual

Early experience with the Circular 212 methodology suggested that capacities were being significantly underestimated. To remedy this, several alterations were introduced, one of which was to reduce size of critical gaps. The approach was also modified to give a "base critical gap" that could be adjusted for some prevailing conditions. Updated critical gaps are shown in Table 4.7.

The practical result of these changes limited the range of critical gap sizes to between 4.0 and 8.5 s, generally lower than those suggested in Circular 212.

4.7.3 The 1994 Update to the 85 HCM

The 1994 update was the first to use both critical gap and follow-up time in the TWSC methodology. Changes to critical gaps and follow-up times were based upon a number of published studies [10–17], as were other changes included in the 1994 update.

Revised values for critical gaps, and new values for follow-up time are shown in Table 4.8.

Table 4.8 represents a retreat from the information provided in the 1985 HCM. YIELD-controlled intersections can no longer be analyzed, and there are no critical

Table 4.7 Critical gap criteria for unsignalized intersections 1985 HCM

Base critical gap for passenger cars (s)

Vehicle maneuver and type of control	Average running speed on major road			
	30 mi/h		55 mi/h	
	Number of lanes on major road			
	2	4	2	4
RT from major road:				
STOP	5.5	5.5	6.5	6.5
YIELD	5.0	5.0	5.5	5.5
LT from major road	5.0	5.5	5.5	6.0
Cross major road:				
STOP	6.0	6.5	7.5	8.0
YIELD	5.5	6.0	6.5	7.0
LT from minor road:				
STOP	6.5	7.0	8.0	8.5
YIELD	6.0	6.5	7.0	7.5

Adjustments and modifications to critical gaps (s)

Condition	Adjustment (s)
RT from minor street: curb radius >50 ft or Turn angle <60°	−0.5
RT from minor street: acceleration lane provided	−1.0
All movements: population >250,000	−0.5
Restricted sight distance*	Up to +1.0

Notes Maximum total decrease in critical gap = 1.0 s; Maximum critical gap = 8.5 s; For values of average running speed between 30 and 55 mi/h, interpolate

*This adjustment is made for the specific movement impacted by restricted sight distance

Source "Highway Capacity Manual," *Special Report 209*, Transportation Research Board, Washington D.C., 1985 [4], Table 10-2, pp. 10–7, used with permission of the National Academy of Science, Courtesy of National Academies Press

gaps or follow-up times for major street approach speeds in excess of 30 mi/h. It is interesting to note that in the 1985 HCM, critical gaps for 55-mi/h approach speeds were shown to be between 1.0 and 1.5 s higher than those for 30-mi/h approach speeds, and critical gaps for YIELD-controlled intersections were uniformly 0.5 s less than those for STOP-controlled intersections.

This was done because the values shown in 1985 for high approach speeds and for YIELD-controlled approaches were always rough estimates at best, without a great deal of data supporting them. By 1994, it was no longer felt that professional judgment was sufficient to provide specific guidance in these areas.

However, there is nothing in the methodology that prevents the analysis of TWSC intersections with high major street speeds, or even YIELD-controlled intersections where local data or defaults for critical gaps and follow-up times are available.

Table 4.8 Critical gap and follow-up time for TWSC intersections 1994 update

Vehicle movement	Critical gap (t_c) (s)		Follow-up time (t_f), (s)
	2-Lane major road	4-Lane major road	
Left turn, major St.	5.0	5.5	2.1
Right turn, minor St.	5.5	5.5	2.6
Through, minor St.	6.0	6.5	3.3
Left turn, minor St.	6.5	7.0	3.4

Note The critical gap and follow-up time values presented in this table reflect data obtained on roadways where the average approach speed of the major street through vehicles approximated 30 mi/h. In cases where no better data are available, these same values may be used to approximate t_g and t_f for roadways with approach speeds other than 30 mi/h

Source "Highway Capacity Manual," *Special Report 209,* Transportation Research Board, Washington D.C., 1994 [19], Table 10-2, pp. 10–9, used with permission of the National Academy of Science, Courtesy of National Academies Press

4.7.4 The 1997 Update to the 85 HCM and the 2000 HCM

There was very little difference between the TWSC intersection methodology of the 1997 update and the 2000 HCM. Essentially, the 1997 version was a preview of the methodology developed for the 2000 HCM. Except for minor editorial changes, the two were virtually identical. The changes from 1994 were, however, significant, and were based upon a targeted study intended to produce an improved model for 2000 [18].

Among other things, base critical gap and follow-up times were again revised, yielding the values shown in Table 4.9. As in the 1994 update, there are no values presented for YIELD-controlled approaches. Major street speeds are not mentioned, and the values of Table 4.9 are intended for use regardless of the major street prevailing speed.

A broader range of adjustments to base values was introduced in 1997. Included were adjustments to base values for heavy vehicles, grades, two-stage situations and certain aspects of intersection geometry. Adjusted values for critical gap are

Table 4.9 Base critical gap and follow-up times: 1997 update and 2000 HCM

Vehicle movement	Base critical gap (s)		Base follow-up time (s)
	2-Lane major St	4-Lane major St	
LT from major street	4.1	4.1	2.2
RT from minor street	6.2	6.9	3.3
TH from minor street	6.5	6.5	4.0
LT from minor street	7.1	7.6	3.5

Source "Highway Capacity Manual," *Special Report 209,* Transportation Research Board, Washington D.C., 1997, Table 10-1, pp. 10–11, used with permission of the National Academy of Science, Courtesy of the National Academies Press

Table 4.10 Adjustment factors for critical gap and follow-up time 1997 update and 2000 HCM

Factor	Values
t_{cHV}	1.0 for two-lane major streets 2.0 for four-lane major streets
t_{cG}	0.1 for movements 9 and 12 (RT from minor street) 0.2 for movements 7, 8, 10, 11 (TH, LT from minor street) 1.0 for all other movements
t_{cT}	1.0 for first or second stage of two-stage process 0.0 for single-stage process
t_{3LT}	0.7 for minor street LT at a T-intersection 0.0 for all other cases
t_{fHV}	0.9 for two-lane major street 1.0 for four-lane major street

computed with Eq. 4.5, while adjusted values for follow-up time are computed with Eq. 4.6.

$$t_c = t_{cb} + t_{cHV} P_{HV} + t_{cG} G - t_{cT} - t_{3LT} \tag{4.5}$$

$$t_f = t_{fb} + t_{fHV} P_{HV} \tag{4.6}$$

where:

t_c critical gap (s),
t_{cb} base critical gap from Table 4.11, (s),
t_{cHV} adjustment to critical gap for heavy vehicles, (s),
t_{cG} adjustment for grade, (s),
t_{cT} adjustment for two-stage STOP situations, (s),
t_{3LT} adjustment for minor street left-turns at T-intersections, (s),
P_{HV} proportion of heavy vehicles in subject movement,
G grade, (decimal),
t_f follow-up time, (s),
t_{fb} base follow-up time from Table 4.11, (s), and
t_{fHV} adjustment to follow-up time for heavy vehicles, (s).

For convenience, the adjustment factors used in Eqs. 4.5 and 4.6 are summarized in Table 4.10.

4.7.5 The 2010 and 2016 Highway Capacity Manuals

There were a number of changes to TWSC procedures in 2010. Most were incremental, and the fundamental nature of the methodology remained unchanged. Of special importance to the determination of critical gaps and follow-up times were that U-turns were considered as a discrete movement, and values were shown to vary depending upon the width of the major street. In 2010, 2-lane, 4-lane, and 6-lane major streets were explicitly treated. The 2016 HCM made no changes to critical gaps and follow-up times.

Critical headways (t_c), formerly referred to as "critical gaps," are still computed using Eq. 4.5. Follow-up times (t_f) are still computed using Eq. 4.6. Base values for both, however, have been revised, and are shown in Tables 4.11 and 4.12 respectively. Adjustments to both remain the same as in 2000, shown previously in Table 4.10.

In Table 4.11, the critical headway for a minor street left turn onto a 6-lane major street seems incongruent. The headway is less than for a similar turn onto a 2-lane or 4-lane major street. Left-turn driver behavior at TWSC intersections on 6-lane major

Table 4.11 Base values of critical headway ($t_{c,base}$)—2010 and 2016 HCMs

Movement	Base critical headway (s) for major streets with:		
	2 Lanes	4 Lanes	6 Lanes
LTs from major St	4.1	4.1	5.3
U-Turns from major St	N/A	6.4 (wide) 6.9 (narrow)	5.6
RTs from minor St	6.2	6.9	7.1
TH from minor St	1-stage: 6.5 Stage I: 5.5 Stage II: 5.5	1-stage: 6.5 Stage I: 5.5 Stage II: 5.5	1-stage: 6.5* Stage I: 5.5* Stage II: 5.5*
LTs from minor St	1-stage: 7.1 Stage I: 6.1 Stage II: 6.1	1-stage: 7.5 Stage I: 6.5 Stage II: 6.5	1-stage: 6.4 Stage I: 7.3 Stage II: 6.7

*Use caution; estimated values

Source "Highway Capacity Manual", Transportation Research Board, Washington D.C., 2010 [6], Exhibit 19-10, pp. 19–15, used with permission of the National Academy of Science, Courtesy of National Academies Press

Table 4.12 Base values of follow-up time ($t_{f,base}$)—2010 and 2016 HCMs

Movement	Base follow-up time (s) for major streets with:		
	2 Lanes	3 Lanes	4 Lanes
LTs from major St	2.2	2.2	3.1
U-Turns from major St	N/A	2.5 (wide) 3.1 (narrow)	2.3
RTs from minor St	3.3	3.3	3.9
TH from minor St	4.0	4.0	4.0
LTs from minor St	3.5	3.5	3.8

Source "Highway Capacity Manual", Transportation Research Board, Washington D.C., 2010 [6], Exhibit 19-11, pp. 19–16, used with permission of the National Academy of Science, Courtesy of National Academies Press

streets has been observed to be extremely aggressive. Because of this, such drivers often accept smaller gaps than what would normally be required.

As in 1997/2000, the issue of two-stage processes exists. Major streets with wide medians capable of storing one or more vehicles create a two-stage process for through and left-turn movements from the minor street. In such cases, drivers cross *each* roadway of the major street as a separate maneuver. The two stages, however, are linked, as some minor street vehicles may have to wait for a storage "slot" to become available in the median. Again, where such a median does not exist, the maneuver is completed in one stage. Critical headways and follow-up times differ based on one-stage or two-stage operation.

4.8 Determining Potential Capacity (C_p)

Once conflicting volumes, critical gaps (headways), and follow-up times have been determined for each movement, the potential capacity of each movement may be determined using gap acceptance modeling. Remember that "potential capacity" assumes that each movement has an exclusive lane, and that no higher-priority movement consumes any gaps in the conflicting traffic stream. One or both of these assumptions is not true in many cases, and subsequent steps must adjust potential capacity to reflect this.

It is in this step that gap acceptance modeling is used. Potential capacities are estimated using an expected distribution of gaps in each conflicting traffic stream, and estimates of critical gap and follow-up time.

4.8.1 Circular 212

In Circular 212, potential capacity is based solely upon the critical gap (t_c). Follow-up times are not used. The circular presents the relationship determining potential capacity in graphic form, as shown in Fig. 4.3.

The red lines illustrate the use of Table 4.4. In the example shown, a conflicting volume of 800 veh/h together with a critical gap of 6.0 s yields a potential capacity (c_p) of approximately 340 pc/h. This value assumes that the subject movement has full use of all available gaps in the 800-veh/h conflicting stream, and that it is made from an exclusive lane. As previously noted, neither of these is necessarily true, and adjustments will have to be made in subsequent computations.

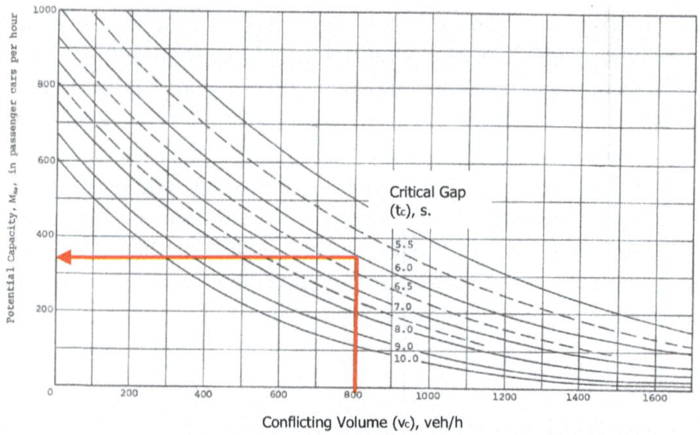

Fig. 4.3 Potential Capacities at Unsignalized Intersections: Circular 212. *Source* "Interim Materials on Highway Capacity—Unsignalized Intersections," *Trasnportation Research Circular 212*, Transportation Research Board, Washington D.C., 1980 [3], Fig. 2, p. 38, used with permission of the National Academy of Science, Courtesy of National Academies Press

4.8.2 The 1985 Highway Capacity Manual

In 1985, a number of revisions to the TWSC methodology were made. Critical gaps were reduced from Circular 212, and the relationship between critical gaps, conflicting volumes, and potential capacity was revised.

Potential capacity was still determined solely using critical gaps; follow-up times were still not explicitly included. The presentation was, like Circular 212, graphic, and is shown in Fig. 4.4.

The red lines in Fig. 4.4 show an example of the potential capacity for a conflicting volume of 800 veh/h and a critical gap of 6.0 s. The result is a potential capacity of 370 pc/h. This is the same example shown in Fig. 4.3 for Circular 212. The result is 30 pc/h higher than in Circular 212 for the same conditions.

Higher potential capacities for the same conflicting volume and critical gap, taken together with generally reduced critical gaps led to sometimes significant increases in potential capacities, which, in general, flow through to similar effects on movement and shared-lane capacities, when compared to the Circular 212 values.

In 1985, a general approach was added which allowed a rough estimate of the effect of platooned major street flow on potential and subsequent capacities. The gap acceptance model of Fig. 4.4 assumes random flow on the major street, which would not exist when a coordinated signal system exists on that street.

This feature is discussed in a separate section—"The Impact of Platoon Flow"—as various editions of the HCM have included the adjustment for this effect in different places—some as an adjustment to potential capacity, others as an adjustment to movement capacity.

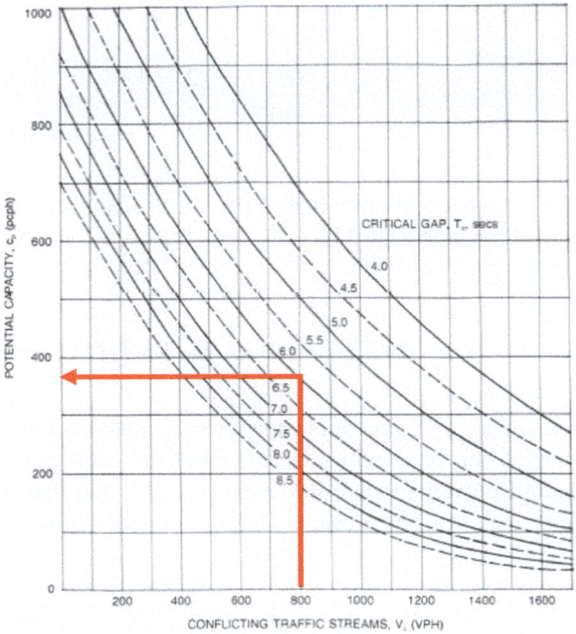

Fig. 4.4 Potential Capacity—Unsignalized Intersections 1985 HCM. *Source* "Highway Capacity Manual," *Special Report 209*, Transportation Research Board, Washington D.C., 1985 [4], Fig. 10-3, pp. 10–7, used with permission of the National Academy of Science, Courtesy of National Academies Press

4.8.3 *The 1994 Update to the 85 HCM*

The priority of movements and the computation of conflicting traffic streams continued to use the values originally presented in Circular 212. The determination of potential capacity, however, was now based on an equation, rather than a graph, and included explicit consideration of both critical gap (t_c) and follow-up time (t_f).

$$c_{px} = \frac{3600}{t_f} e^{-\left(\frac{v_{cx} t_o}{3600}\right)} \tag{4.7}$$

where:

c_{px} potential capacity for movement x, pc/h,
v_{cx} conflicting volume for movement x, veh/h,
t_f follow-up time, s,
t_o $t_c - (t_f/2)$, and
t_c critical gap, s.

The values for the critical gap (t_g) and follow-up time (t_f) for 1994 were shown in Table 4.8.

Examples shown in Figs. 4.3 and 4.4 produced estimates of potential capacity for a critical gap of 6.0 s and a conflicting volume of 800 veh/h. From Table 4.8, a critical gap of 6.0 s is used for through movements from the minor street across a 2-lane major street. The follow-up time for this case is 3.3 s. The potential capacity becomes:

$$t_o = 6.0 - 3.3/2 = 4.35\ \text{s}$$

$$c_p = \frac{3600}{3.3} e^{-\left(\frac{800*4.35}{3600}\right)} = 414.9,\ say\ 415\ \text{pc/h}$$

This reflects another increase in the potential capacity of 340 pc/h (Circular 212) and 370 pc/h (85 HCM).

4.8.4 The 1997 Update to the 85 HCM and the 2000 HCM

The equation for potential capacity of a movement was revised for 1997 and 2000 to more closely align with source documents. "Potential capacity" continues to assume that all movements are in exclusive lanes, and that no higher-priority movements consume any of the available gaps in the conflicting traffic stream.

$$c_{px} = v_{cx}\left[\frac{e^{-(v_{cx}t_{cx}/3600)}}{1 - e^{-(v_{cx}t_{fx}/3600)}}\right] \tag{4.8}$$

where:

c_{px} potential capacity of movement x, pc/h,
v_{cx} conflicting flow rate for movement x, conflicts/h (which now include both vehicle and pedestrian conflicts),
t_{cx} critical gap for movement x, s, and
t_{fx} follow-up time for movement x, s.

Critical gaps and follow-up times for 1997/2000 were shown in Table 4.9, with adjustment factors in Table 4.10.

To continue the comparative example, for a minor street through movement across a 2-lane major street, the base critical gap is given as 6.5 s and the follow-up time as 4.0 s. It is assumed that there are no adjustments. The potential capacity becomes:

$$c_{px} = 800\left[\frac{e^{-(800*6.5/3600)}}{1 - e^{-(800*4.0/3600)}}\right] = 320.4,\ say\ 320\ \text{pc/h}$$

This result is even lower than the Circular 212 estimate. The 1997/2000 version of the TWSC methodology reverses the trend of increasing capacity estimates between 1980 and 1994. This is primarily due to improved studies and field data for U.S. cases that indicated larger critical gaps and follow-up times.

4.8.5 *The 2010 and 2016 Highway Capacity Manuals*

For general cases of random major street flow, Eq. 4.8 remains the model for estimating potential capacity. Input values of critical gap and follow-up time were changed in 2010, but were not further revised in 2016. In some cases, the results would not be the same as the 1997/2000 version.

From Tables 4.11 and 4.12, for a minor street through movement across a 2-lane major street (one stage), the critical headway is 6.5 s and the follow-up time is 4.0 s. The potential capacity then becomes:

$$c_{px} = 800\left[\frac{e^{-(800*6.5/3600)}}{1 - e^{-(800*4.0/3600)}}\right] = 320.4, \textit{say}\ 320\ \text{pc/h}$$

This is the same as the value predicted for the 2000 HCM.

The pattern of predicted potential capacities for a minor street through movement with a conflicting volume of 800 veh/h has varied considerably over the years:

1980:	340 veh/h
1985:	370 veh/h
1994:	415 veh/h
1997/2000:	320 veh/h
2010/2016:	320 veh/h.

The 1994 estimate is the highest, 95 veh/h higher than the lowest estimate from the 1997 through 2016 editions.

4.9 Determination of Movement Capacity—Impedance Effects

Potential capacity assumes that every movement has an exclusive lane, and that there are no higher-priority movements consuming some of the available gaps in the conflicting stream. The model adjusts for each of these assumptions separately.

The first adjustment is to convert potential capacity to movement capacity. In this adjustment, the priority of movements is critical. The impact of higher-priority movements consuming gaps that will no longer be available to lower-priority movements is estimated in this step. The result of this adjustment is called the "movement capacity (c_m)." It is still, however, assumed that each movement is operating out of an exclusive lane.

4.9.1 Circular 212

The Circular 212 approach relies on a strict hierarchy in which available gaps are used. Assuming that a gap appears, the following is the priority in which a waiting subject vehicle would use the gap:

- Right turn from the minor street.
- Left turn from the major street.
- Crossing movement from the minor street.
- Left turn from the minor street.

Thus, if both a left turn from the major street and a crossing movement from the minor street were waiting, the first appropriate gap would be used by the left turn. The crossing (or through) movement would have to wait for a second gap to appear. The priority list is not absolute, however, as some gaps may be used by two subject vehicles, as long as they do not conflict with each other. Thus, a left turn from the major street and a right turn from the minor street may use the same gap.

Each potential capacity must be adjusted to account for the use of some gaps by higher-priority vehicles. This process is referred to as "impedance." There are three fundamental impedance situations that are accounted for in Circular 212:

1. Left turns from the major street and right turns from the minor street are not impeded by any other movements. Thus, in these cases, the movement capacity is equal to the potential capacity.
2. At a t-intersection, left turns from the minor street are impeded by left turns from the major street, as both movements may use the same gaps.
3. At a 4-leg intersection, crossing movements from the minor street are impeded by both left-turn movements from the major street.
4. At a 4-leg intersection, left turns from the minor street are impeded by both left-turn movements on the major street, and the opposing minor street through and right-turn movements.

These impedance situations are illustrated in Fig. 4.5. The variables in Fig. 4.5 are defined below:

c_{pi} potential capacity of subject movement i, pc/h,
c_{mi} movement capacity of subject movement i, pc/h,
P_j impedance factor for impeding movement j, pc/h.

The impedance factor, P_j, is the proportion of time that a given movement is *not* blocked by a higher-priority movement.

Each impeding movement generates an impedance factor (P). In the first case, the 4th priority left turn from the minor street is impeded by 2nd priority left turn from the minor street, approaching from the right. In the second case, the 3rd priority through movement from the minor street is impeded by *both* left turns from the major street, which are both priority 2 movements. In the last case, the 4th priority left turn from the minor street is impeded by *both* left turns from the major street, and by the opposing minor street through and right-turn movements.

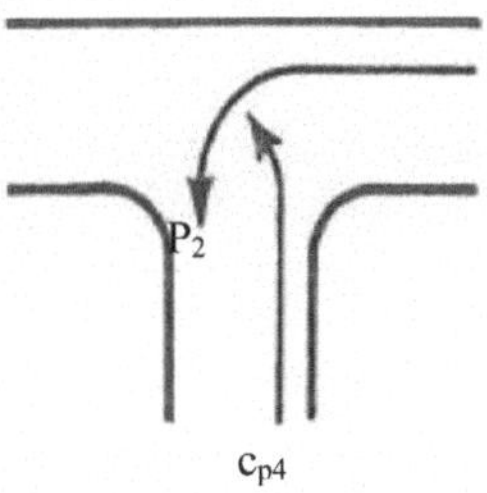

Left Turns from the Minor Street at a T – Intersection:

$c_{m4} = c_{p4} P_2$

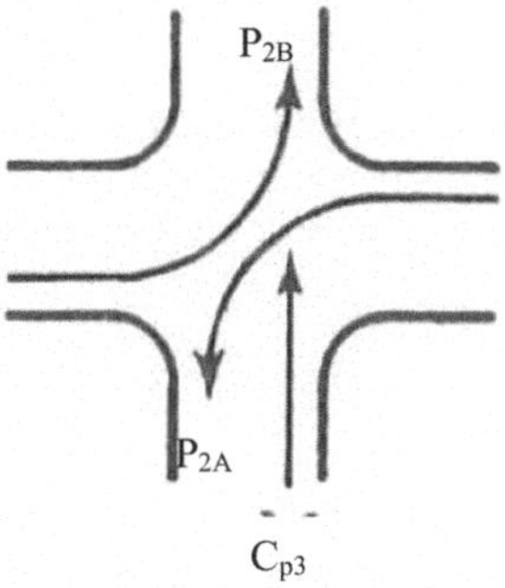

Through Movements from the Minor Street at a 4-Leg Intersection:

$c_{m3} = c_{p3} P_{2A} P_{2B}$

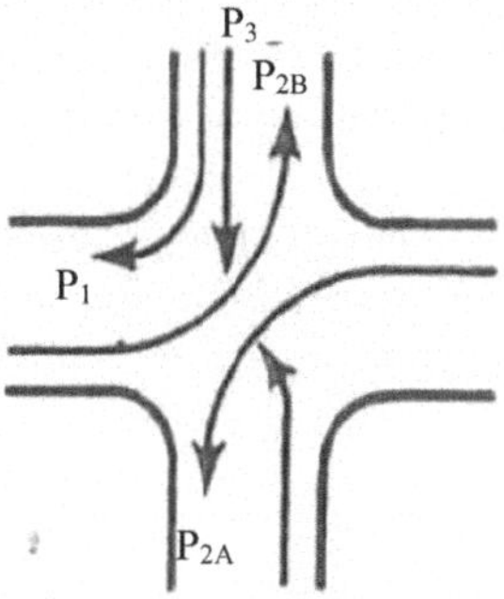

Left Turns from the Minor Street at a 4-Leg Intersection:

$c_{m4} = c_{p4} P_{2A} P_{2B} P_1 P_3$

Fig. 4.5 Impedance Effects at Unsignalized Intersections: Circular 212. *Source* Modified from "Interim Materials on Highway Capacity—Unsignalized Intersections," *Transportation Research Circular 212*, Transportation Research Board, Washington D.C., p. 39, used with permission of the National Academy of Sciences, Courtesy of National Academies Press

The impedance factor, P, for each impeding movement is computed using Fig. 4.6. It is entered with the percent of capacity used by the existing demand, to find each impedance factor. The "percent of capacity used" is defined as:

$$\% \, cap = \frac{v_j}{c_{pj}} * 100 \tag{4.9}$$

where:

v_j demand flow rate for impeding movement j, pc/h,

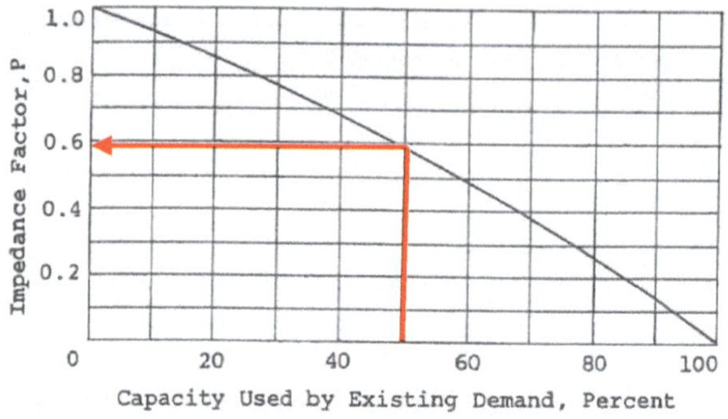

Fig. 4.6 Impedance Factors—Circular 212. *Source* "Interim Materials on Highway Capacity—Unsignalized Intersections," *Transportation Circular 212*, Transportation Research Board, Washington D.C., 1980 [3], Fig. 3, p. 39, used with permission of the National Academy of Science, Courtesy of National Academies Press

c_{pj} potential capacity for impeding movement j, pc/h.

Note that as potential capacities are given in units of pc/h, that the existing demand flow rates must also be converted to pc/h, using the equivalents given in Table 4.1.

The red lines on Fig. 4.6 illustrate its use. If an impeding movement uses 50% of its potential capacity, it would reduce the potential capacity of a lower-priority movement that it impedes by a factor of 0.59. Essentially, the higher-priority movement blocks available gaps 41% of the time.

Once the impedance factor for *each* impeding movement is found, the movement capacity is found as shown in Fig. 4.5. Remember that for Priority 2 movements (right turns from the minor street, left turns from the major street), there are no impeding movements, and $c_{mi} = c_{pi}$.

4.9.2 The 1985 Highway Capacity Manual

The process for determining movement capacities in the 1985 HCM was the same as in Circular 212. An appendix was added to allow a general approach to the effect of platoon flow on the major street. This subject is treated in a later section.

4.9.3 The 1994 Update to the 85 HCM

The impedance effects illustrated in Fig. 4.4 remain in effect. That is:

- Right turns from the minor street (Rank 1) and left turns from the major street (Rank 2) are unimpeded by other movements. For these movements, movement capacity (c_m) = potential capacity (c_p).

- Through movements from the minor street (Ranks 3) are impeded by both left turns from the major street (Rank 2).
- Left turns from the minor street (Rank 4) are impeded by the opposing minor street right turn (Rank 2), the opposing minor street through movement (Rank 3) and both major street left turns (Rank 2).

The graphic approach to determining each impedance factor (P_y) was replaced by equations for the 1994 update. Note that there is an impedance factor for *each* impeding movement.

$$P_y = 1 - \frac{v_y}{c_{my}} \tag{4.10}$$

where:

P_y impedance factor for impeding movement y,
v_y demand flow for impeding movement y, pc/h, and
c_{my} movement capacity for impeding movement y, pc/h.

Note that to find an impedance factor, the movement capacity for the impeding movement must be determined. Thus, computations must proceed in the priority order of the movement involved.

The total adjustment to a potential capacity is the product of the individual impedance factors for each impeding movement, or:

$$f_x = \prod_y P_y \tag{4.11}$$

and:

$$c_{mx} = c_{px} f_x \tag{4.12}$$

where f_x is the total impedance adjustment factor for subject movement x, and all other variables are as previously defined.

The situation is, however, a bit more complicated when the left turn movements from the minor street are considered. Normally, each left-turn movement is impeded by *four* higher-priority movements:

- The opposing minor street right turn (P_{or}),
- The opposing minor street through movement (P_{ot}),
- The two major street left-turn movements (P_{L1}, P_{L2}).

If the methodology were directly applied, the total impedance factor for a left turn movement from the minor street would be:

$$f_x = P_{or} P_{ot} P_{L1} P_{L2}$$

Three of these, however, involve partial overlaps—that is, impeding vehicles from several impeding movements could be present *at the same time.* Thus, the simple product of the individual impedance factors could overstate the total impact on capacity. The three movements that might overlap are the opposing through movement (P_{ot}) and the major street left turns (P_{L1}, P_{L2}). The 1994 update presents a procedure for correcting this overlap, as follows:

1. Let $P'' = P_{or} P_{L1} P_{L2}$
2. Then:

$$P' = 0.65\ P'' - \left(\frac{P''}{P'' + 3}\right) + 0.6\sqrt{P''}$$

3. and:

$$f_x = P_{or}\ P'$$

Using the adjusted value of f_x, the movement capacity of each minor street left turn may now be computed using Eq. 4.12.

It should also be noted that the 1994 update gives an additional impedance factor to account for major street left turns that are not made from an exclusive LT lane (which the methodology assumes). It, however, requires field data collection to establish saturation flow rates, and its use is optional. Consult the 1994 update [20] directly for this modification. By 1997, this adjustment became mandatory, and thus is outlined in the next section.

4.9.4 The 1997 Update to the 85HCM and the 2000 HCM

Movement capacities are still computed using Eqs. 4.11 and 4.12, which are unchanged from 1994.

However, there are significantly more impedances accounted for in the 1997 update and 2000 HCM than in the 1994 version. This is primarily due to inclusion of pedestrians, which not only add to conflicting volumes, but also contribute to impedance by using some available gaps. As a result, by 1997 *all* movements were impacted by impedance, not just those in Priorities 3 and 4. Table 4.13 summarizes the impedances considered.

As shown in Table 4.13, minor street left turns now face up to six impeding flows—4 vehicular and 2 pedestrian. The factor (P_y) for an impeding vehicular flow is the same as in 1994—Eq. 4.10 is used. Where the impeding flow is composed of pedestrians, the factor is computed as:

$$P_y = 1 - \left[\frac{v_j(w/S_p)}{3600}\right] \tag{4.13}$$

Table 4.13 Impeding flows in the 1997 update and 2000 HCM

Subject vehicular flow	Impeding vehicular flows	Impeding pedestrian flows
1 (Major St LT)	None	16
4 (Major St LT)	None	15
7 (Minor St LT)	1,4,11,12	15,13
8 (Minor St TH)	1,4	15,16
9 (Minor St RT)	None	15,14
10 (Minor St LT)	1,4,8,9	16,14
11 (Minor St TH)	1,4	15,16
12 (Minor St RT)	None	16,13

Source Roess, Prassas, and McShane, *Traffic Engineering*, 3rd Edition, Pearson-Prentice Hall, Upper Saddle Brook, NJ, 2004, Table 23.4, p. 672, reprinted with permission of Pearson Education Inc., New York, New York

where:

v_j conflicting pedestrian flow rate, peds/h,
w lane width (for impeded movement), ft, and
S_p walking speed of pedestrians, ft/s.

Where local calibrations are not available, a default walking speed of 4 ft/s was recommended in the 2000 HCM.

For Movements 7 and 10 (minor street left turns), where there are four vehicular impedances, the adjustment for overlap specified in 1994 is still used.

Also, the modification for major street left turns made from shared lanes is also continued, but is made mandatory. The general model assumes that such left turns are made from exclusive LT lanes. Where this is not the case, the impedance factors normally associated with Movements 1 and 4 (the major street left turns) should be modified as follows:

$$P*_{1/4} = 1 - \left[\frac{1 - P_{1/4}}{1 - \left(\frac{v_{mTH}}{s_{mTH}} + \frac{v_{mRT}}{s_{mRT}} \right)} \right] \tag{4.14}$$

where:

$P*_{1/4}$ adjusted impedance factor for impeding movement 1 or 4,
$P_{1/4}$ base value of impedance factor for impeding movement 1 or 4,
v_{mTH} major street TH flow rate in direction of impeding movement, veh/h,
v_{mRT} major street RT flow rate in direction of impeding movement, veh/h,
s_{mTH} saturation flow rate for the major street through movement, veh/h,
s_{mRT} saturation flow rate for the major street RT movement, veh/h.

Note that this adjustment requires a field calibration of the saturation flow rates on the major street (the 2000 HCM provides measurement instructions). Where such data is not available, default rates between 1500 and 1600 veh/h were cited and may be used.

4.9.5 The 2010 and 2016 Highway Capacity Manuals

As in the past, movement capacities reflect the fact that not all gaps are available for the use of every movement. Higher priority movements may consume some gaps that are then no longer available to lower priority movements. This process has been referred to as "impedance" in the past. This terminology was replaced in 2010, and former impedance factors (P) are, in 2010, referred to as adjustment factors.

The 2010 and 2016 methodology for movement capacities is fundamentally the same, although there are some minor revisions in the presentation and in some notation. The only fundamental difference is that the 2010 HCM mandates the consideration of pedestrian impedances, while the 2016 HCM makes it optional, noting that pedestrians most often have a negligible impact on movement capacities.

Applicable impedances that must be taken into account remain the same as in the 2000 HCM, as shown in Table 4.13. The addition of U-turns as an explicit movement, has introduced a few minor changes.

Rank 2 Movements

There are now *three* Rank 2 movements: Right turns from the minor street, left turns from the major street, and U-Turns from the major street. The latter two, however, often operate out of the same lane, creating a shared-lane situation.

As in the past, right turns from the minor street are not impeded by any other movements. Thus, for these turns $c_{mx} = c_{px}$.

The equality also holds for left turns from the major street, as long as they are operating out of an exclusive lane. U-turns, however, may also be operating out of such a lane. The movement capacities for U-turns operating out of an exclusive lane is:

$$
\begin{aligned}
c_{m1U} &= c_{p1U}\left(1 - \frac{v_{12}}{c_{m12}}\right) \\
c_{m4u} &= c_{p4U}\left(1 - \frac{v_9}{c_{m9}}\right)
\end{aligned}
\tag{4.15}
$$

In the most normal case, where a U-turn from the major street operates out of the same lane as left turns from the minor street, a shared-lane situation is created. Thus, a shared-lane capacity (a concept discussed in the next section) for the exclusive LT/U-Turn lane on a major street is computed as:

$$
c_{SH1/1U} = \frac{v_1 + v_{1U}}{\left(\frac{v_1}{c_{m1}}\right) + \left(\frac{v_{1U}}{c_{m1U}}\right)} \tag{4.16}
$$

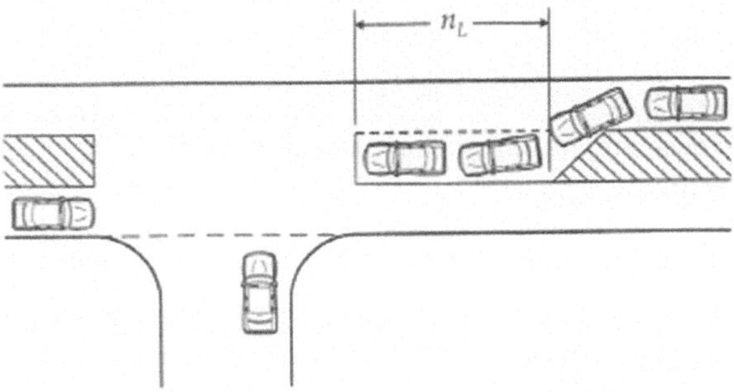

Fig. 4.7 Short left-turn bay on major street at a TWSC intersection 2010 HCM. *Source* "Highway Capacity Manual", Transportation Research Board, Washington D.C., 2010 [6], Exhibit 19-13, pp. 19–20, used with permission of the National Academy of Science, Courtesy of National Academies Press

where all terms are as previously defined.

The 1997 and 2000 methodologies allowed for an adjustment where major street left turns operated out of a shared lane. In 2010, this adjustment was expanded to also account for a short exclusive lane where spillback into the through lane or lanes might occur. This is the situation shown in Fig. 4.7.

In the 2000 HCM, Eq. 4.14 provided a revised impedance factor (P*) for major street left-turn movements that shared a lane with major street through and/or right turn movements. The situation with short exclusive lane with potential spillback was not treated. Equation 4.17 is used in the 2010 HCM:

$$P*_{1/4} = 1 - \left(1 - P_{1/4}\right)\left[(n_L - 1)\sqrt{1 + \frac{X_M^{(n_L-1)}}{1 - X_M}}\right]$$

$$X_M = \left(\frac{v_{mTH}}{s_{mTH}}\right) + \left(\frac{v_{mRT}}{s_{mRT}}\right) \tag{4.17}$$

where:

- $P_{1/4}$ impedance factor for impeding left turns from the major street (Mvts 1 or 4), assuming an exclusive LT lane exists with no spillback,
- $P*_{1/4}$ impedance factor for impeding left turns from the major street (Mvts 1 or 4), adjusted to reflect the impacts of no exclusive LT lane or a short exclusive LT lane with spillback,
- n_L number of vehicles that can be held in the exclusive LT lane on the major street (vehs); $n_L = 0$ where no exclusive LT lane exists,
- X_m combined degree of saturation for major street through and right turn movements (in the impeding direction),
- v_{mTH} through vehicle flow rate on the major street in the impeding direction, veh/h,
- v_{mRT} RT vehicle flow rate on the major street in the impeding direction, veh/h,

s_{mTH} saturation flow rate for through movements on the major street, veh/h/ln; a default value of 1800 veh/h/ln is recommended (a change from the 2000 HCM), and

s_{mRT} saturation flow rate for RT movements on the major street, veh/h/ln; a default value of 1500 veh/h/ln is recommended (a change from the 2000 HCM).

When a shared-lane is used for major street LTs, $n_L = 0$, and Eq. 4.17 becomes Eq. 4.14.

Rank 3 and 4 Movements

The movement capacity computations for Rank 3 (minor street through movements and minor street left turns at T intersections) and Rank 4 (minor street left turns) did not change. The computations used in the 2000 HCM remain in effect.

4.10 Shared-Lane Capacity

At this point in the TWSC methodology, movement capacities have been determined for all affected movements. It is, however, still assumed that each movement occupies an exclusive lane. In most cases of TWSC, this is not true. The final step in determining capacities is, therefore, the capacity of lanes shared by more than one minor street movement. The simple model used for this was introduced in Circular 212 in 1980, and has not changed since. It is still used in the 2016 HCM.

Where more than one movement uses a lane, the capacity of the shared lane is computed as:

$$c_{SH} = \frac{\sum_n v_i}{\sum_n (v_i / c_{mi})} \tag{4.18}$$

where:

c_{SH} capacity of a shared lane, pc/h,
v_i demand flow rate of movement i in the shared lane, pc/h,
c_{mi} movement capacity of movement i in the shared lane, pc/h,
n number of movements sharing the lane, usually 2 or 3.

This is a simple, logical equation, as opposed to one developed using field data. Essentially, if three movements share a lane, and one demand flow uses 25% of its available movement capacity, the second uses 30%, and the third uses 35%, then it is reasonable to assume that when placed in a single lane, the total demand flow would use 25% + 30% + 35% = 90% of the shared capacity. The capacity of the lane is then logically the shared-lane demand divided by 0.90.

It should be noted that each minor street approach lane is evaluated for its shared-lane capacity. In some cases, widening of curb lanes at the intersection may provide space for a defacto right-turn lane where none is marked. The determination of which movements are truly sharing lanes requires careful field observations.

4.11 Delay and Level of Service

As with most capacity analysis procedures, the last step is the determination of level of service. The criteria for assigning a level of service is a critical *policy* decision made by the Highway Capacity and Quality of Service Committee of the Transportation Research Board.

For consistency with signalized intersection criteria, it has always been the objective of the HCQSC to use *control delay* as the service measure for LOS. However, in 1980, when the TWSC intersection methodology was first introduced, *control delay* was a concept not yet fully developed or implemented, even for signalized intersections. It was formally introduced as a service measure for signalized intersections in the 1985 HCM. While the fundamental theory was also applicable to TWSC (and other unsignalized intersections), a delay algorithm for such intersections was not introduced until 1994.

4.11.1 Circular 212

To determine approximate levels of service, the service measure used in Circular 212 was "reserve" or "unused capacity." The reserve (or unused) capacity for a given minor street lane was computed as:

$$c_R = c_{SH} - v_{SH} \tag{4.19}$$

where:

c_R reserve capacity, pc/h,
c_{SH} shared-lane capacity, pc/h,
v_{SH} total demand in shared lane, pc/h.

Table 4.14 shows the criteria for level of service, and the presumed general delay levels expected for each.

The criteria are a bit odd in some ways. There is no clear rationale for relating "reserve capacity" to delay, or any other operational parameter that users might directly experience. The verbal delay descriptors are not quantified in any way, leaving the exact meaning of these terms to the user.

Even the use of the term "failure" is inconsistent with its common use throughout the manual. A negative "reserve capacity," which means that demand exceeds capacity, is termed level of service E, not F. Level of service F is reserved for cases in which the minor street approaches are blocked by "external causes," which presumably means congestion and queueing on the major street.

Part of this is explained by a view of how unsignalized intersections fail. All of the capacity computations assume that critical gaps are stable, and that the rigid priority scheme modeled is followed. When congestion (or excessive delay) is experienced

Table 4.14 Level of service and expected delay for reserve capacity ranges—circular 212

Reserve capacity (pc/h)	Level of service	Expected traffic delay
400 or more	A	Little or no delay
300–399	B	Short traffic delays
200–299	C	Average traffic delays
100–199	D	Long traffic delays
0–99	E	Very long traffic delays
Less than 0	E	Failure—extreme congestion
(Any value)	F	Intersection blocked by external causes

Source "Interim Materials on Highway Capacity—Unsignalized Intersections," *Transportation Research Circular 212*, Transportation Research Board, Washington D.C., 1980 [3], Table 3, p. 40, used with permission of the National Academy of Science, Courtesy of National Academies Press

on a minor street, drivers will often select shorter, even unsafe, gaps through which to execute their desired maneuver. Even the priority order of gap usage can become unglued as frustrated drivers start selecting gaps on the basis of relative aggressiveness, not the strict priority of their movement. Thus, "failure" of an unsignalized intersection may or may not be accompanied by a long queue of vehicles on the minor approaches. Rather, one might see major street vehicles sometimes negatively impacted by minor street drivers "forcing" their way through or into the major street stream. Nevertheless, the LOS criteria of Table 4.16 represent a distinctly different view of "failure" than other methodologies of the HCM, even in 1980.

4.11.2 The 1985 Highway Capacity Manual

Level of service criteria, and the use of "reserve capacity" to determine it were the same as in Circular 212 in the 1985 HCM.

4.11.3 The 1994 Update to the 85 HCM

Perhaps the most important change introduced in 1994 was an algorithm for prediction of delay to minor street movements. The algorithm was similar in structure to that used for signalized intersections, and has the same theoretical basis. With a prediction of delay, level of service could now be related to average delays, and more direct comparisons to signalized alternatives could be made.

Delay is computed for each *lane* (whether exclusive or shared) separately, and is defined as the average time between a vehicle joining the end of a queue (at the STOP sign) and departing from the STOP line.

$$D_x = \frac{3600}{c_{mx}} + 900T\left[\frac{v_x}{c_{mx}} - 1 + \sqrt{\left(\frac{v_x}{c_{mx}} - 1\right)^2 + \frac{\left(\frac{3600}{c_{mx}}\right)\left(\frac{v_x}{c_{mx}}\right)}{450T}}\right] \tag{4.20}$$

where:

D_x average delay to movement x, s/veh,
T analysis period, h (for 15 min, T = 0.25),
c_{mx} movement (or shared-lane) capacity for movement (or shared lane) x, pc/h, and
v_x demand flow rate for movement (or shared lane) x, pc/h.

Where more than one movement shares a lane, c_{mx} is replaced by the shared-lane capacity (c_{SH}) and v_x is replaced by the total shared-lane demand flow rate. The result is then the average delay to all vehicles in the shared lane.

Level of service is, for the first time, determined by delay, using the criteria in Table 4.15. The criteria are different than for signalized intersections (delay thresholds are generally lower than for signalized intersections) due to driver expectations. Further, at signalized intersections, drivers may relax during red intervals, whereas drivers at a STOP-controlled location must be constantly alert and looking for gaps in the conflicting traffic stream.

Where there are multiple lanes on a STOP-controlled approach, an average delay weighted by demand flow rate in each lane may be computed. A weighted average delay for the entire intersection may also be computed, using a delay of 0.0 s/veh for all Priority 1 movements. This latter average may be of limited utility, given that large major street flows will often drive the result to a very low value, which might imply a good LOS. LOS designations for the entire intersection are not recommended because of this.

Table 4.15 Level of service criteria for TWSC intersections 1994 update

Level of service	Average total delay (s/veh)
A	≤ 5
B	$>5 \leq 10$
C	$>10 \leq 20$
D	$>20 \leq 30$
E	$>30 \leq 45$
F	>45

Source "Highway Capacity Manual," *Special Report 209*, Transportation Research Board, Washington D.C., 1994 [19], Table 10-3, pp. 10–12, used with permission of the National Academy of Science, Courtesy of National Academies Press

Table 4.16 Level of service criteria for TWSC and AWSC intersections 1997 update and 2000 HCM

Level of service	Control delay (s/veh)
A	0–10
B	>10–15
C	>15–25
D	>25–35
E	>35–50
F	>50

Source "Highway Capacity Manual", Transportation Research Board, Washington D.C., 2000 [5], Exhibit 17-2, pp. 17–1, used with permission of the National Academy of Science, Courtesy of National Academies Press

4.11.4 The 1997 Update to the 85HCM and the 2000 HCM

The delay equation used in 1997 and 2000 was slightly updated from 1994. The basic equation remains the same, but a constant of "5" has been added to reflect experience showing that the 1994 equation tended to underestimate delay:

$$D_x = \frac{3600}{c_{mx}} + 900T\left[\frac{v_x}{c_{mx}} - 1 + \sqrt{\left(\frac{v_x}{c_{mx}} - 1\right)^2 + \frac{\left(\frac{3600}{c_{mx}}\right)\left(\frac{v_x}{c_{mx}}\right)}{450T}} + 5\right] \quad (4.21)$$

where all terms are as previously defined. Level of service criteria were also revised, and are shown in Table 4.16.

In general, upper delay limits on LOS C, D, and E were increased by 5 s/veh, values which remained less than similar criteria for signalized intersections, but gap was smaller than in 1994.

In cases where the major street left turns are not made from an exclusive lane, the 1997 and 2000 models also introduced an approach to estimate delay to Rank 1 vehicles (which in weighted averages were considered to be "0" in 1994). The inclusion of this was, however, optional, with the decision left to the user. It was made mandatory in 2010.

4.11.5 The 2010 and 2016 Highway Capacity Manuals

The delay to individual exclusive or shared lanes is still computed using Eq. 4.21. The approach remains as in the 2000 HCM. This equation, however, covers delay to vehicles in Ranks 2, 3, and 4. It is assumed that Rank 1 vehicles do not experience delay.

In 2000, the HCM provided a procedure to estimate delay to Rank 1 vehicles where major street left turns operated out of a shared lane. The inclusion of this delay was optional. In 2010 and 2016, its inclusion is mandated. The control delay to Rank 1 vehicles is estimated using Eq. 4.22.

$$D_{Rank1} = \frac{\left(1 - P*_{1/4}\right) D_{m1/4}\left(\frac{v_{3/5}}{N}\right)}{v_{m3/5} + v_{m(1+3)/(4+6)}} \quad for\ N > 1$$
$$D_{Rank1} = \left(1 - P*_{1/4}\right) D_{m1/4} \quad for\ N = 1 \tag{4.22}$$

where:

D_{Rank1} control delay to through and right-turn vehicles on the major street when left turns operate out of a shared lane or short exclusive lane with spillback, s/veh,

$P*_{1/4}$ adjusted impedance factor for movement 1 or 4 operating out of a shared lane or short exclusive lane with spillback; note that $P*_1$ is used to determine the delay to movements 2 and 3; $P*_4$ is used to determine the delay to movements 5 and 6.

$D_{m1/4}$ delay to vehicles in movement 1 or 4 (as appropriate), from Eq. 4.32, s/veh,

$v_{3/5}$ demand flow rate for movement 3 or 5 (as appropriate), veh/h, and

$v_{m(1+3)/(4+6)}$ demand flow rate for the sum of movements 1 + 3 or 4 + 6 (as appropriate), veh/h; these are the left- and right-turning vehicles sharing the through lane(s).

Once again, weighted average delays for each approach can be computed. LOS criteria are the same as in the 2000 HCM (Table 4.18). For 2010 and 2016, however, ANY lane or approach for which v/c > 1.00 is labeled as LOS F, regardless of the delay predicted.

Weighted average delay for the entire intersection may be computed, but it is *not* used to assign a level of service, due to the large major street movements with either no delay or very little delay.

4.12 Two-Stage Movements at a TWSC Intersection

By 1997, it was recognized that there were TWSC intersections, particularly in suburban areas, located on major streets with wide medians—wide enough to store at least one vehicle in the median. In such cases, drivers who had to cross the first major street roadway to make their maneuver (through and left turn movements from the minor street) did so in a two-stage process. They effectively sought to get through the first major street roadway to the median holding area, then find their gap through the second set of conflicting movements.

Drivers were, in effect, sorting out their conflicting volumes into two sets—one of which they had to negotiate in Stage I to get to the median holding area, and the remaining conflicts that they had to negotiate in Stage II across the second major street roadway.

Modifications to the TWSC methodology to account for this were introduced in the 1997 update, and have been referred to previously in the approaches to determining conflicting volumes, and in some cases, critical gaps and follow-up times.

Where the major street includes a median wide enough to store one or more vehicles, the process for minor street through and left-turning vehicles involves two separate movements, or stages, defined as follows:

- The driver initially seeks to move into the median holding area by first crossing the traffic streams approaching from the left.
- Once in the median, the driver then seeks to move into or through the traffic streams approaching from the right.

In essence, the driver is treating each one-way roadway separately. Because storage in the median is limited, however, the directional major street roadways cannot be simply treated as two separate STOP-controlled intersections. In addition, major street left-turning vehicles also enter the median storage area before crossing the opposing major street flow.

It should be noted that such intersections generally involve *two* STOP signs, one at the entry to each major street directional roadway. In some cases, the sign posted in the median will be a YIELD sign, as sight distances are rarely restricted in the median. While the methodology assumes two STOP signs, it can be approximately applied where the median sign is a YIELD sign.

The two-stage process described is illustrated in Fig. 4.8.

The methodology is based upon three capacities for each affected movement:

- Movement capacity of the subject movement, assuming it was a one-stage process (i.e., no median existed).
- Movement capacity of Stage I of the subject movement.
- Movement capacity of Stage II of the subject movement.

Note that each of the potential capacities of each of these is computed using the usual procedures, but that the critical gaps, follow-up gaps, and conflicting volumes will be different for each case. Further, impedances are also modified, as some impeding movements are not involved in both stages of a two-stage process.

The movement capacity at a two-stage intersection is computed as follows:

$$c_{mxT} = \frac{a}{y^{m+1} - 1}\left[y\left(y^{m} - 1\right)(c_{mxII} - v_L) + (y - 1)c_{mx}\right] \quad for\ y \neq 1$$

$$c_{mxT} = \frac{a}{m + 1}[m(c_{mxII} - v_L) + c_{mx}] \quad for\ y = 1 \tag{4.23}$$

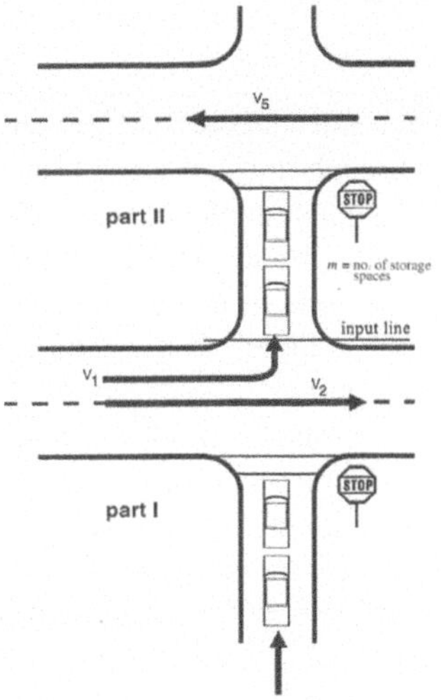

Fig. 4.8 Two-stage process at a TWSC intersection 1997 update and 2000 HCM. *Source* "Highway Capacity Manual," *Special Report 209*, Transportation Research Board, Washington D.C., 1997, Fig. 10-10, pp. 10–20, used with permission of the National Academy of Science, Courtesy of National Academies Press

where:

$$a = 1 - 0.32\, e^{-1.3\sqrt{m}}$$

$$y = \frac{c_{mxI} - c_{mx}}{c_{mxII} - v_L - c_{mx}} \tag{4.24}$$

and:

c_{mxT} movement capacity of two-stage movement x, pc/h,
c_{mxI} movement capacity for stage I of two-stage movement x, pc/h,
c_{mxII} movement capacity for stage II of two-stage movement x pc/h,
c_{mx} movement capacity for two-stage movement x if considered as a one-stage movement, pc/h,
m number of storage spaces in the median,
v_L major street left-turn flow rate (Mvt 1 when considering Mvts 1, 2, or 3, and Mvt 4 when considering Mvts 4, 5, or 6), pc/h, and
a,y intermediate variables.

In remaining parts of the methodology (where a 2-stage process exists), the movement capacity for the two-stage movement is used. This approach remains in effect through the 2016 HCM. The 2016 HCM, however, treats two-stage impacts as an additional "adjustment" to potential capacities, as opposed to a separate analysis step.

4.13 Flared Intersection Approaches

Also introduced in 1997, an approach to considering the impact of a minor street approach that has a flared geometry that might allow one or more vehicles to leave a designated lane to make its desired maneuver was added.

This situation is illustrated in Fig. 4.9. In this case, the flared intersection approach allows one or more right-turning vehicles to be at the STOP line without blocking other vehicles in the lane from also being at the STOP line. In the figure, "n" is the number of vehicles that can be accommodated by the flared approach, and is established by field observations.

The methodology compares the actual storage capacity of the flare with the average queue length for a movement that would normally exist. If the length of the actual storage area is bigger than the maximum expected queue length, the flare creates a separate lane for the right-turn movement. If this is not the case, the existence of the flare will still enhance capacity, but not to the same extent. The enhanced capacity is treated as an interpolation between the shared-lane capacity (without a flare) and the capacity of the movement in a separate lane.

The average queue size for each movement sharing the right lane is found as:

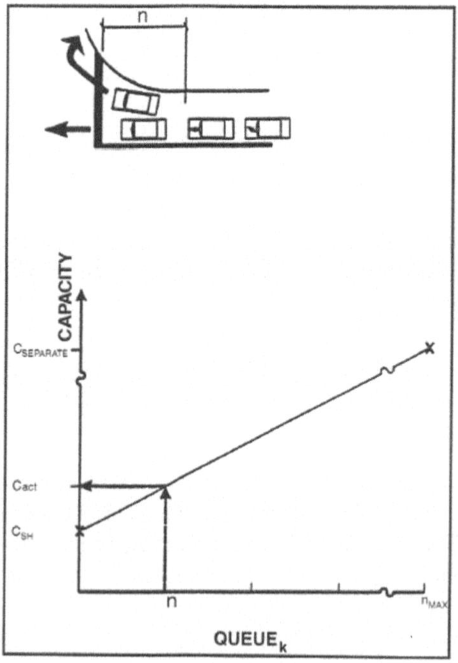

Fig. 4.9 Flared approaches on a TWSC intersection 1997 update and 2000 HCM. *Source* "Highway Capacity Manual," *Special Report 209*, Transportation Research Board, Washington D.C., 1997, Fig. 10-10, pp. 10–20, used with permission of the National Academy of Science, Courtesy of National Academies Press

$$Q_{sepi} = \frac{D_{sepi} v_i}{3600} \tag{4.25}$$

where:

Q_{sepi} average queue size for movement i assuming separate lane operation, pc,
D_{sepi} average control delay for movement i assuming separate lane operation, s/veh, and
v_i demand flow rate for movement i, pc/h.

After computing the appropriate values of Q_{sepi} (there can be up to three—for the right turn, through movement, and left turn sharing the lane), the maximum expected queue size is determined as:

$$n_{\max} = \max_i\left[round\left(Q_{sepi} + 1\right)\right] \tag{4.26}$$

where the "round" function rounds off the value of $Q_{sepi} + 1$ to the nearest integer, and the "max" function selects the maximum value of the result for the various movements "i" that occupy the subject lane.

In Eq. 4.25, a delay estimate is needed. Delay estimations have been discussed in a previously. The capacity of the flared lane is then computed as:

$$c_F = \left(\sum_i c_{sepi} - c_{SH}\right)\frac{n}{n_{\max}} + c_{SH} \quad for\; n \le n_{\max}$$
$$c_F = \sum_i c_{sepi} \quad for\; n > n_{\max} \tag{4.27}$$

where:

c_F shared-lane capacity of the flared lane, pc/h,
c_{SH} shared-lane capacity assuming no flare, pc/h,
c_{sepi} potential capacities of each separate movement assuming a separate lane, pc/h, and all other variables as previously defined.

This portion of the methodology has not fundamentally changed, and continues in the 2016 HCM, although its presentation has been altered.

4.14 The Effect of Platoon Flow on the Major Street

Perhaps the most complex of all of the modifications to the TWSC methodology is the one that deals with platoon flow on the major street. The fundamental gap acceptance approach of the TWSC model assumes that flow on the major street is random. Where a coordinated signal system exists on the major street, organized platoons are formed and discharged at adjacent signalized intersections, producing a pattern of conflicting flows at the TWSC location that is anything but random.

This issue was not dealt with in Circular 212. However, the 1985 HCM added a very generalized approach that could be used to estimate the impact of platoon flow. In 1997 and again in 2010, this modification was greatly enhanced, but was also made extremely difficult to the point where software is necessary to attempt computations. Further, the explanation of the modification is spread across several chapters in the 2010 HCM, making it almost impossible to follow.

4.14.1 The 1985 Highway Capacity Manual

Figure 4.10 illustrates what happens when a controlled vehicle attempts to enter or cross an arterial with platoon flows created by a signal system.

Figure 4.10 shows an idealized case of a STOP- or YIELD-controlled approach facing platoon flow along an arterial. For simplicity, the 60-second cycle has a 50/50 split of green time for all signalized intersections along the arterial. It is further assumed that platoons are near-perfect, and fill the entire 30-s bandwidth along the arterial.

The illustration shows three STOP- or YIELD-controlled approaches located at various distances from the origin signal. Each faces a dramatically different set of conflicting volume patterns, as follows:

- *Vehicle 1*, at 990 ft, faces alternating platoons. Just as the platoon from the left ends, the platoon from the right arrives (and vice versa). There is no point at which a vehicle wishing to enter or cross at this location does not face a platoon

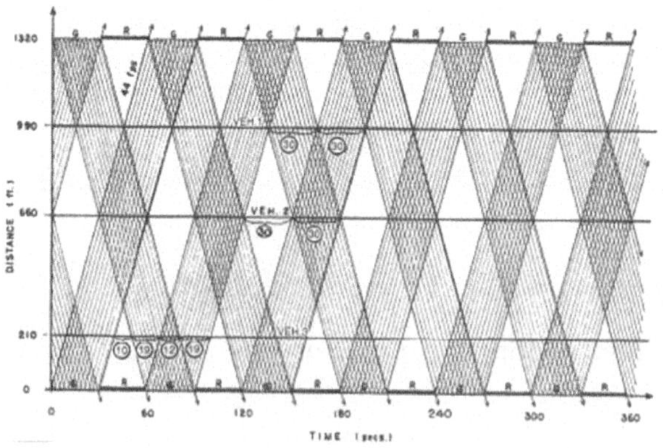

Fig. 4.10 Platoon flow across a TWSC approach 1985 HCM. *Source* "Highway Capacity Manual," *Special Report 209*, Transportation Research Board, Washington D.C., 1985 [4], Fig I10-2, pp. 10–29, used with permission of the National Academy of Science, Courtesy of National Academies Press

of vehicles from one direction. If it is assumed that there are no usable gaps with these platoons, then only left turns from the major street and right turns from the minor street can be made (as in each case the driver deals with only *one* directional platoon.

- *Vehicle 2*, at 660 ft, has a very different situation. Platoons in both directions arrive simultaneously. Thus, for 30 s, platoons in both directions are faced, and for 30 s, no platoons are faced. During every platoon cycle, half the time presents a great opportunity for clear merging or crossing maneuvers.
- *Vehicle 3*, at 210 ft, faces the most complex situation. For some of the time, both platoons are present. For some of the time, only one platoon is present, and for some of the time, neither platoon is present. Only when no platoon is present can left turns and through movements from the minor street be made; left turns from the major street and right turns from the minor street can be made (in one direction) when only one platoon is present. No minor street vehicle moves when *both* platoons are present.

The base methodology assumes random distribution of gaps. Applying the methodology more exactly to a platoon situation requires that time be partitioned into four categories:

- Time when both platoons are present.
- Time when no platoons are present.
- Time when only the Direction 1 platoon is present.
- Time when only the Direction 2 platoon is present.

Each of these must be separately analyzed; total capacity is the sum of the capacities available within each time period. Doing this assumes that the gap distribution is random *within each period*. It also requires that main street demand in each direction be divided into demand occurring within platoons, and demand occurring between platoons. This allows the demand to be divided into the four potential time periods noted above, and allows a separate analysis of each period.

The approach is numerically complex, but simple in concept. It still involves assumptions of randomness within each time period, and that an approximate split of main street demand within and outside of platoons can be established. It also assumes that platoons use the entire available bandwidth. Because of these assumptions, the analysis, while more accurate that simply applying the methodology to an average time period, is still a gross approximation.

4.14.2 The 1994 Update to the 85 HCM

The 1994 update did not include any changes in the general approach to platoon flow presented in the 1985 HCM.

4.14.3 The 1997 Update to the 85 HCM and the 2000 HCM

In 1985 and 1994, the impacts of platoon flow on the major street were treated in a very generalized way. For 1997, this approach was replaced by a more analytic treatment that accounted for specific conditions at the subject location.

The method presented accounted for the impact of upstream signalized intersections on the major street that were located within 0.25 miles of the subject STOP-controlled location. The effect of platoons were considered minor for more distant signals, and were not treated. The model was based primarily on the literature, specifically three studies of platoon dispersal downstream of signalized intersections [20–22].

Like the more general 1994 methodology, the 1997/2000 model recognizes that there are up to four platoon “regimes” that a vehicle from a STOP-controlled approach may face:

- Regime 1: No platoons present.
- Regime 2: Platoon from the left present.
- Regime 3: Platoon from the right present.
- Regime 4: Platoons from both the left and right present.

The analytic approach taken five component steps:

(1) Determining the time for queues to clear at each upstream adjacent signalized intersection within 0.25 miles of the TWSC location.
(2) Determining the proportion of time that the TWSC location is blocked by platoons by each upstream adjacent signalized intersection.
(3) Determining the duration of time that the TWSC location spends in Regimes 1 through 4 as defined.
(4) Determining the conflicting flows during each unblocked period of time.
(5) Determining the weighted capacity for each movement over the four regimes as defined.

Determining Queue Clearance Times

The major street flow faced by STOP-controlled vehicles consists of three components from each adjacent signalized intersection: Through vehicles on the major street (v_T), left turns from the minor street (at the signalized intersection) (v_L), and right turns from the minor street (at the signalized intersection) (v_R). Further, vehicles discharged form two groups at each adjacent signalized intersection: (a) a stable platoon the discharges at the saturation flow rate when the signal goes from RED to GREEN, and (b) random discharges of vehicles arriving after the saturated queue has cleared. The first group includes through vehicles on the major street arriving while the signal is RED, plus those who arrive and join the back of the queue while the initial standing queue is discharging. It may also include a second saturated discharge of left-turning vehicles from the minor street if they are made from an exclusive LT lane with a protected LT phase. The second group includes right turns from the minor street, left turns (that are not operating out of an exclusive lane with a protected LT phase) from the minor street, and through vehicles on the major and minor (in the case of protected LT phases and exclusive LT lane) street arriving after the saturated queue has discharged.

Obviously, the situation is quite complicated. The objective is to find the time necessary to discharge the standing queue of vehicles present when the signal turns GREEN (g_{q1}) and the time necessary to discharge vehicles that join the back of standing queue before it fully discharges (g_{q2}). This must be done for each of the two adjacent signalized intersections (assuming they are both within 0.25 miles of the subject location). In cases where one or both of the upstream signalized intersections have a protected LT phase with an exclusive LT lane on the minor street, there will be two component platoons to examine. Thus, the TWSC analysis may involve up to *four* potentially blocking platoons.

For *each* adjacent upstream signal (where applicable) and *each* blocking movement, the following computational sequence is followed:

$$P_i = R_{pi}(g_i/C) \tag{4.28}$$

where:

P_i proportion of blocking movement i that arrives on GREEN,
R_{pi} platoon ratio for blocking movement i,
g_i effective green time for blocking movement i, s, and
C cycle length, s.

P_i can be measured directly in the field, or estimated using Eq. 4.28.

The platoon ratio (R_{pi}) is based upon the arrival type. Arrival types are numbered 1 through 6, with AT-1 representing very poor progression, with few vehicles arriving on GREEN, and AT-6 representing almost perfect progression, with most vehicles arriving on GREEN. These arrival types are defined in the signalized intersection methodology (for the appropriate edition of the HCM), and platoon ratios are found there.

Remember that there are up to two potential blocking flows that form platoons—the through vehicles from the major street (v_T) and left turns from the minor street (v_L) when controlled by a protected phase and operating out of an exclusive lane.

For each potential blocking movement (up to two for each upstream signalized intersection), the time to discharge vehicles arriving on RED is given by:

$$g_{q1i} = \frac{v_i\, C\,(1 - P_i)}{s} \tag{4.29}$$

and the time to discharge vehicles joining the back of the standing queue during discharge is given by:

$$g_{q2i} = \frac{v_i\, C\, P_i\, g_{q1i}}{s g_i - v_i\, C\, P_i} \tag{4.30}$$

where:

v_i demand flow rate for blocking movement i (either v_T or v_L), veh/h,
s saturation flow rate for discharging vehicles, veh/hg and all other variables as previously defined.

The total time it takes to discharge queues from blocking movement i is the sum g_{q1i} and g_{q2i}.

The saturation flow rate is estimated from the signalized intersection methodology of the appropriate edition of the HCM, or default values may be used. Default values between 1500 and 1700 veh/hg/ln may be used in such cases, but would be multiplied by the number of lanes from which v_i is discharged.

Determining Time That TWSC Intersection is Blocked by Platoons

When a queue is discharged from an upstream signal, a dense or saturated platoon is formed. As that platoon travels downstream, a variety of elements will cause it to disperse. If there are no further downstream signals, the platoon will eventually fully disperse, and resemble random flow. The critical question, therefore, in determining how much time a particular STOP-controlled movement will be blocked by platoon flow on the major street is the degree to which the platoon will have dispersed by the time it reaches the STOP-controlled location. A platoon is determined to be blocking a STOP-controlled movement when the platoon headways are less than the critical gap (t_c). This portion of the model was based upon Ref. 22.

The model is based upon a number of platoon dispersion parameters, defined as follows:

$$\begin{aligned}
&\alpha \quad \text{(Table 4.18)}\\
&\beta = \frac{1}{1+\alpha}\\
&t_a = d/S_{prog}\\
&F = \frac{1}{1+\alpha\beta t_a}\\
&f = v_i/v_c\\
&v_{cMAX} = sf\left[1-(1-F)^{g_q}\right]\\
&v_{cMIN} = 1000\,N
\end{aligned} \tag{4.31}$$

where:

α platoon dispersal factor, Table 4.17,
β intermediate variable,
F intermediate variable,
t_a travel time from the signalized intersection to the TWSC location begin blocked, s,
v_i demand flow rate in platoon movement (either v_T or v_L) blocking the TWSC intersection, veh/h,
v_c conflicting volume for the subject movement under consideration, veh/h,

Table 4.17 Platoon dispersion factor, α

Median type	No. of through lanes per direction (major street)		
	1	2	3
Undivided	0.55	0.50	0.40
Divided, raised curb	0.45	0.40	0.35
TWLTL	0.40	0.35	0.30

Source "Highway Capacity Manual," *Special Report 209*, Transportation Research Board, Washington D.C., 1997, Table 10-4, pp. 10–18, used with permission of the National Academy of Science, Courtesy of National Academies Press

v_{cMAX} maximum platoon flow rate in the conflicting stream, veh/h,
v_{cMIN} minimum platoon flow rate in the conflicting stream, veh/h,
s saturation flow rate for conflicting movement, veh/hg,
g_q queue clearance time for conflicting movement, s, and
N number of lanes (per direction) on major street.

Using these parameters, the duration of the blocked period for either the through movement from the upstream signalized intersection or the left turn (if made from an exclusive LT lane with a protected phase) is found as:

$$t_{pi} = g_q - \frac{\ln\left[\left(1 - \frac{v_{cMIN}}{sf}\right)\left(\frac{v_{cMAX} - v_i R_p f}{v_{cMIN} - v_i R_p f}\right)\right]}{\ln[1 - F]} \quad when\ v_{cMAX} > v_{cMIN}$$

$$t_{pi} = 0 \quad when\ v_{cMAX} \leq v_{cMIN} \tag{4.32}$$

where all terms are as previously defined, and "ln" is the natural logarithm function.

The total time a TWSC location is blocked by an upstream signal discharge is given as:

$$p = \frac{t_{pT} + t_{pL}}{C} \tag{4.33}$$

where:

p proportion of time TWSC location is blocked by discharging platoons from one upstream signalized intersection,
t_{pT} time TWSC location is blocked by the through platoon from the upstream signalized intersection, s,
t_{pL} time TWSC location is blocked by the left-turn platoon (only if made from an exclusive lane with a protected phase) from the upstream signalized intersection, s, and
C cycle length, s.

Of course, there may be *two* upstream signalized intersections, one in each direction from the TWSC location, and *two* values of "p" would be computed in such cases.

Determining Platoon Event Periods
This situation is more complex than it might appear. Once the two blockage proportions, p, have been estimated for the two upstream signalized intersections, determining event periods should be relatively straightforward.

This is complicated by two situations that must be taken into account:

(1) The nature of the blockage changes based upon the subject movement from the TWSC intersection.
(2) Identifying movements becomes difficult, as both intersections can be referred to as "upstream," and platoons from the left on one TWSC approach are from the right on the other. Thus, the convenient movement numbering system for TWSC movements can actually be confusing in this context.

The general principal governing this process is as follows: When either one or two platoons block a subject movement, it cannot move. Only when there is *no* blockage, and the subject movement proceed. Consider the following:

- A through or left-turn movement from the STOP-controlled intersection is effectively blocked if a platoon is present from *either* the left or right (or both) directions.
- The left turn from the major street is blocked with a platoon is present *in the opposing direction.* It is not affected by a platoon in the direction of the major street left turn.
- The right turn from the STOP-controlled street is effectively blocked by a platoon *from the left*, but is not affected by a platoon from the right.

Therefore, it is convenient to adopt the following notation with respect to blocked periods of time (p). Where a subject TWSC movement may be blocked by a platoon approaching from either direction, the larger of the two will be called the *dominant* platoon, and the smaller of the two will be called the *subordinate* platoon and:

p_1 proportion of time blocked by upstream intersection 1,
p_2 proportion of time blocked by upstream intersection 2,
p_{dom} max (p_1, p_2), and
p_{sub} min (p_1, p_2).

For these movements, it doesn't matter which upstream signalized intersection is labeled "1" and which is labeled "2."

For left turns from the major street, the only platoon blockage that matters in the opposing direction. Thus, p_o is defined as the portion of time blocked by the opposing platoon(s).

For right turns from the minor street, the only platoon blockage that matters is the one arriving from the left. Thus, p_L is defined as the portion of time blocked by the platoon arriving from the left. "Left," of course, will change based upon which minor street right turn is under consideration.

At this point, the model uses a simplification: On average, it is assumed that about ½ of the subordinate platoon is present at the same time as the dominant platoon. Based upon this assumption, one of two conditions may exist:

Table 4.18 Proportion of time movement i is unblocked by platoons average case 1997 update and 2000 HCM

Movement that is blocked		Proportion of time Mvt i is unblocked by platoons, p_{xi}	
Mvt No., i	Type	Unconstrained condition	Constrained condition
1	LT from major street	$1 - p_o$	$1 - p_o$
4	LT from major street	$1 - p_o$	$1 - p_o$
7	LT from minor street	$1 - (p_{dom} + p_{sub}/2)$	0
8	TH from minor street	$1 - (p_{dom} + p_{sub}/2)$	0
9	RT from minor street	$1 - p_L$	$1 - p_L$
10	LT from minor street	$1 - (p_{dom} + p_{sub}/2)$	0
11	TH from minor street	$1 - (p_{dom} + p_{sub}/2)$	0
12	RT from minor street	$1 - p_L$	$1 - p_L$

Source Modified from “Highway Capacity Manual,” *Special Report 209*, Transportation Research Board, Washington D.C., Table 10-6, pp. 10–19, used with permission of the National Academy of Science, Courtesy of National Academies Press

- *Constrained*: condition when one or both platoons is always present, which occurs when $p_{dom} + p_{sub}/2 > 1.00$.
- *Unconstrained*: condition when there is some period of time during which neither platoon is present, which occurs when $p_{dom} + p_{sub}/2 \leq 1.00$.

If we return to the original movement-numbering scheme of Fig. 4.1, then the values shown in Table 4.18 estimate the amount of time each movement at a TWSC intersection is *unblocked* by a platoon. This is the only period of time when vehicles from the subject movement can proceed.

Determining Conflicting Flow During Unblocked Periods

The total conflicting flow for each movement i has already been computed, assuming to platooning within these conflicting flows. The base value computed for no platoon blockage must be modified to account for the impacts of platoon flow on the major street. Each conflicting flow is modified as follows:

$$v_{cmi} = \frac{v_{ci} - s(1 - p_{xi})}{p_{xi}} \; for \; v_{ci} > s(1 - p_{xi})$$
$$v_{cmi} = 0 \; for \; all \; other \; cases. \tag{4.34}$$

where:

v_{cmi} modified conflicting flow rate for movement i, veh/h,
v_{ci} conflicting flow rate for movement i, assuming no platoon impacts, veh/h,
s saturation flow rate for movement i, veh/hg, and
p_{xi} proportion of time that movement i is unblocked.

Capacity of the Subject Movement

The movement capacity, c_m, of each movement i, may now be computed as:

$$c_{mi}^{*} = p_{xi} c_{mui} \tag{4.35}$$

where:

c_{mi}* movement capacity of movement i, adjusted for the effects of platoon blockage, veh/h,

c_{mui} capacity of movement i during the unblocked portion of time, veh/h, computed assuming random flow during the unblocked portion of time and using v_{cMi} as a starting point.

This is obviously a difficult model to implement numerically, and hand computations are beyond what would be attempted by a reasonable user. Thus, implementing the TWSC methodology from 1997 onward is dependent upon available software, primarily the HCS Package.

4.14.4 The 2010 Highway Capacity Manual

In the 1997 update and 2000 HCM, adjustments to account for the effect of platoon flows on the major street were applied to movement capacities. In 2010, the adjustments are made to potential capacity values. As previously, the approach relies upon predicting the "unblocked" portion of time for each minor street movement. "Unblocked" is that time when a platoon from the signalized intersection either to the left and/or the right blocks the minor street movement from taking place.

The proportion of time that each movement "x" is blocked (p_x) is shown in Table 4.19.

Table 4.19 Proportion of time that movement x is blocked 2010 HCM

Subject movement (x)	Single-stage movements	2-stage movement	
		Stage I	Stage II
1, 1U	p_{b1}	N/A	N/A
4, 4U	p_{b4}	N/A	N/A
7	p_{b7}	p_{b4}	p_{b1}
8	p_{b8}	p_{b4}	p_{b1}
9	p_{b9}	N/A	N/A
10	p_{b10}	p_{b1}	p_{b4}
11	p_{b11}	p_{b1}	p_{b4}
12	p_{b12}	N/A	N/A

Source "Highway Capacity Manual", Transportation Research Board, Washington D.C., 2010 [6], Exhibit 19-12, pp. 19–17, used with permission of the National Academy of Science, Courtesy of National Academies Press

As noted previously, the estimation of the values shown in Table 4.19 is extremely complex, and the description in the HCM is far from clear. Part of the problem is that the model is described over four different chapters (17, 18, 19, 30, 32—2010 HCM) with cross-references. While similar in framework to the 2000 HCM model, it is more complex, and is not detailed here. Where existing situations are being analyzed, these values can be measured in the field—but only for the signal timing and coordination plan that is in effect when field observations are made.

Once the blocked times are determined, the conflicting flow rates for various subject movements *during the unblocked portion of time* are estimated as follows:

$$v_{cux} = \frac{v_{cx} - (1.5v_{c\min} p_{bx})}{1 - p_{bx}} \quad for\ v_{cx} > 1.5v_{c\min} p_{bx}$$
$$v_{cux} = 0 \quad for\ all\ other\ cases. \tag{4.36}$$

where:

v_{cux} conflicting flow rate for movement x during unblocked portion of time, veh/h,
v_{cx} conflicting flow rate for movement x assuming no signal (platoon) effects on the major street, veh/h,
v_{cmin} minimum conflicting flow rate for movement x, assumed to be 1000 N where N is the number of through lanes per direction on the major street, veh/h, and
p_{bx} blocked portion of time for movement x.

Once the conflicting flow rate during the unblocked period is computed, then the potential capacity of the movement is found as:

$$c_{rx} = v_{cux} \frac{e^{-v_{cux} t_{cx}/3600}}{1 - e^{-v_{cux} t_{fx}/3600}}$$
$$c_{px} = c_{rx}(1 - p_{bx}) \tag{4.37}$$

where c_{rx} is the potential capacity of movement x, assuming random flow during the unblocked portion of time, and assuming that 100% of time is unblocked. This is multiplied by the proportion of time that is unblocked to produce an estimate of potential capacity (c_{px}).

In 2010, the basic approach to platoon impacts was simplified (in HCM Chapter 17), but made more complex to implement, with explanation of the estimation of "blocked times" spread out over four HCM chapters. Again, the user is fully dependent upon software for the implementation of this element.

4.14.5 The 2016 Highway Capacity Manual

The 2016 HCM did not help clarify this element. The TWSC chapter references the Urban Street Segments chapter, which contains additional references to other sections of the manual. As it was virtually impossible to fully document all of the details of the 2010 HCM on the effect of platoon flows, it is just as impossible to do it for the 2016 HCM. From a cursory evaluation of related materials, it does not appear that any fundamental changes were introduced in 2016.

4.14.6 A Controversy

In 2015, a paper by Kyte et al. [23] demonstrated that applying the current adjustment for platoon flow (2010 HCM) did not meaningfully or logically change the results, and questioned whether it should be included at all. There has been no significant study based upon field data to evaluate this potential effect, and it is based primarily upon analytic modeling. The overall reliance on software for computations has allowed increasingly complex models to be implemented.

Little thought, however, has been given to presenting some of the more difficult aspects of these models in written form in a manner that allows professionals to see into the literal black box they are being asked to use.

In the case of the impact of platoons on TWSC intersections, Kyte et al. raises an interesting question: If the adjustment has no significant impact on results, why include it?

4.15 A Final Thought

A fundamental question, often overlooked, is important: do users need all of the detail of the current TWSC methodology to get what they need from a TWSC intersection assessment? Given that the hierarchical nature of movement priorities in the model is rigid, and, in practical terms, can be expected to break down when capacity is approached or delays become significant, can a simpler approach be developed that provides somewhat less complexity and precision with less effort? Particularly for unsignalized intersections, the trade-off between complexity and accuracy is an important one. Relying completely on the proverbial "black box" (software) for implementation carries with it its own risks, particularly the inability to know whether the software is implementing the stated procedure properly, or whether inevitable "bugs" exist.

4.16 Sample Calculations

Appendix I of this book contains a sample problem worked out using each generation of HCM, as well as a comparison and discussion of the relative results.

References

1. Highway Capacity Manual (1950) Bureau of public roads. U.S. Government Printing Office, Washington D.C.
2. Highway Capacity Manual (1965) Special report 87. Transportation Research Board, Washington D.C.
3. Interim Materials on Highway Capacity (1980) Circular 212. Transportation Research Board, Washington D.C.
4. Highway Capacity Manual (1985) Special report 209. Transportation Research Board, Washington D.C.
5. Highway Capacity Manual (2000) Transportation research board. 4th edn. Washington D.C.
6. Highway Capacity Manual (2010) Transportation research board, 5th edn. Washington D.C.
7. Highway Capacity Manual (2016) A guide for multimodal mobility analysis, 6th edn. Transportation Research Board, Washington D.C.
8. Capacity of at-grade junctions. Organization for Economic Co-operation and Development, Paris, France, 1974.
9. Merkblatt for Lichtsignalanlagen an Landstrassen (1972) Ausgabe 1972. Forschungshasellschaft fur das Strassebwesen, Koln, Germany
10. Siegloch W (1973) Die Leistungsermittlung an Knotenpunkten ohne Lichtsignalsteurung, Schriftenreihe Strassenbau und Strassenverkehrstechnik, Heft 154. Bonn, Germany
11. Harders J (1968) Die Leistungsfahigkeit nicht Signalgeregelter Stadatischer Verkehrsknoten, Schriftenreihe Strassenbau und Strassenverkehrstechnik, Heft 76. Bonn, Germany
12. Brilon W, Grossman M (1991) Akatualisiertes Berechnungsverfahren fur Knotenpunkte ohne Lichsignalanlagen, Schriftenreihe Forschung Strassenbau und Strassenverkehrstechnic, Heft 596, Bonn, Germany
13. Kittelson W, Vandehey M (1991) Delay effects on driver gap acceptance characteristics at two-way STOP-controlled intersections. Transportation research record 1320. Transportation Research Board, Washington D.C.
14. Troutbeck R (1992) Estimating the critical acceptance gap from traffic movements, research report 92-5. Queensland University of Technology, Australia
15. Kyte M, Lall K, Mahfood N (1992) An empirical method to estimate capacity and delay on the minor street approach of a two-way STOP-controlled intersection. Transportation Research Record 1365. Transportation Research Board, Washington D.C.
16. Wu N (1994) An approximation for the distribution of queue lengths at unsignalized intersections. In: Proceedings of the second international symposium on highway capacity, vol 2. Australian Road Research Board Ltd., Australia
17. Interim Materials on Unsignalized Intersection Capacity (1991) Circular 373. Transportation Research Board, Washington D.C.
18. Kyte M, Tian Z, Mir Z, Hameedmansoor Z, Kittleson W, Vandehey M, Robinson B, Bondzio L, Wu N, Troutbeck R (1996) Capacity and level of service at unsignalized intersections: volume 1-two way stop controlled intersections. Final Report, National Cooperative Highway Research Program Project 3–46, Transportation Research Board, Washington D.C.
19. Highway Capacity Manual (1994) Special report 209. Transportation research board, Washington D.C.
20. Robertson DI (1983) Coordinating traffic signals to reduce fuel consumption. In: Proceedings of the royal society, vol 387A, England, May 9, 1983
21. Yu L, Van Aerde M (1995) Implementing TRANSYT's macroscopic platoon dispersion in microscopic traffic simulation models. In: 75th annual meeting of the transportation research board. Transportation Research Board, Washington D.C.
22. Bonneson J, Fitts J (1997) Effect of an upstream signal on non-priority movement capacity and delay. Transportation research record 1572, Transportation Research Board, Washington D.C.
23. Kyte M, Enas A, Troutbeck R (2015) The effect of upstream signals on the capacity of a downstream TWSC Intersection, Transportation Research Record 2483. Transportation Research Board, Washington D.C.

Chapter 5
Unsignalized Intersections: All-Way STOP Control (AWSC)

All-way STOP-controlled intersections have a STOP sign on every approach to the intersection. The 1965 HCM referred to these as "four-way STOP-controlled" intersections, but three-leg intersections and intersections of one-way streets, which have only two approaches, also qualify if fully STOP controlled. YIELD signs, of course, may not be used in such situations.

5.1 The 1965 Highway Capacity Manual: A Meager Beginning

Based upon a single study [1], the 1965 HCM [2] included capacity criteria for multiway STOP control, as shown in Tables 5.1 and 5.2. The guidance offered is general, and no references to level of service are included.

The 1965 HCM put all-way STOP-controlled intersections on the map. While it did not provide any comprehensive insight to or modeling of these intersections, it was an important statement that such intersections were worthy of interest and future work, which would not come for some time.

The use of all-way STOP control is a subject of debate in the profession. Because of this, the Manual on Uniform Traffic Control Devices [3] provides only general criteria for their use, which is never mandated by any warrant. Some traffic engineers love them, others avoid them at all costs. The fundamental negative of AWSC is that drivers really don't know how to use them, and right-of-way confusion leads to a lot of uncertain movements. On the other hand, since everyone stops, the potential for high-speed collisions is virtually eliminated.

All-way STOP signs are sometimes installed as an interim measure at intersections that are awaiting signalization. In other cases, unique safety or efficiency considerations can make such an installation desirable.

E. S. Prassas and R. P. Roess, *The Highway Capacity Manual: A Conceptual and Research History Volume 2*, Springer Tracts on Transportation and Traffic 12,
https://doi.org/10.1007/978-3-030-34480-1_5

Table 5.1 Total capacity (All Legs) of a four-way STOP controlled intersection with balanced demand (50/50 on Two Intersecting Streets)—1965 HCM

Number of lanes per approach (Street 1)	Number of lanes per approach (Street 2)	Total capacity (veh/h)
1	1	1900
1	2	2800
2	2	3600

Source "Highway Capacity Manual," *Special Report 87*, Transportation Research Board, Washington D.C., 1965, Table 6.7, Pg. 158, used with permission of the National Academy of Science, Courtesy of National Academies Press

Table 5.2 Total capacity of a four-way STOP controlled intersection with one lane on each approach for varying demand splits—1965 HCM

Demand split (Street 1/Street 2)	Total capacity (veh/h)
50/50	1900
55/45	1800
60/40	1700
65/35	1600
70/30	1550

Source "Highway Capacity Manual" *Special Report 87*, Transportation Research Board, Washington D.C., Table 6.8, Pg. 158, used with permission of the National Academy of Science, Courtesy of National Academies Press

5.2 The 1985 Highway Capacity Manual

The next appearance of all-way STOP-controlled intersections in the HCM occurs in 1985, but is not a significant improvement to the information available in 1965. It repeats Tables 5.1 and 5.2 from the 1965 HCM with minor reformatting. Table 5.3 reflects some new material from a 1981 study [4]. It shows approximate LOS C service volumes (full hour, in veh/h) for various demand splits and lane configurations.

If anything, the 1985 HCM highlighted the dearth of research on the subject of all-way STOP-controlled intersection operations. It spurred the Committee to actively seek new funding and research on the subject.

5.3 The 1994 Update to the 85 HCM

The 1994 update is the first time a formal methodology for the analysis of multiway STOP-controlled intersections is presented. Specifically, it is a model for 4-way STOP-controlled intersections, but the structural model can be applied to cases with

Table 5.3 Approximate LOS C service volumes (veh/h) for 4-way STOP-controlled intersections—1985 HCM

Demand split	Number of lanes		
	2 by 2	2 by 4	4 by 4
50/50	1200	1800	2200
55/45	1140	1720	2070
60/40	1080	1660	1970
65/35	1010	1630	1880
70/30	960	1610	1820

Source "Highway Capacity Manual" *Special Report 209*, Transportation Research Board, Washington D.C., 1985, Table 10–6, Pg. 10–14, used with permission of the National Academy of Science, Courtesy of National Academies Press

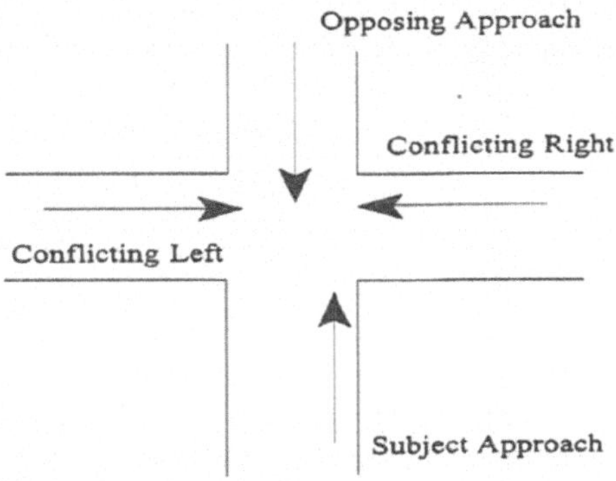

Fig. 5.1 Approaches at a 4-Way STOP-controlled intersection. *Source* "Highway Capacity Manual" *Special Report 209*, Transportation Research Board, Washington D.C., 1997, Fig. 10.21, Pg 10–59, used with permission of the National Academy of Science, Courtesy of the National Academies Press

fewer than four approaches, all of which are STOP-controlled. The methodology is based upon U.S research detailed in Ref [5].

The model treats each approach separately. A capacity and an average delay are estimated for each approach, and a LOS assigned.

The operation of a multiway STOP-controlled intersection is quite complex. A vehicle on one approach may face four different scenarios:

(1) No vehicles present on any other approach.
(2) Vehicles present on the *opposing* approach only.
(3) Vehicles present on one or two *conflicting* approaches only.
(4) Vehicles present on both the *opposing* approach and on one or two *conflicting* approaches.

Each case presents a different set of decisions for the controlled driver to make, and a different set of priorities. While, technically, the basic rules-of-the-road apply to drivers at a multiway STOP intersection, it is very difficult for drivers to apply it, given that all vehicles must stop before proceeding.

Figure 5.1 illustrates the designation of approaches. The approach under consideration is the *subject approach.* Approaches from the right and left are *conflicting approaches*, and the *opposing approach* is opposite the subject approach. The model was developed using a two-phase signalized analogy for cases with single-lane approaches, and a four-phase signalized analogy for cases with multiple lanes on approaches. This framework was verified using field data, and produces some interesting results.

Where there are no vehicles on the opposing or conflicting approaches, adding lanes to the subject approach proportionately increases capacity. However, when vehicles are present on other approaches, capacity is not significantly increased by adding lanes. This is due to the additional complexity confronting the driver in multilane cases, which makes the use of multiple lanes far less efficient than a single lane. Therefore, the addition of lanes on a multiway STOP-controlled approach does not add significant capacity to that approach.

The methodology consists of two regression-based equations, one which predicts the capacity of the subject approach, and one which predicts the average delay:

$$c_s = 1000\,P_s + 700P_o + 200N_s - 100N_o - 300P_{LTo} + 200P_{RTo} - 300P_{LTc} + 300P_{RTc} \quad (5.1)$$

$$D_x = e^{3.8\,(v_s/c_s)} \quad (5.2)$$

The variables in these equations are based entirely upon demand flow rates by movement on each approach. The demand flow rates should be for the peak 15 min within the hour, which can be measured directly, or can be estimated using a peak hour factor (PHF). It should be noted that for many cases, the PHF is taken to be 1.00 unless significant variation in flow throughout the analysis hour is observed. The variables in Eqs. (5.1) and (5.2), along with demand flow rate definitions are shown in Table 5.4. Note that symbols for these variables have been changed from the 1994 update to reflect consistency with other methodologies presented herein.

Note that there is no accounting for heavy vehicles or pedestrians in the model. The 1994 update warns users that significant presence of either could decrease capacities and increase delays. The data used in calibration included some trucks and some pedestrians, but they were not a significant presence at the study sites.

Level of service is determined directly from the delay estimate of Eq. (5.2). The definitions are the same as those for TWSC intersections in the 1994 HCM, and are shown in Table 5.5.

5.4 The 1997 Update, and the 2000–2016 HCMs

The AWSC methodology was replaced in 1997 with a new model developed under NCHRP Project 3–46 [6]. While it was based on some of the same principles as the

Table 5.4 Variables used in Eqs. (5.1) and (5.2) 1994 Update

Variable	Definition	Equation
v_{SL}	Left-turn flow rate, subject approach, veh/h	
v_{ST}	Through flow rate, subject approach, veh/h	
v_{SR}	Right-turn flow rate, subject approach, veh/h	
v_S	Total flow rate, subject approach, veh/h	$v_S = v_{SL} + v_{ST} + v_{SR}$
v_{OL}	Left-turn flow rate, opposing approach, veh/h	
v_{OT}	Through flow rate, opposing approach, veh/h	
v_{OR}	Right-turn flow rate, opposing approach, veh/h	
v_O	Total flow rate, opposing approach, veh/h	$v_O = v_{OL} + v_{TL} + v_{OR}$
v_{C1L}	Left-turn flow rate, conflicting approach 1, veh/h	
v_{C1T}	Through flow rate, conflicting approach 1, veh/h	
v_{C1R}	Right-turn flow rate, conflicting approach 1, veh/h	
v_{C2L}	Left-turn flow rate, conflicting approach 2, veh/h	
v_{C2T}	Through flow rate, conflicting approach 2, veh/h	
v_{C2R}	Right-turn flow rate, conflicting approach 2, veh/h	
v_C	Total flow rate, both conflicting approaches, veh/h	$v_C = v_{C1L} + v_{C1T} + v_{C1R} +$ $v_{C2L} + v_{c2T} + v_{c2R}$
N_s	Number of lanes, subject approach	
N_o	Number of lanes, opposing approach	
P_{LTo}	Proportion of left turns in opposing flow rate	$P_{LTo} = v_{LTo}/v_O$
P_{RTo}	Proportion of right turns in opposing flow rate	$P_{RTo} = v_{RTo}/v_O$
P_{LTs}	Proportion of left turns in subject flow rate	$P_{LTs} = v_{LTs}/v_S$
P_{RTs}	Proportion of right turns in subject flow rate	$P_{RTs} = v_{RTs}/v_S$
c_s	Capacity of subject approach, veh/h	Equation (5.1)
D_x	Average delay on subject approach, s/veh	Equation (5.1)

Table 5.5 Levels of service—1994 update

Level of service	Average total delay (s/veh)
A	≤5
B	>5 and ≤ 10
C	>10 and ≤ 20
D	>20 and ≤ 30
E	>30 and ≤ 45
F	>45

Source "Highway Capacity Manual" *Special Report 209*, Transportation Research Board, Washington D.C., 1994, Table 10–3, Pg 10–12, used with permission of the National Academy of Science, Courtesy National Academies Press

1994 methodology, it recognized the inter-dependency of many of the underlying variables, creating an iterative approach that sometimes took a large number of iterations to close to a result. While there were no fundamental changes between 1997 and 2000, the presentation and explanation of the methodology was significantly revised for the 2000 HCM [7]. There were no significant updates in the 2010 or 2016 HCMs.

The model continues to use the movement designations of the 1994 methodology, illustrated in Fig. 5.1. The number of conflict situations that a subject vehicle may face has, however, been increased to five:

- *Case 1*: Vehicles on subject approach only.
- *Case 2*: Vehicles on subject and opposing approaches only.
- *Case 3*: Vehicles on subject and one conflicting approach only.
- *Case 4*: Vehicles on subject and two other approaches (either opposing or conflicting) only.
- *Case 5*: Vehicles on all approaches.

Saturation headways and service times are dependent upon which conflict case is encountered, with such times increasing as the conflict case becomes more complicated. Case 1 is the least complicated; each subsequent case is more complicated than the previous case. A key part of the methodology is determining the probability of each case occurring to determine representative values of saturation headway and service time.

The methodology continues to focus on each approach individually. There are four subject approaches, each with a capacity and a delay that will be estimated by the model. A number of new variables are introduced, as defined in Table 5.6.

The departure headway, h_d, is defined as the departure time of a vehicle in queue after the departure of the previous vehicle in queue. It is, in effect, the saturation flow rate for the subject approach when a continuous queue is present on the subject approach. This departure headway is the key parameter used in the methodology to estimate capacity. Research [7] has shown that this parameter is dependent on several factors:

- The degree of conflict faced by a subject driver, measured by the presence of vehicles on the opposing and conflicting approaches,
- The intersection geometry, specifically the number of lanes on each approach,
- The directional actions (LT, TH, RT) of interacting vehicles, and
- The mix of vehicle types, specifically heavy vehicles.

Due to the need to iterate, methodology becomes extremely complex, even in relatively simple applications. The need to iterate involves the following interdependency: the intensity of traffic demand on an approach is based upon its capacity; its capacity is based on the intensity of traffic demand on the opposing and conflicting approaches. Iterations (sometimes numbering into the hundreds) are required to reach a stable estimate of the departure headway and service time in each approach lane of the intersection. The methodology allows for the detailed analysis of cases

Table 5.6 Variables used in AWSC methodology—1997–2016

Variable	Definition
c	Capacity of subject approach (veh/h)
D	Control delay on subject approach (s/veh)
h_{is}	Saturation headway for conflict case i (s)
h_d	Departure headway for subject movement (s)
h_{adj}	Departure headway adjustment for proportion of left turns, right turns, and heavy vehicles (s)
P_{LT}	Proportion of left turns
P_{RT}	Proportion of right turns
P_{HV}	Proportion of heavy vehicles
P_{adj}	Adjusted probability of degree of conflict
$P[C_i]$	Probability of degree-of-conflict case i
V	Volume (veh/h)
v	Flow rate (pc/h) = V/PHF
X	Degree of utilization, or $vh_d/3600$
α	Constant to adjust degree-of-conflict case probability to account for serial correlation between forecast headways

having two lanes on each approach – meaning that each subject approach must consider opposing and conflicting flows in up to 6 other lanes. Needless to say, like TWSC intersections, implementation of the methodology for AWSC intersections (since 1997) is almost exclusively through the application of software, specifically the HCS package.

Like the 1994 methodology, the model is based on an analogy to a two-phase signal (for the intersection of two one-way streets) or a four-phase signal (for the intersection of two two-way streets). In each case, each approach is assumed to have an "exclusive" phase. This is because at an AWSC location, vehicles generally move from one approach at a time (a conservative approach).

The 2000 HCM contains very detailed presentations and extensive sample problems illustrating this methodology. Herein, the fundamental steps in the model are presented analytically to hopefully simplify understanding of the methodology.

The basic approach can be applied to the simplest case of two one-way streets, each with one lane per approach, to the most complex with two two-way streets each having two or three lanes.

Table 5.7 Geometry groups for AWSC intersections—2000 HCM

Geometry group	Intersection configuration	Number of lanes on:		
		Subject approach	Opposing approach	Conflicting approaches
1	4-leg or T	1	1	1
2	4-leg or T	1	1	2
3a/4a	4 leg or T	1	2	1
3b	T	1	2	2
4b	4-leg	1	2	2
5	4-leg or T	2	1 or 2	1 or 2
5	4-leg or T	3	1[a]	1[a]
6	4-leg or T	3	3	3

[a]If the number of lanes on the subject approach is 3 and the number of lanes on either the opposing or conflicting approaches is 1, the geometry group is 5. Otherwise, if the number of lanes on the subject approach is 3, the geometry group is 6

NOTE When the stem of a T-intersection is the subject approach, the opposing approach obviously has 0 opposing lanes

Source "Highway Capacity Manual" Transportation Research Board, Washington D.C., 2000, Exhibit 17–32, Pg 17–42, used with permission of the National Academy of Science, Courtesy of National Academies Press

5.4.1 Geometry Groups

In each situation, the geometry (with respect to the number of lanes on each approach) must be defined. The methodology classifies intersections into one of eight groups (or subgroups) as shown in Table 5.7.

STOP control of any kind on a three-lane approach is extremely rare (or should be). STOP control on an approach of 4 or more lanes is not recommended under any circumstances.

5.4.2 Saturation Headways

The model considers each subject approach in the context of how many different situations might be encountered on opposing and conflicting approaches. It explicitly allows for up to 6 lanes on the opposing and conflicting approaches (2 for each), although the concept could clearly be extended for more lanes. Such situations, as noted, are relatively rare. The saturation headway is estimated for each of the potential opposing and conflicting scenarios that might occur. Then, the methodology computes a departure headway based upon the probability of each of these scenarios occurring. Unfortunately, in the case of 6 lanes on the opposing and conflicting approaches, there are up to 64 different scenarios that might exist. This, of course, greatly simplifies if there is only one lane on each approach. In that case, only 5 scenarios exist, corresponding to the five conflict cases defined previously.

The model, however, is structured to determine the departure headway (h_d) of *each approach* separately. In each case, there are up to 64 opposing/conflicting scenarios that might be encountered. The scenarios include the five basic opposing/conflict cases defined, and the number and position of each opposing/conflicting vehicle that might be present. Where 6 opposing and conflicting lanes exist, there are many combinations of vehicle positions that might occur. Table 5.8 shows the 64 cases that are generated by such a situation.

For *each* scenario present (up to 64), the saturation headway (h_{si}) must be determined. The saturation headway for scenario i is computed as:

$$h_{si} = h_{basei} + h_{adj} \tag{5.3}$$

where:

h_{si} saturation headway for scenario i, s/veh,
h_{basei} base saturation headway for scenario i, s/veh, and
h_{adj} adjustment to the base saturation headway, s/veh.

The base saturation headways are shown in Table 5.9. The adjustment to the base saturation headway is computed using Eq. (5.4) and the values are shown in Table 5.10.

$$h_{adj} = h_{LTadj} P_{LT} + h_{RTadj} P_{RT} + h_{HVadj} P_{HV} \tag{5.4}$$

where

h_{adj} adjustment to base saturation headway, s/veh,
h_{adjLT} adjustment factor for left turns,
h_{adjRT} adjustment factor for right turns,
h_{adjHV} adjustment factor for heavy vehicles,
P_{RT} proportion of right turns in subject movement,
P_{LT} proportion of left turns in subject movement, and
P_{HV} proportion of heavy vehicles in subject movement.

5.4.3 Departure Headways

This is conceptually simple, but computationally very complex. At this point, the saturation headways for up to 64 scenarios (of vehicles that may be on opposing or conflicting approaches) has been determined. Now, the probability that each of these scenarios will exist must be determined. The departure headway is simply the sum of the products of (up to) 64 saturation headways and the probability that each scenario exists.

Table 5.8 Possible Degree of conflict cases for multilane AWSC intersections (Two lanes on each approach) 1997–2016

Scenario	DOC Case	No. of Opp/Con Vehs	Opposing approach		Conflicting left app		Conflicting right app	
			L1	L2	L1	L2	L1	L2
1	1	0	0	0	0	0	0	0
2	2	1	1	0	0	0	0	0
3			0	1	0	0	0	0
4	2	2	1	1	0	0	0	0
5	3	1	0	0	0	1	0	0
6			0	0	1	0	0	0
7			0	0	0	0	1	0
8			0	0	0	0	0	1
9	3	2	0	0	1	1	0	0
10			0	0	0	0	1	1
11	4	2	0	0	0	1	0	1
12			0	0	1	0	0	1
13			0	0	1	0	1	0
14			0	0	0	1	1	0
15			0	1	0	1	0	0
16			1	0	1	0	0	0
17			0	1	0	0	1	0
18			1	0	0	1	0	0
19			0	0	1	0	0	0
20			0	1	0	0	0	1
21			1	0	0	0	1	0
22			1	0	0	0	0	1
23	4	3	0	0	0	1	1	1
24			0	0	1	1	0	1
25			0	0	1	1	1	0
26			1	0	1	1	0	0
27			1	1	1	0	0	0
28			1	1	0	0	1	0
29			1	1	0	0	0	1
30			0	1	1	1	0	0
31			1	0	0	0	1	1
32			0	0	1	0	1	1
33			1	1	0	1	0	0
34			0	1	0	0	1	1
35	4	4	1	1	0	0	1	1
36			0	0	1	1	1	1
37			1	1	1	1	0	0

(continued)

Table 5.8 (continued)

Scenario	DOC Case	No. of Opp/Con Vehs	Opposing approach		Conflicting left app		Conflicting right app	
			L1	L2	L1	L2	L1	L2
38	5	3	0	1	0	1	0	1
39			1	0	0	1	1	0
40			0	1	1	0	1	0
41			0	1	0	1	1	0
42			0	1	1	0	0	1
43			1	0	1	0	0	1
44			1	0	0	1	0	1
45			1	0	1	0	1	0
46	5	4	1	0	0	1	1	1
47			0	1	1	1	1	0
48			0	1	1	1	0	1
49			1	0	1	0	1	1
50			1	0	1	1	1	0
51			0	1	0	1	1	1
52			1	1	1	0	0	1
53			1	0	1	1	0	1
54			0	1	1	0	1	1
55			1	1	0	1	1	0
56			1	1	0	1	0	1
57			1	1	1	0	1	0
58	5	5	1	0	1	1	1	1
59			1	1	0	1	1	1
60			1	1	1	0	1	1
61			0	1	1	1	1	1
62			1	1	1	1	1	0
63			1	1	1	1	0	1
64	5	6	1	1	1	1	1	1

Source "Highway Capacity Manual" Transportation Research Board, Washington D.C., 2000 [7] Exhibit 17–30, pg 17–40, used with permission of the National Academy of Science, Courtesy of National Academies Press

Unfortunately, this becomes iterative, as the probabilities for each of the (up to) 64 scenarios depend on the degree of utilization (X) for each opposing and conflicting lane. That value, however, depends upon knowing the departure headways for these lanes, which requires the probabilities to be computed.

There are up to 6 lanes on opposing and conflicting approaches. In each case, there is a probability that there will be a vehicle present, and a probability the there is not vehicle present. If a_j is the number of vehicles present on opposing or conflicting lane j, its value can be 1 or 0. The probability of given scenario existing is the product of the probabilities that a vehicle is present in the defined locations for that scenario, and that no vehicle is present in other opposing or conflicting lanes.

Table 5.9 Base Saturation Headways for AWSC intersections 1997–2016

DOC*	No. of Veh**	Base saturation flow headway (s) for geometry group							
		1	2	3a	3b	4a	4b	5	6
1	0	3.9	3.9	4.0	4.3	4.0	4.5	4.5	4.5
2	1 2 ≥3	4.7	4.7	4.8	5.1	4.8	5.3	5.0 6.2	6.0 6.8 7.4
3	1 2 ≥3	5.8	5.8	5.9	6.2	5.9	6.4	6.4 7.2	6.6 7.3 7.8
4	2 3 4 ≥5	7.0	7.0	7.1	7.4	7.1	7.6	7.6 7.8 9.0	8.1 8.7 9.6 12.3
5	3 4 5 ≥6	9.6	9.6	9.7	10.0	9.7	10.2	9.7 9.7 10.0 11.5	10.0 11.1 11.4 13.3

*Degree of conflict case. **No. of vehicles on opposing and conflicting approaches

Source "Highway Capacity Manual" Transportation Research Board, Washington D.C., 2000 [7] Exhibit 17–35, Pg 17–44, used with permission of the National Academy of Science, Courtesy of National Academies Press

Table 5.10 Adjustments to base saturation headways at AWSC intersections 1997–2016

Adj Factors	Saturation headway adjustments (s) for geometry group							
	1	2	3a	3b	4a	4b	5	6
h_{adjLT}	0.2	0.2	0.2	0.2	0.2	0.2	0.5	0.5
h_{adjRT}	−0.6	−0.6	−0.6	−0.6	−0.6	−0.6	−0.7	−0.7
h_{adjHV}	1.7	1.7	1.7	1.7	1.7	1.7	1.7	1.7

Source "Highway Capacity Manual" Transportation Research Board, Washington D.C., 2000 [7] Exhibit 17–33, Pg 17–42, used with permission of the National Academy of Science, Courtesy of National Academies Press

Table 5.11 Probability of a_j for AWSC intersections 1997–2016

a_j	V_j	$P(a_j)$
1	0	0
0	0	1
1	>0	X_j
0	>0	$1–X_j$

Source "Highway Capacity Manual" Transportation Research Board, Washington D.C., 2000 [7] Exhibit 17–34, Pg 17–43, used with permission of the National Academy of Science, Courtesy of National Academies Press

The probability a_j (being either 1 or 0) in any given case is shown in Table 5.10. "v_j" is the demand flow rate in opposing or conflicting lane j, and $P(a_j)$ is the probability that the stated value of a_j will exist (Table 5.11).

The first two rows of the table are obvious: If there is no demand flow on opposing/conflicting lane j, then the probability that a vehicle is there is 0, and the probability that no vehicle is there is 1. Where there is a non-zero demand flow present, then the probabilities are based upon X, the degree of utilization (or saturation), as computed by Eq. (5.5).

$$X_j = \frac{v_j h_{dj}}{3600} \tag{5.5}$$

where

X_j degree of utilization for opposing/conflicting lane j,
v_j demand flow rate for opposing/conflicting lane j, and
h_{dj} departure headway for opposing/conflicting lane j.

As noted, the departure headways are needed for this computation, which is part of the process for determining departure headways. Thus, iteration becomes necessary. Computations start with an assumed value of *3.2 s/veh for all departure headways.*

Consider Scenario 2 from Table 5.8. In this case, there is only one vehicle on one opposing lane. There are no vehicles on any of the other five opposing/conflicting lanes. Let us assume, for illustration, that the utilization factor, X, for each opposing/conflicting lane is 0.20. For Scenario 2 to exist, the first opposing lane must have a vehicle present. The probability of this occurring is 0.20. Each of the other five opposing/conflicting lanes must have no vehicles present. The probability of each lane having no vehicle present is $1–0.20 = 0.80$. The probability of Scenario 2 existing is, therefore, the product of these 6 probabilities: $0.20 \times 0.80 \times 0.80 \times 0.80 \times 0.80 \times 0.80 = 0.0655$. In general terms, this is described by Eq. (5.6).

$$P(i) = \prod_j P(a_j) \tag{5.6}$$

where P_i is the probability of Scenario i existing. Using this procedure, up to 64 such probabilities have to be computed. Of course, in many cases there will be fewer scenarios, as not all conflicting and/or opposing approaches will have two lanes. The 2000 HCM details some of the simpler combinations that might exist. In general, however, there may be as few as *one* value of $P(a_j)$ where the intersection of two one-way streets with one lane on each approach exists. In other cases, there will be between one and six values depending upon the geometry present. Where the geometry is simpler, there will be, of course, fewer scenarios to sort through.

Equation (5.6) assumes that there are no statistical correlations among the up to 64 scenario probabilities. These probabilities can be adjusted to account for interdependencies. First, the probability of each Degree of Conflict occurring must be computed. Note that in Table 5.7, DOC 1 exists only in Scenario 1; DOC 2 exists in Scenarios 2 through 4; DOC 3 exists in Scenarios 5 through 10; DOC 4 exists in Scenarios 11 through 37; DOC 5 exists in Scenarios 38 through 64. Therefore:

$$P(C_1) = P(1)$$

$$P(C_2) = \sum_{i=2}^{4} P(i)$$

$$P(C_3) = \sum_{i=5}^{10} P(i)$$

$$P(C_4) = \sum_{i=11}^{37} P(i)$$

$$P(C_5) = \sum_{i=38}^{64} P(i) \tag{5.7}$$

where all variables are as previously defined.

Each of the 64 scenario probabilities should be adjusted by the following factors:

$$Adj\,P(1) = \alpha[P(C_2) + 2\,P(C_3) + 3\,P(C_4) + 4\,P(C_5)]$$

$$AdjP(2-4) = \frac{\alpha[P(C_3) + 2\,P(C_4) + 3\,P(C_5) - P(C_2)]}{3}$$
$$AdjP(5-10) = \frac{\alpha[P(C_4) + 2\,P(C_5) - 3\,P(C_3)]}{6}$$
$$AdjP(11-37) = \frac{\alpha[P(C_5) - 6\,P(C_4)]}{27}$$
$$AdjP(38-64) = \frac{-\alpha[10\,P(C_5)]}{27} \qquad (5.8)$$

where AdjP(i) is the adjustment to probability of i. The adjustment is added to the probabilities computed in Equation (5.6).

$$P'(i) = P(i) + AdjP(i) \qquad (5.9)$$

where P′(i) is the adjusted probability that Scenario i will occur.

Once these probabilities have been established, the departure headway for the subject movement may be estimated as:

$$h_d = \sum_{i=1}^{63} P'(i) * h_{si} \qquad (5.10)$$

where all terms are as previously defined.

These computations begin with all values of h_d assumed to be 3.2 s/veh. Obviously, when h_d is estimated in Eq. (5.10), the results must be compared. If the result is more than 0.01 different from the assumed value, an additional iteration must be done. Iterations proceed as follows:

- Departure headways for each approach are initially set as 3.2 s/veh.
- Departure headways for each approach are computed using the methodology described.
- Resulting departure headways for each approach are compared to the assumed value.
- If an additional iteration is necessary, all departure headways are set to the results of the previous computation.
- Iterations continue until the computed and assumed values are within 0.01 of each other.

The process is, obviously, computationally complex, requiring software for implementation. The analytic framework has been described here using the most complex case of multilane approaches on a 4-leg AWSC intersection. For simpler geometries, the number of scenarios (Table 5.7) is greatly reduced. The 2000 HCM shows details for several simpler cases, but the fundamental analytic approach is same in all cases.

5.4.4 Capacity

Once departure headways for all approaches is determined, another set of iterations is necessary to determine capacity. Capacity for a subject approach is the maximum throughput from the subject approach with fixed volumes on all opposing and conflicting approaches (lanes). This value is found iteratively using the same procedure outlined for determining departure headways—except that in this case, the subject demand flow rate is incrementally increased with each iteration, until the degree of utilization on the subject approach (X) is 1.00. Note that capacity is actually determined for a given lane of the subject approach. If there are two lanes, the approach capacity is the sum of the lane capacities on the subject approach.

Once again, this is computationally complex, as each increased demand flow causes a full set of iterations to determine final values of h_d. Software must be used to implement this computation.

5.4.5 Delay and Level of Service

Control delay for an AWSC lane (or approach) is computed as:

$$D_i = t_{si} + 900\,T\left[(X_i - 1) + \sqrt{(X_i - 1)^2 + \frac{h_{di} X_i}{450\,T}}\right] + 5 \tag{5.11}$$

where

- D_i control delay for subject approach or lane i, s/veh,
- t_{si} service time for approach or lane i, s/veh,
- T length of analysis period, h,
- X_i degree of utilization for approach or lane i, and
- h_{di} departure headway for approach or lane i, s/veh.

The service time, t_{si}, is estimated as the departure headway, h_{di}, minus the move-up time (m_i), where the move-up times are 2.0 s for all movements in Geometry Groups 1 through 4, or 2.3 s for all movements in Geometry Groups 5 and 6.

T is the analysis period in hours. For the standard 15-minute analysis period, $T = 0.25$ h. When the degree of utilization on an approach nears 1.00, the length of the analysis period may have a significant impact on delay. Most analyses use analysis periods of between 0.50 and 1.00 h in such cases.

Level of service is based directly on control delay. Criteria are shown in Table 5.12.

Where multiple lanes exist on an approach, a weighted average (by demand flow in each lane) delay may be computed. An aggregate weighted average delay for the intersection as a whole may also be computed. Levels of service may be assigned to individual lanes, approaches, and/or to the intersection as a whole.

Table 5.12 Levels of service for AWSC intersections: 1997–2016

Level of service	Control delay (s/veh)
A	0–10
B	>10–15
C	>15–25
D	>25–35
E	>35–50
F	>50

Source "Highway Capacity Manual" Transportation Research Board, Washington D.C., 2000, Exhibit 17–22, pg 17–32, used with permission of the National Academy of Science, Courtesy of National Academies Press

5.5 Sample Calculation

A sample calculation demonstrating the computations of various HCMs for AWSC intersections is found in Appendix B.

References

1. Herbert J (1963) A study of four-way stop intersection capacities. Highw Res Rec vol 27, Transportation Research Board, Washington DC
2. Transportation Research Board (1965) Highway capacity manual. Special Report 87, Transportation Research Board, Washington DC
3. Manual on Uniform Traffic Control Devices (2009) Federal Highway Administration, Washington DC. As updated through May 2012
4. Energy savings traffic operations project manual (1981) Barton-Aschmann Associates Inc., Evanston IL
5. Interim Materials on Unsignalized Intersection Capacity (1991) Circular 373. Transportation Research Board, Washington DC
6. Kyte M, Tian Z, Mir Z, Hameedmansoor Z, Kittleson W, Vandehey M, Robinson B, Bondzio L, Wu N, and Troutbeck R (1996) Capacity and level of service at unsignalized intersections: volume 1-two way stop controlled intersections. Final report, National Cooperative Highway Research Program Project 3–46, Transportation Research Board, Washington DC
7. Transportation Research Board (2000) Highway capacity manual, 4th edn. Transportation Research Board, Washington DC

Chapter 6
Unsignalized Intersections: Roundabouts

Europe has made extensive use of roundabouts for many decades, and traffic circles, often around significant monuments, have been around for centuries. While some traffic circles and roundabouts have always existed in the U.S., interest in their ability to reduce delays (compared to signalized intersections) and create safer and more efficient flow began to increase in the 1980s and 1990s. A formal methodology for analysis of roundabouts first appeared in the 1997 update to the 1985 HCM [1]. *It was somewhat expanded (editorially) for 2000, and significantly expanded in 2010, with minor changes in 2016.*

For the purposes of this book, we begin with 2000 HCM [2], with the understanding that it is fundamentally the same methodology introduced in the 1997 update.

It is also important to note that "traffic circles" and "roundabouts" operate with different control strategies. In a traditional "traffic circle," normal right-of-way rules are in place, i.e., the driver on the left yields to the driver on the right. In this situation, drivers entering the traffic circle are *always* on the right, and therefore have preference over vehicles already circulating through the circle. In a roundabout, entering vehicles must yield to circulating traffic. As this differs from the normal right-of-way rules, YIELD signs must be placed on all entering roadways. All methodologies in the HCM deal with roundabouts, NOT traffic circles.

In 1995, there were only 12 true roundabouts operating in the U.S. By 2016, the number was approaching 5000, with no signs of slowing down.

6.1 The 2000 Highway Capacity Manual

The methodology of the 2000 HCM focused on estimating the capacity of each approach to the roundabout. It was a first step in providing an analysis methodology to a rapidly-growing use of roundabouts at significant intersections throughout the U.S.

E. S. Prassas and R. P. Roess, *The Highway Capacity Manual: A Conceptual and Research History Volume 2*, Springer Tracts on Transportation and Traffic 12,
https://doi.org/10.1007/978-3-030-34480-1_6

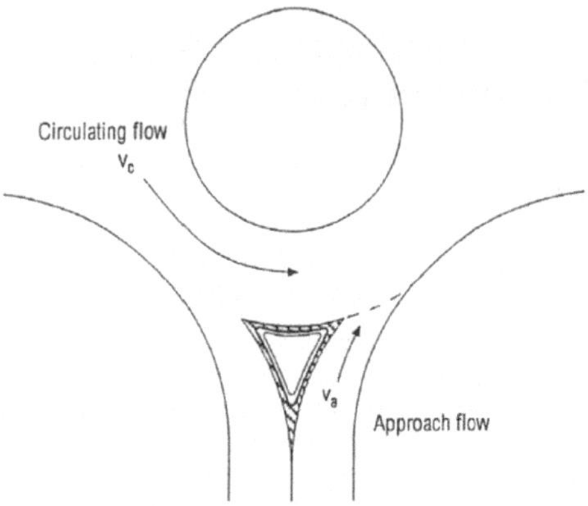

Fig. 6.1 Critical conflict at a roundabout entrance. *Source* "Highway Capacity Manual", Transportation Research Board, Washington D.C., 2000 [4], Exhibit 17-36, pp. 17–46, used with permission of the National Academy of Science, Courtesy of National Academies Press

While a start on a complex issue, the methodology was limited:

- Only single-lane roundabouts with single-lane approaches were addressed.
- While capacity was estimated, no level of service criteria were provided.

The methodology was developed from a number of studies [3–7] that addressed various aspects, including gap acceptance, at single-lane roundabouts in Europe, Australia, and the U.S.

Figure 6.1 illustrates the controlling operational feature limiting capacity at one-lane roundabouts. Because vehicles entering the roundabout are performing a maneuver similar to that of a right-turning vehicle at a TWSC intersection, it is assumed that their gap acceptance behavior will be similar to such vehicles.

The capacity for approach "a" (a one-lane approach to a one-lane roundabout) is estimated as:

$$c_a = \frac{v_c\, e^{-v_c t_c/3600}}{1 - e^{-v_c t_f/3600}} \tag{6.1}$$

where:

c_a capacity of approach a, veh/h,
v_c demand flow rate of circulating traffic, veh/h,
t_c critical gap, s/veh, and
t_f follow-up time, s/veh.

For 1997 and 2000, specific values for t_c and t_f could not be established. Rather, upper and lower bounds were given. Thus, the methodology established a range of capacities for each approach, rather than a specific value. Critical gaps (t_c) ranged from 4.1 to 4.6 s/veh, while follow-up times (t_f) ranged from 2.6 to 3.1 s/veh.

The most critical part of the analysis is determining the circulating demand flow rate (v_c) for each approach to the roundabout. For a 4-approach roundabout, the situation is illustrated in Fig. 6.2. Table 6.1 shows the flows comprising v_c for each approach.

In each case, the conflicting flow rate involves through and left-turn flows from the adjacent approach on the left, and the left turn flow from the opposing approach. Table 6.1 assumes that there are no U-turning vehicles involved. If there are, these would create a fourth movement on each approach. The U-turn flow rate would be added to the circulating flow rate for all approaches (except the subject approach).

It should also be noted that the methodology does not account for heavy vehicles, and is carried out in mixed veh/h. Demands should, however, be flow rates for the worst 15-min of the analysis hour. Where hourly volumes are given, they would be divided by the PHF to convert to flow rates.

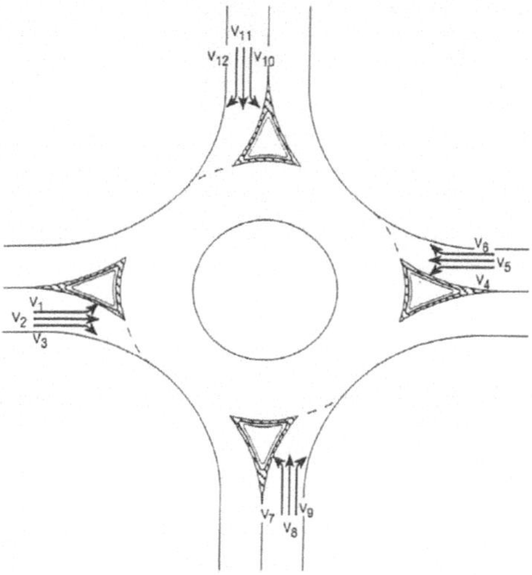

Fig. 6.2 Flows at a 4-Approach Roundabout. *Source* "Highway Capacity Manual", Transportation Research Board, Washington D.C., 2000 [2], Exhibit 17-39, pp. 17–47, used with permission of the National Academy of Science, Courtesy of National Academies Press

Table 6.1 Conflicting flows (v_c) in a roundabout

Approach and served flows	Conflicting flow rate (veh/h)
A: 7,8,9	$v_1 + v_2 + v_{10}$
B: 4,5,6	$v_7 + v_8 + v_1$
C: 10,11,12	$v_4 + v_5 + v_7$
D: 1,2,3	$v_{10} + v_{11} + v_4$

6.2 The 2010 Highway Capacity Manual

The 2000 HCM introduced the subject of roundabouts in a significant way, but was limited to addressing capacity of single-lane roundabouts. While the 2010 HCM was based primarily on studies completed prior to 2010 [8–11], by 2010 a far more sophisticated and inclusive methodology had been developed.

The methodology continues to focus on the capacity of entry roadways, and adds a level of service measure based upon delay. Three basic variables are treated:

v_e the entry flow rate, pc/h,
v_c the circulating flow rate in the roundabout, pc/h, and
v_{ex} the exit flow rate, pc/h.

Of course, each of these values is different for each entering/exiting roadway.

The method is no longer restricted to one-lane roundabouts with one-lane entry roadways. Four different scenarios may be analyzed:

- A one-lane roundabout with one-lane entry roadways.
- A two-lane entry roadway entering a one-lane circulating roadway.
- A one-lane entry roadway entering a two-lane circulating roadway.
- A two-lane entry roadways entering a two-lane circulating roadway.

In addition, the model also deals with by-pass lanes for right turns from an entering roadway. Figure 6.3 illustrates the four basic cases treated in the 2010 HCM.

6.2.1 Converting Demand Volumes to Flow Rates in Pce's

If entering demand volumes are given in full-hour veh/h, they must be converted to flow rates (for the peak 15 min of the analysis hour) and passenger-car equivalents (pce's). This is generally done as a two-step process:

$$v_i = \frac{V_i}{PHF} \tag{6.2}$$

where:

V_i demand volume, veh/h,
v_i demand flow rate, veh/h, and
PHF peak hour factor.

Then:

$$v_{ipce} = \frac{v_i}{f_{HV}}$$
$$f_{HV} = \frac{1}{1 + P_{HV}(E_{HV} - 1)} \tag{6.3}$$

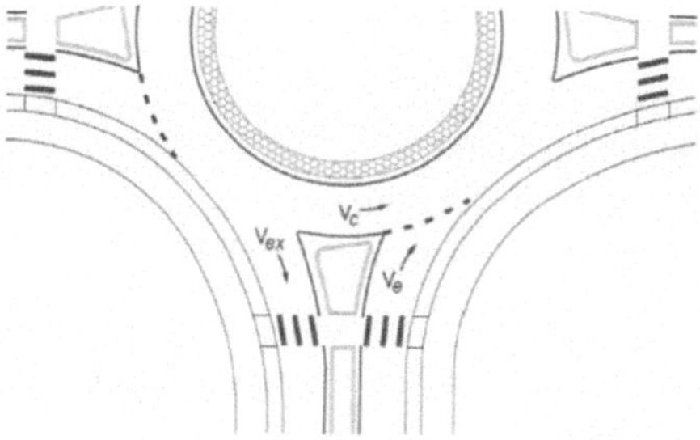

(a) One-Lane Entry; One-Lane Circulating Roadway

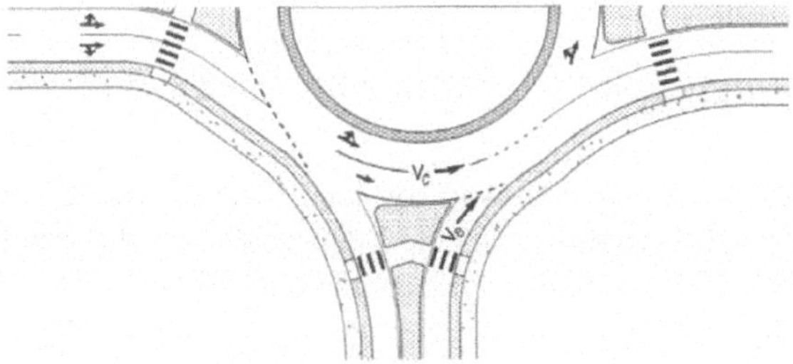

(b) One-Lane Entry; Two-Lane Circulating Roadway

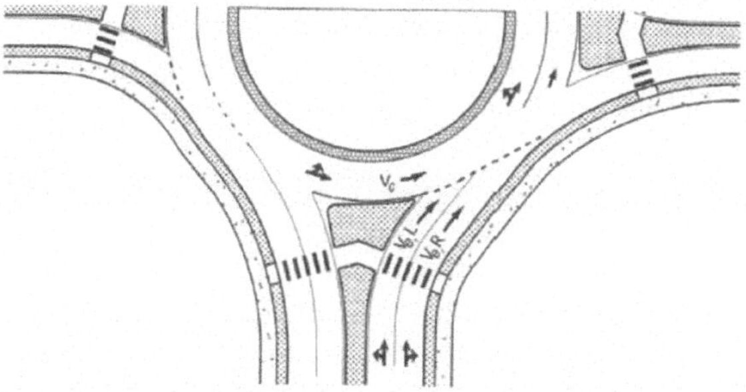

(c) Two-Lane Entry; One-Lane Circulating Roadway

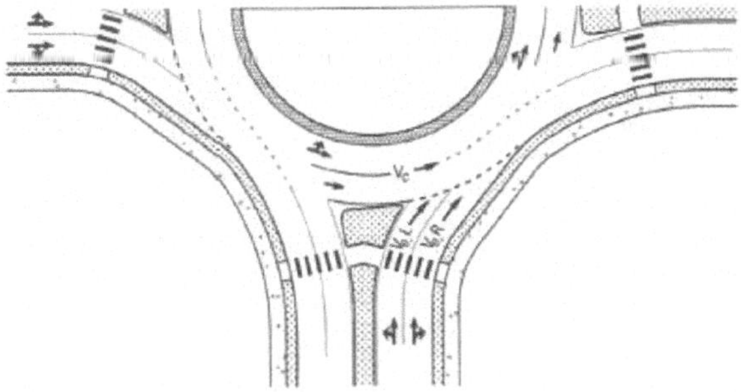

(d) Two-Lane Entry; Two-Lane Circulating Roadway

Fig. 6.3 Roundabouts Included in the 2010 HCM Methodology. *Source* "Highway Capacity Manual", Transportation Research Board, Washington D.C., 2010 [12], Exhibits 21-3, 21-4, 21-5, 21-6, pp. 21-6–21-8, used with permission of the National Academy of Science, Courtesy of National Academies Press

where:

v_{ipce} demand flow rate, pc/h,
f_{HV} heavy vehicle adjustment factor,
P_{HV} proportion of heavy vehicles in demand flow, and
E_{HV} heavy vehicle equivalent (1 for pc, 2 for HV).

For this conversion, all vehicles with more than 4 wheels on the ground are considered to be heavy vehicles. Most subsequent computations in the methodology use flow rates in pc/h as inputs.

6.2.2 Estimating Capacity of Entering Roadways

The capacity of each entering roadway (by lane) is computed using the equations shown in Table 6.2. The equation selected depends upon the number of entering lanes, and the number of circulating lanes at the entry point.

The 2010 HCM also gives some guidance on the effect of by-pass lanes for entering right-turning vehicles. In general, two types of by-pass lanes are in use, as illustrated in Fig. 6.4.

Where either type of by-pass is present, the right-turning demand is removed from the normal computations for entry capacity. The capacity of the "normal" lane or lanes is handled using the equations of Table 6.2, without the right-turn entry flow included.

Where a YIELDING type by-pass lane is present, its capacity is estimated using Eq. 6.4.

$$c_{bypass,pce} = 1,130\, e^{(-1*10^{13})\, v_{\exp ce}} \quad for\ 1\ conflicting\ exit\ lane$$
$$c_{bypass,pce} = 1,130\, e^{(-0.7*10^{-3})\, v_{\exp ce}} \quad for\ 2\ conflicting\ exit\ lanes \tag{6.4}$$

Table 6.2 Estimating capacity of entry roadway lanes at a roundabout 2010 HCM

Entry lanes	Circ. lanes	Equation
1	1	$c_{epce} = 1130\, e^{(-1*10^{-3})\, v_{cpce}}$
1	2	$c_{epce} = 1130\, e^{(-0.07*10^{-3})v_{cpce}}$
2	1	$c_{eRpce} = 1130\, e^{(-1*10^{-3})\, v_{cpce}}$ $c_{eLpce} = 1130\, e^{(-1*10^{-3})\, v_{cpce}}$
2	2	$c_{eRpce} = 1130\, e^{(-0.7*10^{-3})\, v_{cpce}}$ $c_{eLpce} = 1130\, e^{(-0.75*10^{-3})\, v_{cpce}}$

c_{epce} = capacity of entry roadway, pc/h; c_{eRpce} = capacity of right lane of entry roadway, pc/h
c_{eLpce} = capacity of left lane of entry roadway, pc/h
v_{cpce} = total circulating flow rate at point of entry, pc/h

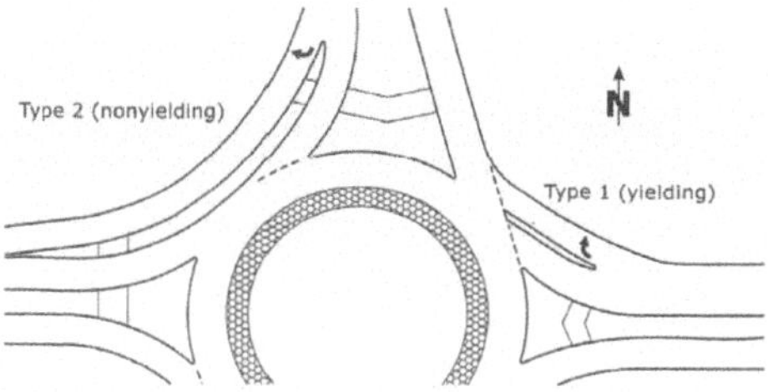

Fig. 6.4 By-pass lanes for right turns at a roundabout 2010 HCM. *Source* "Highway Capacity Manual", Transportation Research Board, Washington D.C., 2010 [12], Exhibit 21-8, pp. 21-10, used with permission of the National Academy of Science, Courtesy of National Academies Press

where:

$c_{bypass,pce}$ capacity of the by-pass lane, pc/h, and
v_{expce} demand flow rate on the conflicting exit roadway, pc/h.

The 2010 HCM does not give any guidance on the capacity of NON-YIELDING by-pass lanes. It notes that the capacity of such lanes could be substantial. The capacity of such a lane is almost entirely dependent on the merge that must take place when the by-pass lane joins the exit roadway.

The methodology focuses on determining the capacity of entry lanes to the roundabout. The capacity of exit lanes is also of some interest, but only general guidance is given. It is suggested that exit capacities may be in the range of 1200 veh/h/ln to 1300 veh/h/ln based upon European and U.S. observations.

6.2.3 Determining Circulating and Exit Demand Flows

The capacity computations described require that the conflicting circulating flow and exit flow rates at each entry point be determined. This is done in a manner similar to the 2000 HCM, and simply follows each demand flow to its intended destination. The 2010 HCM separately identifies U-turn vehicles, and provides illustrations of the process. Note that right turns *never* become conflicting circulating vehicles to any other movement, but do show up in conflicting exit flows. Conflicting exit flows are only used in the estimation of the capacity of a by-pass lane.

Figure 6.5 illustrates the computation of a circulating conflicting flow rate, and Fig. 6.6 illustrates the computation of a conflicting exit flow rate.

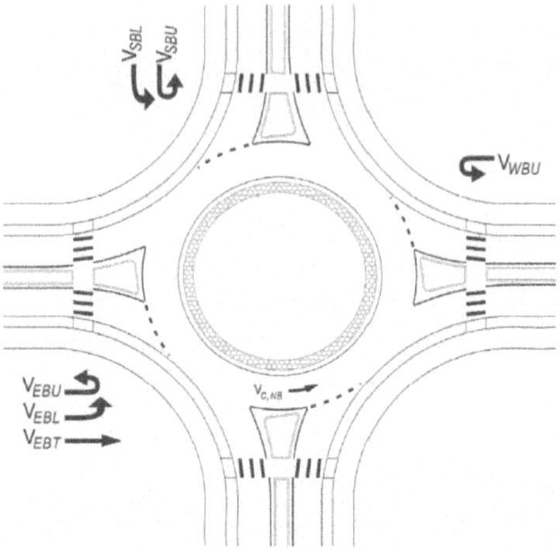

$$v_{c,NB,pce} = v_{WBU,pce} + v_{SBL,pce} + v_{SBU,pce} + v_{EBT,pce} + v_{EBL,pce} + v_{EBU,pce}$$

Fig. 6.5 Computation of conflicting circulating flows illustrated—2010 HCM. *Source* "Highway Capacity Manual", Transportation Research Board, Washington D.C., 2010 [12], Exhibit 21-11, pp. 21–13, used with permission of the National Academy of Science, Courtesy of National Academies Press

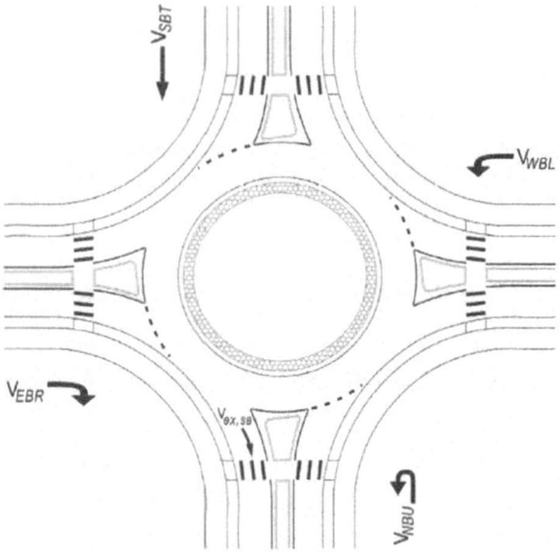

$$v_{ex,pce} = v_{NBU,pce} + v_{WBL,pce} + v_{SBT,pce} + v_{EBR,pce} - v_{EBR,pce,bypass}$$

Fig. 6.6 Computation of conflicting exit flows illustrated—2010 HCM. *Source* "Highway Capacity Manual", Transportation Research Board, Washington D.C., 2010 [12], Exhibit 21-11, pp. 21–13, used with permission of the National Academy of Science, Courtesy of National Academies Press

6.2.4 Determining Demand Flow Rates by Lane

Because the roundabout methodology deals with each entry lane individually, it is necessary to assign entry flows to the appropriate lane when two-lane entry roadways exist. This is primarily a logical assignment based upon movements and the lanes they would most likely be using. The 2010 HCM does give some guidance for doing this, but field observations and knowledge of local driving habits are equally important.

There are three simple guidelines for assigning entry flow rates to lanes:

- Where a right-turn by-pass lane is provided, all right turns are assigned to this lane; none are assigned to other lanes on the entry roadway.
- Where only one lane is available for a given movement (either because only one lane exists, or lane-use controls prohibit the use of certain lanes for certain movements), it is assigned to that lane.
- Other flows are distributed across all available lanes more-or-less uniformly, unless this is not possible due to physical or control restrictions, or as evidenced by observed lane use imbalances.

6.2.5 Pedestrian Interference

The 2010 roundabout model allows for the capacity of entry roadway lanes to be reduced due to pedestrian interference. The model is based upon pedestrians having an absolute right-of-way that drivers scrupulously observe. In reality, when vehicular queues exist on entry roadways, pedestrians may pass between vehicles, and have very little impact on operating conditions. The adjustment factors, shown in Table 6.3, depend upon whether the entry roadway has one or two lanes.

6.2.6 Capacities, Demand Flow Rates, and V/C Ratios

At this point, the analysis model has provided capacities and demand flow rates in pc/h. Both are converted to vehicles per hour; then v/c ratios (X) are computed for each approach lane, as follows:

$$
\begin{aligned}
v_i &= v_{ipce} f_{HV} \\
c_i &= c_{ipce} f_{HV} f_{ped} \\
X_i &= v_i / c_i
\end{aligned}
\tag{6.5}
$$

where all variables are as previously defined.

Table 6.3 Capacity adjustment factors (f_{ped}) for pedestrian interference at a roundabout—2010 HCM

Case	Equation
One-lane entry roadways	
$v_{cpce} > 881$	$f_p = 1.00$
$v_{cpce} \leq 881$ & $n_{ped} \leq 101$	$f_p = 1 - 0.000137\, n_{ped}$
$v_{cpce} \leq 881$ & $n_{ped} > 101$	$f_p = \frac{1119.5 - 0.715\, v_{cpce} - 0.644\, n_{ped} + 0.00073\, v_{cpce}\, n_{ped}}{1068.6 - 0.654\, v_{cpce}}$
Two-lane entry roadways	
$n_{ped} < 100$	$f_{ped} = \min\left[1 - \frac{n_{ped}}{100}\left(1 - \frac{1260.6 - 0.329\, v_{cpce} - 38.1}{1{,}380 - 0.50\, v_{cpce}}\right),\ 1\right]$
$n_{ped} \geq 100$	$f_{ped} = \min\left[\left(\frac{1260.6 - 0.329\, v_{cpce} - 0.381\, n_{ped}}{1380 - 0.50\, v_{cpce}}\right),\ 1\right]$

Note n_{ped} = pedestrian flow rate per hour, peds/h
Source "Highway Capacity Manual", Transportation Research Board, Washington D.C., 2010 [12], Exhibits 21-17 and 21-19, pp. 21–16 and 21–17, used with permission of National Academy of Science, Courtesy of National Academies Press

6.2.7 Delay and Level of Service

The delay for any given entry lane is computed as:

$$D_i = \frac{3600}{c_i} + 900\,T\left[X_i - 1 + \sqrt{(X_i - 1)^2 + \frac{\left(\frac{3600}{c_i}\right)X_i}{450T}}\right] + 5 * \min[X_i, 1] \tag{6.6}$$

where all terms are as previously defined. Weighted average (by demand flow rate) delays may be computed for two-lane approaches and for the roundabout as a whole.

Level of service criteria are the same as for TWSC and AWSC intersections, as shown in Table 6.4.

6.3 The 2016 HCM

There were no substantial changes to the roundabout methodology in the 2016 HCM. Editorial improvements in presentation were made.

Table 6.4 Levels of service for roundabouts—2010 HCM

Control delay (s/veh)	LOS by v/c Ratio (X)	
	v/c ≤ 1.00	v/c > 1.00
0–10	A	F
>10–15	B	F
>15–25	C	F
>25–35	D	F
>35–50	E	F
>50	F	F

Note For approaches and intersection-wide assessment, LOS is defined solely by control delay
Source "Highway Capacity Manual", Transportation Research Board, Washington D.C., 2010 [12], Exhibit 21-1, pp. 21–1, used with permission of the National Academy of Science, Courtesy of National Academies Press

6.4 Sample Calculation

A sample roundabout calculation illustrating and comparing computations for various versions of the HCM is found in Appendix C.

6.5 Closing Comments for Unsignalized Intersections (TWSC, AWSC, and Roundabouts)

With the 2010 and 2016 HCMs, a complete set of materials for signalized and unsignalized intersection analysis now exists. All use a common approach to level of service based upon control delay. While direct comparison of signalized and unsignalized approaches to control solely on the basis of delay would ignore other factors, like safety, driver familiarity, and others, it gives analysts a somewhat level playing field to compare outcomes.

Obviously, there are many factors involved in the decision to signalize an intersection. These are covered in traditional signal warrants of the Manual on Uniform Traffic Control Devices (MUTCD), available on-line through the FHWA web site.

Care must be taken when comparing overall intersection delay, as TWSC intersections include many major street vehicles experiencing no, or negligible delay, while all vehicles at AWSC, roundabouts, or signalized intersections experience delay. Remember that the TWSC methodology does *not* recommend assigning an overall intersection LOS because of this.

References

1. Highway Capacity Manual (1885) Special Report 209, Transportation Research Board, Washington D.C., as updated in 1997
2. Highway Capacity Manual (2000) Transportation research board, 4th edn. Washington D.C.
3. Brilon, W., Stuwe, B. (1994) Capacity and design of roundabouts in germany. In: Transportation research record 1398, Transportation Research Board, Washington D.C.
4. Akcelik R, Besley M (1996) SIDRA-5 users guide, Australian Road Research Board, Transport Research Limited
5. Flannery A, Datta T (1997) Operational performance measures of american roundabouts. Transportation research record 1572, Transportation Research Board, Washington D.C.
6. Troutbeck R (1996) The analysis of the performance of roundabouts, unpublished technical memorandum to the committee on highway capacity and quality of service, Transportation Research Board, Washington D.C.
7. Rodegerdts L, Blogg M, Wemple E, Myers E, Kyte M, Dixon M, List G, Flannery A, Troutbeck R, Brilon W, Wu N, Persaud C, Lyon D, Harkey L, Carter D (2008) Roundabouts in the United States. NCHRP Report 572, Transportation Research Board, Washington D.C.
8. Flannery A, Datta T (1997) Operational performance measures of american roundabouts. In: Transportation research record 1572. Transportation Research Board, Washington D.C.
9. Rodegerdts L, Blogg M, Wemple E, Myers E, Kyte M, Dixon M, List G, Flannery A, Troutbeck R, Brilon W, Wu N., Persaud C, Lyon D, Harkey L, Carter D (2007) Roundabouts in the United States. In: NCHRP Report 572. Transportation Research Board, Washington D.C.
10. Rodegerdts L, Bansen J, Tiesler C, Knudsen J, Myers E, Johnson M, Moule M, Persaud B, Lyon C, Hallmark S, Isebrands H, Crown R, Guichet B, O'Brien A (2010) Roundabouts: an informational guide. In: NCHRP Report 672, 2nd edn. Transportation Research Board, Washington
11. Robinson B, Rodegerdts L, Scarbrough W, Kittelson W, Troutbeck R, Brilon W, Bondzio L, Courage K, Kyte M, Mason J, Flannery A, Myers E, Bunker J, Jacquemart G (2000) Roundabouts: an informational guide, report FHWA-RD-00-067. Federal Highway Administration, Washington D.C.
12. Highway Capacity Manual (2010) 5th edn. Transportation Research Board, Washington D.C.

Chapter 7
Signalized Intersections: 1950 Through 1980

Perhaps the most complex point in any traffic network is the signalized intersection. At a typical signalized intersection with four entry approaches on two two-way streets, there are 12 vehicular movements (4 approaches, each with a left-turn, through, and right-turn movement), four crosswalk pedestrian flows, and bicycles that must be taken into consideration. The intricacies of signal timing have a significant impact on operations, as does the physical design of each approach. Moreover, the "intersection" does not operate as a whole: each approach can have different demands and experience different operating conditions. The development of capacity and level of service analysis methodologies for signalized intersections reflects the complexity of operations in the field, and the number of underlying factors that influence them. This chapter details signalized intersection analysis methodologies from their beginnings in the 1950 HCM through the 1980 Interim Procedures for Highway Capacity.

7.1 Fundamental Concepts

Chapter 2 presented fundamental concepts and terminology with respect to signalized intersections that are used herein. Chapter 3 provided an overview of the general development of the concepts of capacity and level of service as applied to interrupted flow in general, and to signalized intersections in particular. These should be kept in mind when considering the material in this chapter. Further discussion of fundamental concepts of interrupted flow can be found in the textbook *Traffic Engineering* by Roess, Prassas, and McShane [1].

E. S. Prassas and R. P. Roess, *The Highway Capacity Manual: A Conceptual and Research History Volume 2*, Springer Tracts on Transportation and Traffic 12,
https://doi.org/10.1007/978-3-030-34480-1_7

7.2 The Early Development of Traffic Signals

The earliest "signals" used in conjunction with moving conveyances were early railroad signals, which had become common on English railroads by the 1860s. The first traffic signal appeared in 1868 at the intersection of Bridge Street, Great George Street, and Parliament Street in London [2]. The design combined semaphore arms with gas-lit red and green lights for nighttime visibility. It was manually operated by a police officer using a lever, which turned the appropriate light towards the appropriate street. The experiment was deemed successful, but was discontinued after a small gas explosion erupted in the lines supplying the gas, injuring several policemen.

Some other key dates and accomplishments [3] include:

- Ernest Sirrine patented the first automatic street traffic system in 1910. His signals used the words "STOP" and "PROCEED."
- Lester Wire of Salt Lake City invented an electric traffic light in 1912 that used red and green colored lights.
- James Hoge patented manually-controlled traffic lights in 1913 using the illuminated words "STOP" and "MOVE." One year later they were installed in Cleveland, Ohio by the American Traffic Signal Company.
- William Ghiglieri of San Francisco invented the first automatic signal using colored red and green lights in 1917. His signal could be manually or automatically operated.
- Around 1920, William Potts invented several automatic electric signal light systems, including an overhanging, four-way, red, green, and yellow light system.

The first interconnected system of traffic signals was introduced in Salt Lake City in 1917. Six intersections were interconnected using a manually-operated switch. Automated operation of interconnected systems was introduced in Houston, Texas in 1922.

7.3 The 1950 Highway Capacity Manual

Suffice it to say that by the time of the 1950 HCM [4], traffic signals were key control feature of all urban and downtown street networks throughout the nation. At the same time, signal systems were still relatively primitive (by today's standards), and all signals were pre-timed, and many were still not interconnected.

The 1950 HCM dealt with signalized intersections in a relatively unsophisticated way. It started with an estimate of the saturation flow rate based upon the curb-to-curb width of the street, parking conditions, and directional distribution. These basic values are shown in Fig. 7.1.

The base conditions for Fig. 7.1 include 10% commercial vehicles, 10% right turns, 10% left turns, and no street cars (not a consideration in most places these days). Two relationships are presented, one for intersections in Downtown Areas,

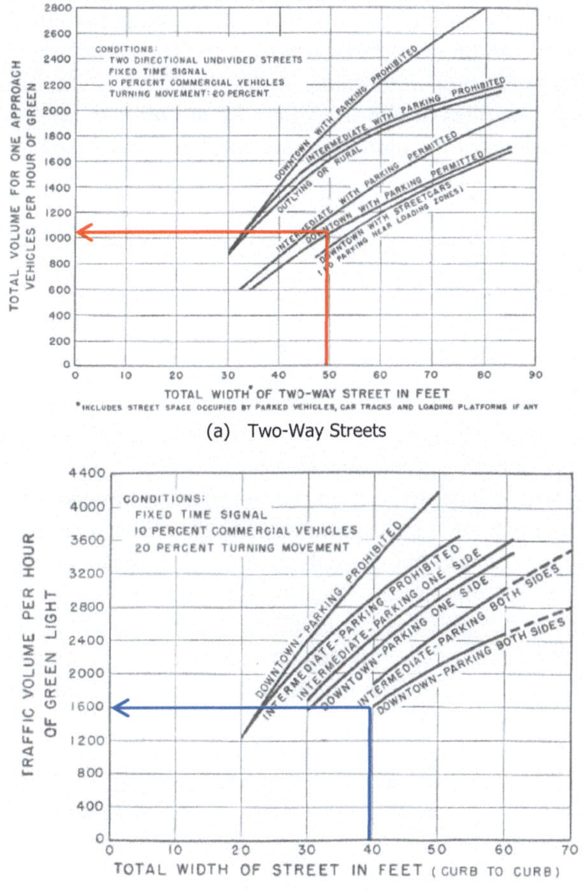

Fig. 7.1 Saturation flow rates for signalized intersections under base conditions—1950 HCM. *Source* "Highway Capacity Manual", Bureau of Public Roads, U.S. Government Printing Office, Washington D.C., 1950 [4], Figs. 24 and 26, pp. 79 and 84, used with permission of the National Academy of Science. Courtesy of National Academies Press

another for intersections in Intermediate Areas. The definitions of these are fuzzy at best.

The "saturation flow rates" given, however, are somewhat crude, in that they do not take into account the difference between actual green time (G) and effective green time (g). For two-way streets, a single-approach saturation flow rate is given based upon the total street width (in both directions). It assumes an equal distribution of travel width in both directions.

Two examples are shown in Fig. 7.1. The first (red) considers a two-way street with parking in a downtown area with a 50-ft curb-to-curb width. The saturation flow rate is seen to be 1050 veh/hg under base conditions.

The second (blue) shows a one-way street with parking on both sides in a downtown area. The street width is 40 feet, curb-to-curb. The saturation flow rate is seen to be 1600 veh/hg under base conditions.

The saturation flow rates drawn from Fig. 7.1 are *total* saturation flow rates for the approach, regardless of the number of lanes. They are for the base conditions of the figure, which are:

- Fixed-time signal,
- 10% commercial vehicles (trucks),
- 10% right-turning vehicles,
- 10% left-turning vehicles, and
- No street cars (trolleys) or buses.

For clarity, this saturation flow rate under base conditions is given the symbol "s_{TB}"—total approach saturation flow rate under base conditions, and is given in units of vehicles per hour of green time (veh/hg).

Values of s_{TB} would have to be adjusted to reflect the impact of conditions that do not conform to the defined base conditions. There is also an adjustment to reflect whether the computation is to find "possible capacity" or "practical capacity," the two primary capacity levels that are predicted by the 1950 HCM. The 1950 HCM also defines "basic capacity," which is essentially the possible capacity when conditions are ideal, i.e., no trucks or turning movements. All capacity computations also have to be adjusted by the green-to-cycle length ratio, to convert from vehicles per hour of green time to vehicles per hour.

The base conditions of Fig. 7.1 also include no special turning lanes or protected turning phases. When either or one of these exists, there are some special rules to be applied when using Fig. 7.1 to determine the saturation flow rate (s_{TB}):

For two-way streets with added turning lane(s), but no protected turning phases:

1. Enter Fig. 7.1 with a street width equaling 2 times the width of the through lanes in the subject direction.
2. Increase the resulting saturation flow rate by 5% for an added right-turn lane, and 10% for an added left-turn lane.
3. For a right-turning lane: Increase the result after step 2 by the number of right-turning vehicles, not to exceed 600 * (G/C) veh/h or N * (D-20)/25 veh/h, where D is the length of the right-turning lane (ft) and N is the number of signal cycles per hour.
4. For a left-turning lane: Increase the result after step 2 by the number of left-turning vehicles, but not to exceed the capacity of the left-turning lane, which is estimated as 1200-V_o, where V_o is the opposing volume. The capacity of the left-turning lane cannot, however, be less than 2 vehicles per signal cycle.

Step 2 increases the saturation flow rate to account for the fact that turning vehicles from a turning lane *will not* be blocking through vehicles on the approach. Steps 3 and 4 increase the saturation flow rate to account for the additional vehicles being served by the turning lane(s).

For two-way streets with added turning lane(s) AND protected turning phase(s):

1. Use steps 1 and 2 as described for two-way streets with added turning lanes and no turning phases.
2. For a right-turning lane with protected phase: Increase the resulting saturation flow rate from step 1 by the number of right-turning vehicles, not to exceed 800 vehicles per 10 ft of width of turning lane per hour of protected green provided.
3. For a left-turning lane with protected phase: Increase the resulting saturation flow rate from step 1 by the number of left-turning vehicles, not to exceed 800 vehicles per 10 ft of width of turning lane per hour of protected greet provided.

For one-way streets with added turning lanes:

1. Enter Fig. 7.1 with the "normal" street width, not including the width of turning lanes.
2. Increase the saturation flow rate obtained from step 1 by 5% for an added right-turn lane and/or 10% for an added left-turn lane.
3. For a right-turn or left-turn lane: add the number of vehicles turning right or left, not to exceed either 600 * G/C vehicles per hour or N * (D-20)/25 vehicles per hour. This is done separately for each right-turn and left-turn lane.

The 1950 HCM also gives very detailed instructions for additional modifications for "high-type intersections," which are not included here. "High-type intersections" include those with high-type design and high approach speeds. Unfortunately, more specific criteria were not given, although the manual includes many illustrated examples.

Figure 7.1, with application of the additional rules regarding turning lanes on signalized intersection approaches, yields a value of total saturation flow rate under base conditions (s_{TB}). The methodology now converts this to a capacity, as follows:

$$c = s_{TB} * f_c * f_{HV} * f_{RT} * f_{LT} * f_{LB} * f_P * \left(\frac{G}{C}\right) \tag{7.1}$$

where:

c capacity of the intersection approach (veh/h),
s_{TB} total saturation flow rate under base conditions (veh/hg),
f_c adjustment for type of capacity being estimated (decimal),
f_{HV} adjustment for heavy vehicles (decimal),
f_{RT} adjustment for right turns (decimal),
f_{LT} adjustment for left turns (decimal),
f_{LB} adjustment for local buses (decimal),
f_P adjustment for parking conditions (decimal),
G green time for subject approach (secs), and
C cycle length (secs).

It should be noted that this equation does not exist in the 1950 HCM. The adjustments are described in paragraph form in the manual. The equation is shown here to clarify the types of adjustments that are made, and to conform with the symbology used in later editions of the manual.

Because the manual itself describes the adjustments in verbal form, for clarity, they have been converted to equation or tabular form herein. Equations 7.2 through 7.5 define adjustments for all conditions *except* local buses and parking:

$$f_c = 1.10\ for\ possible\ capacity$$
$$f_c = 0.90\ for\ practical\ capacity \qquad (7.2)$$

$$f_{HV} = 1.00 - (P_{HV} - 0.10) \qquad (7.3)$$

$$f_{RT} = 1.00 - \left(\frac{P_{RT} - 0.10}{2}\right) \qquad (7.4)$$

$$f_{LT} = 1.00 - (P_{LT} - 0.10) \qquad (7.5)$$

where:

P_{HV} proportion of heavy vehicles in the traffic stream (decimal),
P_{RT} proportion of right-turns in the traffic stream (decimal), and
P_{LT} proportion of left-turns in the traffic stream (decimal).

The difference between "heavy vehicles" and "local buses" is interesting, although the 1950 HCM is not quite explicit in its definition. A "local bus" is one that stops within the confines of the intersection (either on the near side or far side of the intersection) to pick-up and/or discharge passengers. For the 1950 HCM, a "heavy vehicle" is any commercial vehicle, *except local buses.* Thus, at an intersection with local buses, but no bus stop, local buses are heavy vehicles.

The 1950 HCM treats the local bus and parking adjustments in a coordinated way, although the discussion is, at best, convoluted. Three basic cases are considered:

- Bus stops on streets where parking is prohibited. On such streets, buses are stopping in a travel lane.
- Bus stops on streets with parking permitted *except* at the bus stop. On such streets, buses can pull out of the moving traffic lane to pick up and discharge passengers. When unoccupied by a bus, the bus stop becomes, in effect, a right-turn lane.
- Streets with parking permitted, but no bus stop.

Local bus and parking adjustment factors for these cases are summarized in Table 7.1.

Equation 7.1 allows the estimation of possible and practical capacities for signalized intersection approaches. The 1950 HCM, however, does not offer much description of the operating conditions when actual demands are at or close to these values. It does cite the results of a delay study made at a rural intersection: at possible capacity, average delays of 2.84 min/vehicle were observed, while at practical capacity, average delays of 1.04 min per vehicle were observed. At least in this case, the average delay at practical capacity was approximately 37% of the delay at possible capacity.

Table 7.1 Adjustments for local buses (f_{LB}) and parking (f_P)—1950 HCM

Parking and bus stop condition	f_{LB}	f_P
Parking prohibited w/bus stop		
Near side bus stop	0.90	1.00
Far side bus stop (downtown)	0.97	1.00
Far side bus stop (Intermediate)	0.85	1.00
Heavy bus volumes	See footnote 1	1.00
Parking permitted w/bus stop		
Near side bus stop	$1.00 + P_{RT} + P_{LT}$, max = 1.06	1.00
Far side bus stop	1.00	1.00
Heavy bus volumes	See footnote 2	1.00
Parking permitted, no bus stop	1.00	$1.00 - [(P_{RT} + P_{LT})/4]$ min = 0.94 See footnotes 3, 4

The 1950 HCM was an important step in the modeling of signalized intersections. It clearly established the concept of saturation flow and saturation flow rates, although it did not recognize the difference between actual and effective green time. It defined the need for critical adjustments that consider commercial vehicles, turning movements, local buses, and parking. It established a framework that would be built on in subsequent manuals, even though there was a departure from this approach in the 1965 HCM.

1. Subtract 12 ft from street width when entering Fig. 7.1, then add the number of buses/h to s_{TP}.
2. Subtract 6 ft from street width when entering Fig. 7.1, then add the number of buses/h to s_{TP}.
3. If parking is eliminated >20 ft from crosswalks, ADD $(P_{RT} + P_{LT}) * [(D\text{-}20)/5G]$, where D is the distance from the crosswalk that parking is eliminated, and $D_{max} = 5G + 30$.
4. If parking is eliminated from *both* sides of the intersection for a distance D > 5G, or the street widens leaving the intersection by one or more lanes: use the upper curve in Fig. 7.1 (for parking prohibited).

7.4 The 1965 Highway Capacity Manual

The 1950 HCM methodology for signalized intersections was developed directly by the members of the Highway Capacity Committee, working with data collected from a variety of traffic agencies across the country. Nevertheless, the need for additional work in this new and ever-important area was evident.

In 1954 and 1955, the Committee again requested information from traffic agencies, this time with significant specificity as to the format and type of data needed. The data was returned to the Bureau of Public Roads for analysis. The effort provided data on 1600 intersection approaches, with data provided on a cycle-by-cycle basis for between 1 and 2 h. The "degree of utilization" for each of these was recorded in a manner consistent with the Committee request. It was this data, and the resulting analysis, that led to the 1965 HCM methodology [5].

Initially, the data was analyzed and studied for the Bureau of Public Roads by a consultant—C-E-I-R Inc.—working under the supervision of O.K. Normann [6]. The consultant conducted a wide variety of graphic and regression analyses to identify critical variables in the data and key relationships among them. It was, however, O.K. Normann himself that developed the final methodology for the 1965 HCM [7]. The 1965 HCM clearly recognizes the importance of Normann's work:

> It is, therefore, fitting that the procedures described in this chapter be designated collectively as the 'Normann Method' of computing signalized intersection capacity and levels of service. [5, p. 112]

A key difference between the theoretical work of the consultant [6] and the procedures finally developed by Normann [7] is apparent from two statements in the consultant report:

> These equations are not sufficiently close to serve as prediction equations for obtaining a precise estimate of the magnitude of flow to be expected under each set of conditions .... [6, p 60]

and:

> For the purpose of predicting the general magnitude of expected flow (within these errors of estimate) and for predicting the probable effect of changing one variable or another, the equations will serve readily for indicating the relative importance of each facility improvement type and the general effect of a change in a number of them. [6, p 61]

In other words, the calibrated relationships were good enough to predict *changes* in operating characteristics resulting from changes in underlying conditions, but were *not* good enough to predict absolute values of capacity or service volume. Normann, of course, ignored this distinction, recognizing that the profession needed to get absolute as well as relative results, and that the available data, despite any shortcomings, had produced something that was far better than the 1950 HCM that was already in widespread use.

Historically, this kind of judgment is made by the Committee in every edition of the manual. Everyone would love to be able to predict everything to within ±5% of reality. However, sometimes predicting something to within ±50% of reality is better than what you've got at any given time. Unfortunately, the "hows" and "whys" of many of these judgments do not get well documented, which becomes a problem for future Committees and researchers.

7.4.1 Signalized Intersection Levels of Service in the 1965 HCM

As was discussed in Chap. 3, despite the inconsistency with the concept definition, the principal measure used to define level of service in the 1965 HCM was the v/c ratio—often referred to as "capacity utilization," a term still in common use today.

A similar measure, however, did not exist for signalized intersections. The Committee, therefore, had to develop one. It did so early in the process, so that the relevant information could be collected as part of the data collected in 1954 and 1955. The variable defined for this purpose was "load factor."

The load factor is formally defined as the *proportion of fully-loaded cycles during the analysis hour*. The analysis hour was almost always the peak hour. It also depends on the definition of "fully loaded cycle." Note that the load factor is a measure for a specific intersection approach, thus, the term "cycle" means the green phase for the approach within the cycle. A cycle is described as "fully loaded" if:

> (1) there are vehicles ready to enter the intersection in all lanes when the signal turns green, and (2) they continue to be available to enter in all lanes during the entire phase with no unused time or exceedingly long spacings between vehicles at any time due to lack of traffic, whether resulting from lack of demand (as often found toward the end of the phase) or from frictional interferences upstream. [5, pp 115 and 116]

Additional guidance is provided as well. Essentially, a cycle is not fully loaded if at any point there is green time available, but no demand present to use it. The 1965 HCM related level of service directly to the load factor, as shown in Table 7.2.

Load factor almost immediately became an interpretation nightmare. To be fully loaded, cycles had to have almost constant demand present on all lanes during the particular cycle. The interpretation of any gaps in the availability of demand were difficult to make, which made it difficult to get uniform observations where multiple people were involved in data collection—which happens virtually all of the time.

Table 7.2 Levels of service for signalized intersections—1965 HCM

Level of service	Description of traffic flow	Load factor
A	Free flow	0.0
B	Stable flow	≤0.1
C	Stable flow	≤0.3
D	Approaching unstable flow	≤0.7
E[a]	Unstable flow	≤1.00
F	Forced flow	–[b]

[a]Capacity, [b]Not applicable

Source "Highway Capacity Manual," *Special Report 87*, Transportation Research Board, Washington D.C., 1965 [5], Table 6.3, p. 131, used with permission of the National Academy of Science, Courtesy of National Academies Press

For example, if there were 60 cycles in the peak hour at a particular intersection approach, and 54 of them were fully loaded, the load factor would be 54/60 = 0.90. On the other hand, if each individual cycle were 90% utilized, none of them would be fully utilized, and the load factor would be 0.00. In other words, because it was based upon use of individual cycles, *not* "capacity utilization," the load factor was fundamentally different from the v/c ratio.

7.4.2 Base Saturation Flow Rates in the 1965 HCM

The basis for the Normann Method was a set of six nomographs used to determine the base saturation flow rate for six fundamental types of signalized intersection approaches:

- One-way streets with no parking.
- One-way streets with parking permitted on one side.
- One-way streets with parking permitted on both sides.
- Two-way streets with no parking.
- Two-way streets with parking.
- Rural intersection approaches with no parking.

The nomographs are all based upon the approach width, although this concept was simplified from the 1950 HCM. For one-way streets, the approach width remained the curb-to-curb width in feet. For two-way streets, however, the approach width was defined as the curb-to-division line width in feet, not including exclusive turning lanes. Thus, like the 1950 manual, the 1965 manual treated signalized intersection approach capacity as dependent upon total width, *not* the number of lanes. Subsequent manuals would treat signalized intersection capacity and service volumes as per-lane characteristics. The six basic nomographs are shown in Fig. 7.2.

Note that the "load factor" lines in Fig. 7.2 coincide with the definitions of levels of service A through E. Thus, the figure yields a saturation flow rate that can be used to estimate the service volume for any given level of service.

Figure 7.2 also *includes* two adjustment factors on each nomograph:

- An adjustment factor for Peak Hour Factor and metropolitan area population (f_{POP}), and
- An adjustment factor for location within a metropolitan area (f_L).

The base conditions for the saturation flow rates drawn from Fig. 7.2 include 5% trucks and through buses, 10% right turns, 10% left turns, and no local buses.

An example in the use of Fig. 7.2 is shown on Fig. 7.2e. Saturation flow rates are found for a two-way street with parking with a curb-to-division line width of 40 ft. The resulting total saturation flow rates are:

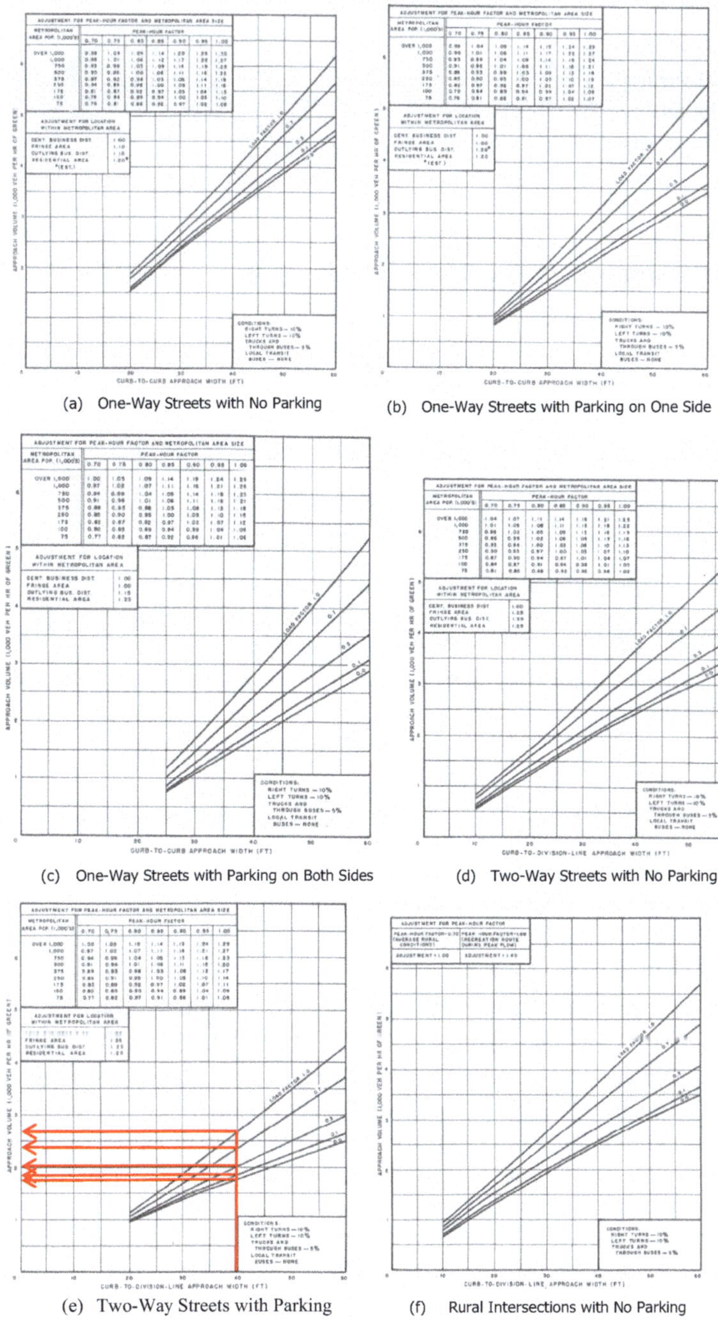

(a) One-Way Streets with No Parking
(b) One-Way Streets with Parking on One Side
(c) One-Way Streets with Parking on Both Sides
(d) Two-Way Streets with No Parking
(e) Two-Way Streets with Parking
(f) Rural Intersections with No Parking

Fig. 7.2 Base saturation flow rates for signalized intersections—1965 HCM. *Source* "Highway Capacity Manual," *Special Report 87*, Transportation Research Board, Washington D.C., pp. 134–136 [5], used with permission of the National Academy of Science, Courtesy of National Academies Press

$s_{TB,E}$ 2700 veh/hg
$s_{TB,D}$ 2400 veh/hg
$s_{TB,C}$ 2020 veh/hg
$s_{TB,D}$ 1800 veh/hg
$s_{TB,E}$ 1750 veh/hg.

7.4.3 Adjustments in the 1965 HCM

These values, of course, must be modified for conditions that do not conform to the base conditions of Fig. 7.2, using Eq. 7.6:

$$SV_i = s_{TBi} * f_{POP} * f_L * f_{HV} * f_{LB} * f_{RT} * f_{LT} * \left(\frac{G}{C}\right) \tag{7.6}$$

where:

SV_i service volume for level of service i (veh/h),
s_{TBi} total saturation flow rate under base conditions for level of service i (veh/hg), obtained from Fig. 7.2,
f_{POP} adjustment factor for metropolitan area population and PHF (decimal), obtained from Fig. 7.2,
f_L adjustment factor for location within a metropolitan area (decimal), obtained from Fig. 7.2,
f_{HV} adjustment for heavy vehicles (decimal),
f_{LB} adjustment for local buses (decimal),
f_{RT} adjustment for right turns (decimal),
f_{LT} adjustment for left turns (decimal),
G actual green time (secs), and
C cycle length (secs).

Once again, this equation does not appear anywhere in the 1965 HCM, but is included here to clarify the application of the many adjustment factors used for signalized intersections. Note that adjustment factors for type of capacity (f_c) and parking (f_P) from the 1950 HCM are gone. Types of capacity are replaced by service volumes, and there is only one capacity—synonymous with the service volume at level of service E. The parking adjustment is baked into Fig. 7.2, and does not have to be separately applied.

Heavy vehicle adjustment factors are given in Table 7.3. They adjust for truck and through bus presence that do not conform to the 5% base condition.

The adjustment for local buses (f_{LB}) is obtained from four nomographs, illustrated in Fig. 7.3. The four nomographs cover near-side and far-side bus stops on streets with and without parking.

Table 7.3 Heavy—vehicle adjustment factors (f_{HV}) for signalized intersections—1965 HCM

% Trucks/Th. buses	Factor (f_{HV})	% Trucks/Th. buses	Factor (f_{HV})	% Trucks/Th. buses	Factor (f_{HV})
0	1.05	7	0.98	14	0.91
1	1.04	8	0.97	15	0.90
2	1.03	9	0.96	16	0.89
3	1.02	10	0.95	17	0.88
4	1.01	11	0.94	18	0.87
5	1.00	12	0.93	19	0.86
6	0.9	13	0.8	20	0.8

Source "Highway Capacity Manual," *Special Report 87*, Transportation Research Board, Washington D.C., 1965 [5], Table 6.6, p. 142, used with permission of the National Academy of Science, Courtesy of National Academies Press

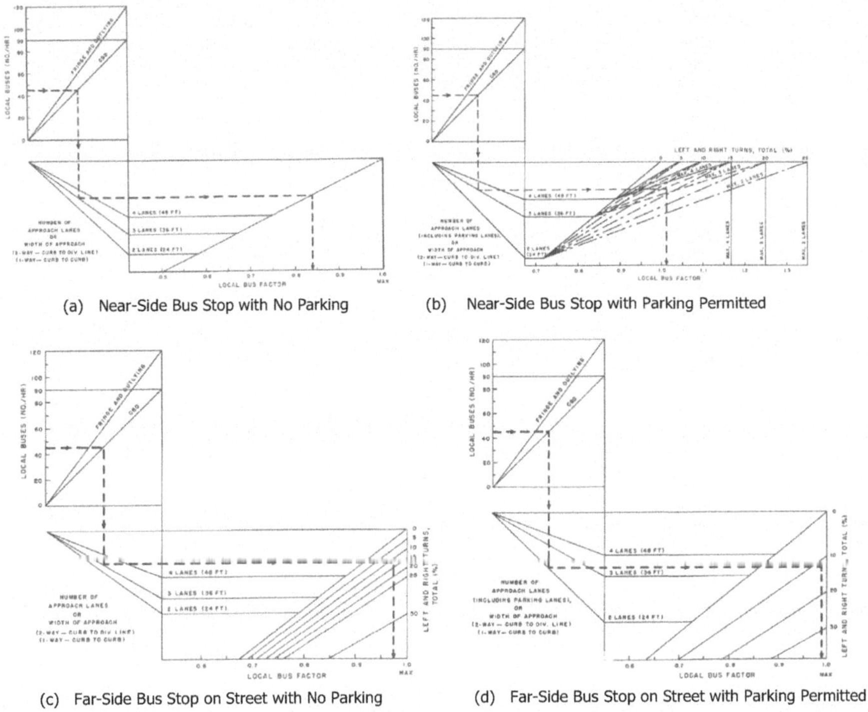

Fig. 7.3 Local bus adjustment factor—1965 HCM. *Source* "Highway Capacity Manual," *Special Report 87*, Transportation Research Board, Washington D.C., 1965 [5], Figs. 6.11–6.14, pp. 143–145, used with permission of the National Academy of Science, Courtesy of National Academies Press

Adjustment factors for right turns (f_{RT}) and left turns (f_{LT}) are shown in Tables 7.4 and 7.5. The tables give factors for ordinary cases in which there are no exclusive turning lanes and no exclusive turning phases. As was the case in the 1950 HCM, such cases are handled with a set of special steps.

Table 7.4 gives turn factors for right turns and for left turns on one-way streets. In effect, on a one-way street, the left turn is analogous to a right turn from any street. However, even on a one-way street, right-turn and left-turn adjustments are handled separately, and are *never* combined. Table 7.5 gives adjustment factors for left turns on a two-way street.

Table 7.4 Adjustment factors for right turns on two-way streets[a], right turns on one-way streets[a], and left turns on one-way streets[a]—1965 HCM

% Turns[b]	Approach width (ft) Streets with no parking[c]			Approach width (ft) Streets with parking[d]		
	≤15	16–24	25–34	≤20	21–29	30–39
0	1.20	1.050	1.025	1.20	1.050	1.025
1	1.18	1.045	1.020	1.18	1.045	1.020
2	1.16	1.040	1.020	1.16	1.040	1.020
3	1.14	1.035	1.015	1.14	1.035	1.015
4	1.12	1.030	1.015	1.12	1.030	1.015
5	1.10	1.025	1.010	1.10	1.025	1.010
6	1.08	1.020	1.005	1.08	1.020	1.010
7	1.06	1.015	1.005	1.06	1.015	1.005
8	1.04	1.010	1.005	1.04	1.010	1.005
9	1.02	1.005	1.000	1.02	1.005	1.000
10	1.00	1.000	1.000	1.00	1.000	1.000
11	0.99	0.995	1.000	0.99	0.995	1.000
12	0.98	0.990	0.995	0.98	0.990	0.995
13	0.97	0.985	0.995	0.97	0.985	0.995
14	0.96	0.080	0.990	0.96	0.080	0.990
15	0.95	0.975	0.990	0.95	0.975	0.990
20	0.90	0.950	0.975	0.90	0.950	0.975
30+	0.85	0.900	1.000	0.90	0.900	1.000

[a]No separate turning lanes or separate signal indication
[b]Handle right turns and left turns separately in all computations; do not sum
[c]For approach width >35 ft, the adjustment factor is 1.000
[d]For approach width >39 ft, the adjustment factor is 1.000
Source Excerpts from "Highway Capacity Manual," *Special Report 87*, Transportation Research Board, Washington D.C., 1965 [5], Table 6.4, p. 140, used with permission of the National Academy of Science, Courtesy of National Academies Press

Table 7.5 Adjustment factors for left turns on two-way streets[a]—1965 HCM

% Turns	Approach width (ft) Streets with no parking			Approach width (ft) Streets with parking		
	≤15	16–34	≥35	≤20	21–39	≥40
0	1.30	1.10	1.050	1.30	1.10	1.050
1	1.27	1.09	1.045	1.27	1.09	1.045
2	1.24	1.08	1.040	1.24	1.08	1.040
3	1.21	1.07	1.035	1.21	1.07	1.035
4	1.18	1.06	1.030	1.18	1.06	1.030
5	1.15	1.05	1.025	1.15	1.05	1.025
6	1.12	1.04	1.020	1.12	1.04	1.020
7	1.09	1.03	1.015	1.09	1.03	1.015
8	1.06	1.02	1.010	1.06	1.02	1.010
9	1.03	1.01	1.005	1.03	1.01	1.005
10	1.00	1.00	1.000	1.00	1.00	1.000
11	0.98	0.99	0.995	0.98	0.99	0.995
12	0.96	0.98	0.990	0.96	0.98	0.990
13	0.94	0.97	0.985	0.94	0.97	0.985
14	0.92	0.96	0.980	0.92	0.96	0.980
15	0.90	0.95	0.975	0.90	0.95	0.975
20	0.85	0.90	0.950	0.85	0.90	0.950
30+	0.80	0.85	0.900	0.80	0.85	0.900

[a]No separate turning lanes or separate signal indications
Source Excerpts from "Highway Capacity Manual," *Special Report 87*, Transportation Research Board, Washington D.C., 1965 [5], Table 6.5, p. 145, used with permission of the National Academy of Science, Courtesy of National Academies Press

As noted, where exclusive turn lanes and/or signal phases exist, the factors in Tables 7.4 and 7.5 do not apply. In such cases, the following special procedures are applied:

Separate Turning Lanes and Separate Turning Signals

- Deduct the width of the exclusive lane(s) from the total approach width. Compute the service volumes of the remaining width using the basic methodology, but with 0% assigned to the turning movement(s) using the exclusive lane(s).
- Each special turning lane will have the following saturation flow rates per 10 ft of approach width:

LOS	Veh/Hg	Assumed truck percentage (%)
A, B, C	800	5
D	1000	5
E	1200	5

Where two or more turning lanes are provided to handle a particular movement, each additional lane should be assigned a saturation flow rate of 0.8 times the above values. Apply the appropriate G/C ratio for the separate signal indication and adjust for trucks as appropriate.

- Add the service volumes computed in steps 1 and 2 to obtain the total for the approach.

Separate Turning Lanes w/o Separate Signalization

- Deduct the width of the exclusive lane(s) from the total approach width. Compute the service volumes of the remaining width using the basic methodology, but with 0% assigned to the turning movement(s) using the exclusive lane(s).
- For a right-turn lane: For any LOS, use 600 * G/C vehicles per hour assuming 5% trucks. If there are no pedestrians present, use the values for the signal-controlled case. Adjust for trucks as appropriate.
- For a left-turn lane: For any level, consider the service volume in pc/h as 1200-V_0, but not less than 2 vehicles per cycle.
- Add the volumes for through lanes, right-turn lane(s) and left-turn lane(s) to obtain the approach service volume.

Separate Turning Signal w/o Separate Turning Lane(s)

This situation exists where there is protected plus permitted signal phasing with no separate left-turn lanes. The green phase must be divided into portions during which opposing flows are present and those in which they are not present.

- For the portion of green time where the opposing flow is present, apply the basic procedure.
- For the portion of green time where the opposing flow is NOT present, apply the basic procedure, but consider left turns as if they were left turns from a one-way street.
- Add the results of the two service volumes for component portions of green time to obtain the full approach service volume.

7.4.4 Comments on the 1965 Intersection Methodology

The 1965 HCM moves forward with the concept of saturation flow rate, but continues to treat it as a variable based upon approach width in feet, as opposed to a lane-based parameter. It deals with the G/C ratio as did the 1950 HCM, but does not address the difference between actual green time and effective green time.

The most glaring "missing" piece from the 1965 HCM is the lack of a specific adjustment related to pedestrians. Ambient levels of pedestrian activity are built into the six base nomographs, and pedestrian interference is part of what is reflected in the adjustment for population and PHF, and the adjustment for location within the urban area. However, the procedure provides no way of dealing with pedestrian levels that are clearly beyond the ordinary or average conditions, such as exists in all major central cities like New York, Boston, Chicago, etc. Later editions of the HCM would deal with pedestrians in a much more specific fashion, but would require input data on pedestrian volumes.

Because the base conditions of the nomographs are NOT "ideal" conditions, some adjustment factors are greater than 1.00. Because the nomographs have a base truck percentage of 5%, actual demands with less than this would have a positive adjustment. The same is true of right turn and left turn adjustments, where the base case is 10% for each turn.

The tying of the level of service to load factor was always problematic. "Load factor" was not analogous to a v/c ratio, nor did it provide any specific description of what operating conditions were like. After publication of the 1965 HCM, there were many attempts to find new service measures to describe intersection operating quality, which led to the adoption of various "delay" measures beginning with the 1985 HCM.

7.5 Circular 212: Interim Materials

By 1970, work was already underway to produce a third edition of the *Highway Capacity Manual*. In 1980, recognizing that an update was sorely needed, but that fundamental research aimed at finalizing methodologies would not be ready for another 5 years, the HCQSC published TRB Circular 212 [8], entitled *Interim Materials on Highway Capacity*. The methodologies of this publication were not formal replacements for material in the 1965 HCM, but represented a major step on the road to the 1985 HCM. Many of the methodologies, though intended for interim and experimental use, were widely used and adopted.

7.5.1 Background and General Approach

Perhaps the most popular methodology of the Circular 212 was a new method for analysis of signalized intersections based upon "critical lane analysis" or "critical movement analysis" as a fundamental approach. Circular 212 presented both a *planning methodology* and a *design and operations methodology*. The former was as close to a "back of the envelope" procedure as could be imagined, and relied heavily on the use of average conditions and characteristics. There were few adjustments for prevailing conditions that differed greatly from the norm. The design and operations procedure included a number of adjustments, but was still a relatively straightforward approach.

The interim methodologies were developed primarily by William Reilly of JHK & Associates as part of an overall effort to prepare materials for a new manual [9]. He drew heavily on previous work in the area by Messer et al. [10], as well as similar work in Australia [11], and earlier work by Don Berry [12].

The basic concept of critical movement analysis (CMA) was not new. When Bruce Greenshields studied discharge headways and lost times in the late 1940s, he was essentially creating critical movement analysis models. Nevertheless, the focus of highway capacity analysis models in the 1950 and 1965 HCMs was on total saturation flow based upon total approach width. In critical movement analysis, the focus is on flow in a single lane.

Critical movement analysis is based upon a number of fundamental concepts:

1. Both the planning and design/operations methodology used a base saturation flow rate of 1800 passenger cars per hour of green per lane (pc/hg/ln), based upon a variety of studies and past work.
2. The methodology essentially analyzes the demand flows and the signal phasing to identify those lane movements that will require the most time to be processed. These are the "critical movements."
3. All critical movements are seeking to use the same space—i.e., the intersection area. While the signal separates the most difficult conflicting movements in time, during any signal cycle, all critical movements must pass through the same space. Thus, the critical demand variable is the *sum of the critical lane volumes*.
4. The 1950 and 1965 HCM used adjustment factors to modify saturation flow rates under ideal conditions to reflect prevailing conditions. In Circular 212, adjustments are used to convert prevailing demand flows to equivalent flows under ideal conditions.

The methodology, in the simplest terms, identifies the critical movements, then uses the sum of the critical movement flows as a service measure to determine level of service. Table 7.6 shows the relationship between the sum of critical lane volumes and the level of service.

Note that for planning applications, the limiting values are given in terms of mixed veh/h. This is because the planning methodology does not adjust for a variety of factors, including truck presence. The design/operations application, however, does convert for a number of prevailing conditions, and the criteria are stated in adjusted pc/h.

The limiting values for sum of critical lane volumes goes down as the number of phases increases. This is to take into account the fact that each phase adds lost time to the signal, and reduces the amount of effective green time available for movement of critical flows. However, cycle length also affects this, and it is not taken into account in the methodology.

Table 7.6 Level of service versus sum of critical lane volumes—circular 212

Level of service	Planning applications Max. sum of critical lane volumes (veh/h)		
	2-Phase signals	3-Phase signals	4+ Phase signals
A	900	855	825
B	1050	1000	965
C	1200	1140	1100
D	1350	1275	1225
E	1500	1425	1375
F	>1500	>1425	>1375
Level of service	Design and operations applications Max. sum of critical lane volumes (pc/h)		
	2-Phase signals	3-Phase signals	4+ Phase signals
A	1000	950	900
B	1200	1140	1080
C	1400	1340	1270
D	1600	1530	1460
E	1800	1720	1650
F	>1800	>1720	>1650

Source "Interim Materials on Highway Capacity—Critical Movement Analysis," *Transportation Research Circular 212*, Transportation Research Board, Washington D.C., 1980 [8] Table 6, p. 11, originally from W.R. Reilly (NCHRP Project 3-28) and Messer et al. [10]. Used with permission of the National Academy of Science, Courtesy of National Academies Press

It should also be noted that while saturation flow rates and v/c ratios are never explicitly computed by the Circular 212 methodologies, they are implicit in the level of service criteria of Table 7.6. In particular, Circular 212 takes effective green time to be equal to the actual green time *plus* yellow and all-red time. This explains why for a 2-phase signal (design/operations applications), the sum of critical lane volumes is permitted to be as high as 1800 pc/h—even though that is the assumed base, or ideal, saturation flow rate. Implicit in this is the assumption that there is *no* lost time in the two-phase cycle. If that were truly the case, however, there is no reason to reduce the sums for 3- and 4+ phase signals. This is a clear inconsistency in the methodology.

The LOS criteria of Table 7.6 do not result in an explicit prediction of a service quality measure. Circular 212 does, however, provide a general relationship between the various levels of service and the typical expected v/c ratios and average stopped-time delays. This relationship is shown in Table 7.7.

As noted, the planning methodology does not adjust for anything. The design/operations methodology adjusts demand volumes for the following conditions:

- Trucks
- Local buses

Table 7.7 Relationship between delay and level of service

Level of service	Typical v/c range	Delay range[a] (secs/veh)
A	0.00–0.60	0.0–16.0
B	0.61–0.70	16.1–22.0
C	0.71–0.80	22.1–28.0
D	0.81–0.90	28.1–35.0
E	0.91–1.00	35.1–40.0
F	Varies	40.1 or greater

[a]Measured as stopped delay. Delay values relate to the man stopped delay incurred by all vehicles entering the intersection. Note that traffic coordination conditions are not considered and could drastically alter the delay range for a given v/c ratio

Source "Interim Materials on Highway Capacity—Critical Movement Analysis," *Transportation Research Circular 212*, Transportation Research Board, Washington D.C., 1980 [8], Table 7, p. 12. Originally from W.R. Reilly (NCHRP Project 3-28), based on a synthesis of various data. Used with permission of the National Academy of Science, Courtesy of National Academies Press

- Lane width
- Lane utilization
- Left turns
- Right turns.

The adjustments shift the demand volume from veh/h under prevailing conditions to through passenger cars per hour under ideal or base conditions. Because adjustments are made to demand volume, not to capacities or saturation flow rates, they do not use the standard symbology where "f_i" is the multiplicative adjustment factor for prevailing condition i.

Table 7.8 summarizes passenger car equivalent values (E) for various types of vehicles. Note that the approach to local buses ignores the details of whether the bus stop is near side or far side, and whether the street has parking or not.

Table 7.9 summarizes through passenger car equivalents for left turns and right turns at a signalized intersection. Equivalents for left turns (E_{LT}) depend upon whether the turn is opposed or unopposed, the opposing volume in veh/h (if it is opposed),

Table 7.8 Passenger car equivalents versus vehicle types—circular 212

Type of vehicle	Equivalent	PCE
Passenger car/motorcycle	E_{PC}	1.0
Truck or through bus	E_T	2.0
Local bus	E_{LB}	5.0

Source Extracted from "Interim Materials on Highway Capacity—Critical Movement Analysis," *Transportation Research Circular 212*, Transportation Research Board, Washington D.C., 1980 [8], Table 8, p. 12. Originally from W.R. Reilly (NCHRP Project 3-28), based on a synthesis of various data. Used with permission of the National Academy of Science, Courtesy of National Academies Press

Table 7.9 Through passenger car equivalents for left turns and right turns—circular 212

Case	Opposing volume (veh/h)			
	0–299	300–599	600–999	1000+
E_{LT} for left turns made from shared left-though lanes				
No turn phase	1.0	2.0	4.0	6.0
With turn phase	1.2	1.2	1.2	1.2
E_{LT} for left turns made from left turn lanes				
No turn phase	1.0	2.0	4.0	6.0
With turn phase	1.05	1.05	1.05	1.05
Case	Pedestrian volume in conflicting crosswalk (peds/h)			
	0–99	100–599	600–1199	1200+
E_{RT} for right turns made from shared or exclusive lanes				
All	1.00	1.25	1.50	2.00 or more[a]

[a]As determined from local conditions

Source "Interim Materials on Highway Capacity—Critical Movement Analysis," *Transportation Research Circular 212*, Transportation Research Board, Washington D.C., 1980 [8], Table 3, p. 9 and Table 4, p. 10. Originally from W.R. Reilly (NCHRP Project 3-28), based on a synthesis of various data, and Messer et al. [10]. Used with permission of the National Academy of Science, Courtesy of National Academies Press

and whether the turn is made from a separate lane or a mixed lane. Equivalents for right turns depend upon pedestrian volumes in the conflicting crosswalk.

Adjustment factors for lane width (W) and lane utilization (U) are shown in Table 7.10. The lane width factor *increases* the demand volume to reflect narrow lanes. The lane utilization factor adjusts for the fact that vehicles will not use multiple lanes equally. It increases demand volume so that the average lane volume will reflect the volume expected in the most heavily-used lane.

Table 7.10 Lane width (W) and Lane utilization (U) factors for critical movement analysis—circular 212

Lane width factor (W)		Lane utilization factor (U)	
Lane width (ft)	W	No. of lanes	U
8.0–9.9	1.1	1	1.00
10.0–12.9	1.0	2	1.05
13.0–15.9	0.9	3	1.10

Source Extracted from "Interim Materials on Highway Capacity—Critical Movement Analysis," *Transportation Research Circular 212*, Transportation Research Board, Washington D.C., 1980 [8], Table 8, p. 12, Originally from W.R. Reilly (NCHRP Project 3-28), based on a synthesis of various data. Used with permission of the National Academy of Science, Courtesy of National Academies Press

7.5.2 The Planning Application

The steps in the planning application are relatively straightforward, and involve little adjustment. Worksheets are provided in Circular 212, as are numerous example problems, to help guide the process. The planning application steps are as follows:

1. Identify the geometrics of the intersection, including lane usage (which may be restricted by design or through lane-use controls).
2. Identify all movement volumes in vehicles per hour.
3. Identify the signal phasing, particularly noting which movements are permitted on which phases.
4. Perform a *left-turn check*:

   Where a left turn is made either partially or completely on a permitted basis (i.e., without a protected left-turn phase), a simple check is made to see if the number of left turns can be handled with the signalization that exists. To do this, the *minimum* number of left turns that can be made must be estimated, as well as the *maximum* number of left turns that be made.

   The *minimum* number of left-turns that can be made (V_{Lmin}) is equal to 2 vehicles per signal cycle. Thus, if the signal has a 60-s cycle length, a minimum of $2 \times (3600/60) = 120$ left-turns/h can be made.

   The *maximum* number of left-turns (V_{Lmax}) that can be made is dependent upon the opposing volume, which includes the total opposing through plus right-turn traffic:

$$V_{L\mathrm{max}} = 1200(G/C) - V_o \tag{7.7}$$

   If $V_{L\mathrm{min}} > V_{L\mathrm{max}}$, then $V_{L\mathrm{max}} = V_{L\mathrm{min}}$. If $V_L > V_{L\mathrm{max}}$, the intersection is not likely to function without breakdown. Changes in signalization and/or geometric design are needed.

5. Assign lane volumes:

   Volumes are assigned to lanes. If a left-turn lane or lanes exist, all left-turn volume is assigned to them. If a right-turn lane or lanes exist, all right-turn volume is assigned to them. Where two turning lanes exist for a given movement, 55% of the volume is assigned to one lane.

   If no left-turn lane is provided, Table 7.9 is used to convert the left-turn volume in veh/h to an equivalent in pc/h. Remaining through and right-turning vehicles are then assumed to be in pc/h. The sum of all pc/h volumes is divided equally amongst available lanes. In doing so, *all* left-turning vehicles must be assigned to the left lane, and *all* right-turning vehicles must be assigned to the right lane. When this assignment is completed, left-turn volumes in pc/h are converted back to veh/h.
6. Select critical lane volumes:

   For a two-phase signal, the critical volume for each phase is the highest total of a through lane volume plus its opposing left-turn volume.

For a multilphase (three- or more phase) signal, phases are listed in the sequence in which they occur. In the case of an actuated signal, the phases are listed in the sequence in which they are *most likely* to occur.

For each phase in which there are *no permitted left turns*, the critical volume is the highest lane volume that is allowed to move in that phase.

For each phase in which there is an *opposing left turn*, the critical volume is the highest total of a through lane volume and its opposing left-turn volume.

In the case of a protected plus permitted left-turn phase, V_{Lmax} (as computed in the left-turn check), is assigned to the permitted portion of the phase; remaining left turns are assigned to the protected portion of the phase.

7. Obtain the sum of critical lane volumes: After being identified, the sum of critical lane volumes is computed.
8. Determine the level of service: Enter Table 7.6 with the sum of critical lane volumes to determine the expected level of service.

The planning methodology is relatively straightforward, with the possible exception of assigning left-turn volumes to the appropriate lanes and signal phases. It allows for a quick assessment of likely overall intersection level of service. It does *not*, like previous and subsequent methodologies, allow for an estimate of the level of service for a specific intersection approach.

7.5.3 The Design and Operations Application

The design/operations methodology in Circular 212 is slightly more complex than the planning method. It converts all demand volumes to flow rates in pc/h, and allows for a more detailed consideration of signalization details. The steps in the design/operations methodology are as follows:

1. Identify lane geometry (same as planning method).
2. Identify hourly volumes (same as planning method).
3. Identify phasing (same as planning method).
4. Perform a *left-turn check* (same as planning method).
5. Convert all movement volumes to equivalent flow rates in through passenger cars per hour. For *each* movement:

$$v_{pci} = \frac{(V_{PCi} + 2V_{Ti} + 5V_{LBi}) * E_M * W^*U}{PHF} \quad (7.8)$$

where:

v_{pci} flow rate for movement i in pc/h,
V_{PCi} passenger car volume for movement i (veh/h),
V_{Ti} truck and through bus volume for movement i (veh/h),
V_{LBi} local bus volume for movement i (veh/h),

E_i passenger car equivalent for movement i, see Table 7.9 for E_{LT} and E_{RT}; for through movements, $E_M = 1.00$.
W lane width adjustment (Table 7.10),
U lane utilization adjustment (Table 7.10), and
PHF peak hour factor.

6. Calculate lane volumes:

   Where there are separate lanes for left- and or right-turns, all such turns are assigned to these lanes. Where left- and/or right-turns are sharing lanes with through vehicles, the equivalent volumes for all movements sharing a set of lanes are added. The total volume using any set of lanes is divided by the number of lanes available to get the lane volume (which is actually a flow rate).

   As in the planning methodology, where shared lanes are involved, all left turns must be assigned to the left lane and all right turns must be assigned to the right lane. Unlike the planning method, left turns are *not* converted back to veh/h.
7. Identify and sum the critical lane volumes (flow rates):

   Figure 7.4 is used to identify and sum the critical lane volumes for the intersection. The figure covers all combinations of signal phases that are likely to be encountered.
8. Determine the intersection level of service: Enter Table 7.6 (design/operations methodology) with the sum of the critical lane volumes to determine the expected level of service. Like the planning method, the LOS determined applies to the overall intersection, not to any specific intersection approach.

The selection of critical lane volumes as shown in Table 7.6 can be somewhat confusing. Unlike the planning application, critical sums do *not* involve adding through volumes to opposing left-turn volumes. This is because the left-turn volumes have already been adjusted to pc/h considering the opposing through volume through the use of the left-turn equivalent. Adding left-turn and opposing through volumes would, in effect, be double-counting the effect of the conflict.

Where fully-protected left-turn phases are in place, the highest through lane volume is added to the highest left-turn lane volume, regardless of the approach from which they originate. Because left turns and through vehicles move on separate phases, the only concern is for the maximum volume during each phase—the volume that would consume the most green time.

For overlapping (protected plus permitted) phasing, the design/operations application is somewhat more approximate. In this case, opposing left-turn volumes *are* added to through movement volumes to reflect the fact that they at least partially move at the same time. Remember that adjusting left-turn volumes in such cases involves a very rough estimate of how the left-turn volume splits between the permitted and protected portions of the approach.

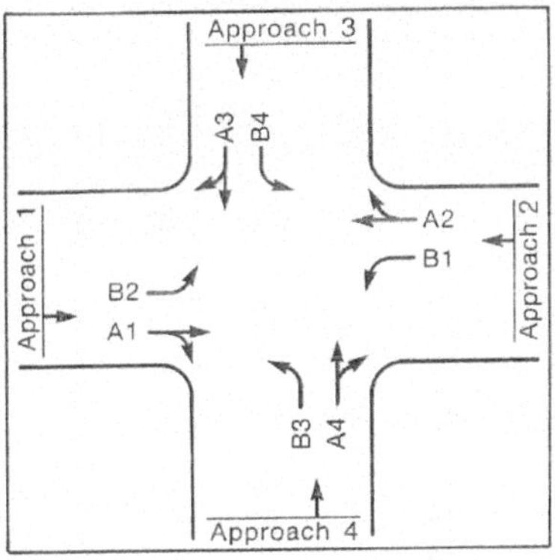

Signal Phasing for Street and Intersection Geometry	Approaches	Critical Movements[1]
One Phase, No LT Bay	1 and 2 3 and 4	A1+B2 or A2+B1 A3+B4 or A4+B3
One Phase, With LT Bay	1 and 2 3 and 4	A1 or A2 or B1 or B2 A3 or A4 or B3 or B4
Two Phases, No Overlap, With LT Bay[2]		
Leading or lagging LTs from both directions	1 and 2 3 and 4	A1 or A2+B1 or B2[4] A3 or A4+B3 or B4[4]
Leading or lagging LTs from one direction	1 and 2 3 and 4	B1+A1 or A2[5] B3+A3 or A4[5]
Two Phases, With Overlap, With LT Bay[3]		
Leading or lagging LTs from both directions	1 and 2 3 and 4	A1+B1 or A2+B2 A3+B3 or A4+B4
Leading or lagging LTs from one direction	1 and 2 3 and 4	B1+A1 or A2[5] B3+A3 or A4[5]

1. By approach, on a per-lane basis. Select maximum of the alternatives shown.
2. Exclusive LT phase in one or both directions.
3. Protected plus permitted phase in one or both directions.
4. Note that the critical volume on a given street is the single highest volume. Combining through flows and opposing left turns *is not* done in the design/operations application.
5. Assume arrow indication is for movements B1 and B3. Other combinations are possible, depending upon intersection configuration.

(Source: modified from "Interim Materials on Highway Capacity – Critical Movement Analysis," *Transportation Research Circular 212*, Transportation Research Board, Washington D.C., 1980, Figure 4 and Table 9, pg. 26, Originally from W.R. Reilly (NCHRP Project 3-28), based on Messer, C.J., Fambro, D.B., and Peterson, S.G. "Critical Lane Analysis for Intersection Design," *Transportation Research Record 644*, 1977, pgs 26-35. Used with permission of the National Academy of Science, Courtesy of National Academies Press.)

Fig. 7.4 Determining sum of critical lane volumes: design and operations method

Signal phasing for street and intersection geometry	Approaches	Critical movements[a]
One phase, no LT Bay	1 and 2 3 and 4	A1 + B2 or A2 + B1 A3 + B4 or A4 + B3
One phase, with LT Bay	1 and 2 3 and 4	A1 or A2 or B1 or B2 A3 or A4 or B3 or B4
Two phases, no overlap, with LT Bay[b]		
Leading or lagging LTs from both directions	1 and 2 3 and 4	A1 or A2 + B1 or B2[d] A3 or A4 + B3 or B4[d]
Leading or lagging LTs from one direction	1 and 2 3 and 4	B1 + A1 or A2[e] B3 + A3 or A4[e]
Two phases, with overlap, with LT Bay[c]		
Leading or lagging LTs from both directions	1 and 2 3 and 4	A1 + B1 or A2 + B2 A3 + B3 or A4 + B4
Leading or lagging LTs from one direction	1 and 2 3 and 4	B1 + A1 or A2[e] B3 + A3 or A4[e]

[a]By approach, on a per-lane basis. Select maximum of the alternatives shown
[b]Exclusive LT phase in one or both directions
[c]Protected plus permitted phase in one or both directions
[d]Note that the critical volume on a given street is the single highest volume. Combining through flows and opposing left turns *is not* done in the design/operations application
[e]Assume arrow indication is for movements B1 and B3. Other combinations are possible, depending upon intersection configuration
Source Modified from "Interim Materials on Highway Capacity—Critical Movement Analysis," *Transportation Research Circular 212,* Transportation Research Board, Washington D.C., 1980 [8], Fig. 4 and Table 9, pg. 26, Originally from W.R. Reilly (NCHRP Project 3-28), based on Messer et al. [10]. Used with permission of the National Academy of Science, Courtesy of National Academies Press

7.5.4 Comments on Critical Movement Analysis

The introduction of the critical movement analysis method in Circular 212 was a major step in the evolution of signalized intersection analysis. Compared to the methodologies that would follow, it was a very simple way to generate a general assessment of how such an intersection would perform. Obviously, its simplicity avoided dealing directly with the myriad and complex details of the operation of a signalized intersection.

The methodology lacked the specificity of identifying operating characteristics in specific terms beyond generalized delay ranges. Further, it yielded an overall view of the intersection as a whole, not individual intersection approaches, or discrete lane groups. The details of signal timing were generally avoided, except during the left-turn check.

Because of its simplicity, and because of the weaknesses of the 1965 HCM approach, CMA became a very popular method of analysis. In fact, a small number of major state highway agencies still use this approach instead of the more complex methodologies in subsequent editions of the HCM.

In any event, it clearly set the stage for all subsequent methodologies in that it (a) focused on the lane as the primary unit of width, (b) used the identification of critical movements as the centerpiece of analysis, and (c) at least in the most general sense, associated levels of service with delay.

Sample calculations using and comparing these and other signalized intersection methodologies are found in Appendix IV.

References

1. Roess RP, Prassas ES, McShane WR (2019) Traffic engineering, 5th edn. Pearson-Prentice Hall, Upper Saddle River, NJ
2. The Westminster Road Semaphore. www.victoriacountyhistory.ac.uk/explore/items/westminster-road-semaphore
3. History of Traffic Lights. History of roads. www.inventors.about.com/od/rstartinventions/a/History-Of-Roads_3.htm
4. U.S. Government Printing Office (1950) Highway capacity manual. Bureau of Public Roads, U.S. Government Printing Office, Washington DC
5. Transportation Research Board (1965) Highway capacity manual, special report 87. Transportation Research Board, Washington, DC
6. Bureau of public roads (Dec 1960) Intersection traffic flow. Final report to the Bureau of public roads. CEIR Inc., Arlington, VA
7. Normann OK (1962) Variations in flow at intersections related to the size of city, type of facility, and capacity utilization. In: Highway research bulletin, vol 352. Transportation Research Board, Washington, DC
8. Transportation Research Board (1980) Interim materials on highway capacity. In: Circular 212. Transportation Research Board, Washington, DC
9. Reilly et al (1982) Signalized intersection capacity study. Final report. NCHRP project 3, vol 28, no 2, JHK & Associates
10. Messer CJ, Fambro DB, Peterson SG (1977) Critical lane analysis for intersection design. Transp Res Rec 644:26–35
11. Signalized Intersections (1978) Capacity Guide, Australian Road Research Board, ARR 79, 1978
12. Berry DS (1997) Other Methods for Computing Capacity of Signalized Intersections. Presented at the 56th Annual Meeting of the Transportation Research Board, Washington DC

Chapter 8
Signalized Intersections: 1985 Through 2000

Between 1980 and 1985, the Highway Capacity and Quality of Service Committee moved to adopt a signalized intersection methodology based upon critical movements. This approach to timing and analysis of signalized intersections had been used by Bruce Greenshields in the early 1950s, and had been developed by Don Berry and others. The methodology incorporated into the 1985 HCM [1] *was based on the work of Carroll Messer and Dan Fambro* [2], *which was modified and calibrated under an NCHRP contract by William Reilly of JHK and Associates* [3].

The basic framework of the 1985 HCM signalized intersection methodology is still in use in the current HCM. The basic framework included worksheets that could theoretically (but not practically) be done by hand. The worksheets were and are useful for understanding and teaching the methodology. This worksheet format remained the same in the 1994 [4], 1997 [5], and 2000 [6] versions of the HCM. The changes in each version were made to the models within the same general framework.

In the 2010 and 2016 HCMs [7, 8], however, the worksheet format was discarded. The 2010 HCM has many unique aspects that make it very different from previous versions. Because of these differences, the 2010 and 2016 HCM methodologies will be discussed in a separate chapter (Chap. 9) of this book.

8.1 The Basic Framework—1985 Through 2000

Starting with a base of critical movement analysis, the 1985 HCM presented an organized framework for analysis of signalized intersections, from the most straightforward pre-timed, two-phase signals to the most complex actuated, multiphase signals. It established a model organized into five distinct analysis modules, as shown in Fig. 8.1, which remained the same through the 2000 HCM. In summary, each module contains a set of self-contained analysis steps, as follows:

E. S. Prassas and R. P. Roess, *The Highway Capacity Manual: A Conceptual and Research History Volume 2*, Springer Tracts on Transportation and Traffic 12,
https://doi.org/10.1007/978-3-030-34480-1_8

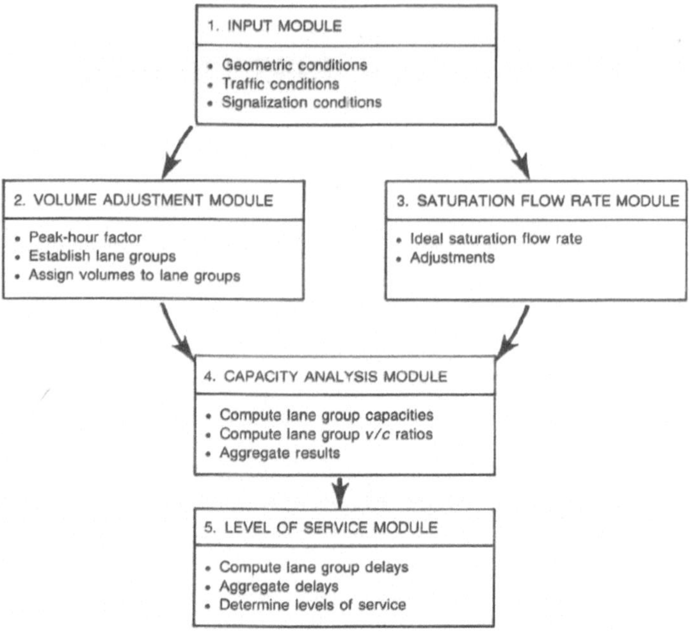

Fig. 8.1 Signalized intersection analysis modules in the 1985 HCM. *Source* "Highway Capacity Manual," *Special Report 209*, Transportation Research Board, Washington D.C., 1985 [1], Fig. 9-1, pp. 9–6, used with permission of the National Academy of Science, Courtesy of National Academies Press

- *Input Module*: The Input Module is used to specify a wide range of prevailing conditions that will impact the results of an analysis.
- *Volume Adjustment Module*: In the Volume Adjustment Module, volumes are adjusted to reflect flow rates, and are assigned to specific lane groups.
- *Saturation Flow Rate Module*: In the Saturation Flow Rate Module, the total saturation flow rate for established lane groups is determined for a range of prevailing conditions.
- *Capacity Analysis Module*: The results of the Volume Adjustment and Saturation Flow Rate Modules are compared and combined to produce v/c ratios for each discrete lane group in the intersection.
- *Level of Service Module*: In the Level of Service Module, specific values of expected delays are computed for each lane group under prevailing conditions.

The 1985 HCM introduced a number of concepts to the methodology, including:

- The basic physical unit for analysis is the *lane group*. Simply defined, a lane group is a collection of adjacent lanes that serve a set of movements that have reached an equilibrium in terms of lane usage. All movements on an approach are assumed to share all approach lanes *unless* there are lane-use controls in place, such as exclusive left-turn and/or exclusive right-turn lanes.

- Worksheets were added for clarification of each step of the analysis. Although the procedure was almost exclusively done using software (the HCS Package was originally created for the 1985 HCM), it was possible to do the entire methodology by hand and even the computer outputs of most vendors displayed the steps and results through the worksheets. This helped users to understand the results and what went into the calculations.

In the following sections, the five modules will be discussed and the changes that were made in each consecutive edition will be detailed for HCM editions from 1985 through 2000.

8.2 The Input Module

8.2.1 The 1985 Highway Capacity Manual

The Input Module summarizes all of the data that are needed to conduct an analysis of a signalized intersection. Because the methodology is detailed, and addresses many characteristics of modern intersections, the list of characteristics that must be specified to begin an analysis is rather long. Table 8.1 indicates the characteristics of the intersection that must be specified.

Parking lanes are noted whenever there is curb parking permitted within 250 feet of the STOP line. When reporting "parking activity," only movements into and out of curb parking spaces within 250 feet of the STOP line were included.

One of the most critical prevailing conditions to be reported is the "arrival type." Arrival type is a general measure of how well the subject intersection is coordinated with the nearest downstream signal, and has a significant impact on delay. Five arrival types were originally defined for the 1985 HCM as follows:

- Type 1: Dense platoons arrive at the subject intersection at the beginning of the RED phase. This is the worst possible platoon condition.
- Type 2: Dense platoons arrive during the middle of the RED phase, or a dispersed platoon arrives throughout the RED phase. This is a more typical situation with poor signal progression.
- Type 3: This condition represents random arrivals, i.e., vehicles are equally likely to arrive at any time during the signal cycle. This is representative of an isolated intersection or a signal that is uncoordinated with its downstream neighbor because the distance is too large for cohesive platoons. It is also the base case for delay estimation.
- Type 4: Dense platoons arrive during the middle of the GREEN phase, or dispersed platoons arrive throughout the GREEN phase. This is a typical situation with relatively good signal progression.
- Type 5: Dense platoons arrive at the beginning of the GREEN phase. This arrival type indicates excellent, nearly ideal, signal progression.

Table 8.1 Input data required for each lane group for signalized intersection analysis—1985 HCM

Type of condition	Parameter	Symbol
Geometric conditions	Area type	CBD or other
	Number of lanes	N
	Lane widths (ft)	W
	Grades (%)	+ (upgrade) − (downgrade)
	Existence of exclusive LT or RT lanes	Y or N
	Length of storage bay, LT or RT lanes	L_s
	Parking conditions	Y or N
Traffic conditions	Volumes by movement (veh/h)	V_i
	Peak hour factor (PHF)	PHF
	Proportion heavy vehicles	P_{HV}
	Conflicting pedestrian flow rate (peds/h)	PEDS N_B
	Number of local buses stopping within the intersection (buses/h)	N_m
	Parking activity (pkg maneuvers/h) Arrival type	1–5
Signalization conditions	Cycle length (s)	C
	Green times (s)	G_i
	Actuated versus pre-timed operation	A or P
	Pedestrian push-button ????	Y or N
	Minimum pedestrian green (s)	G_p
	Phase plan	

Source "Highway Capacity Manual," *Special Report 209*, Transportation Research Board, Washington D.C., Table 9-2, pp 9–7, used with permission of the National Academy of Science, Courtesy of National Academies Press

When originally drafted, the 1985 HCM signalized intersection methodology provided only verbal descriptions of arrival types, with the expectation that users would know how to evaluate this through field observations. As an afterthought, however, numerical criteria were added based upon the ratio of the percent of vehicles arriving during the green and the percentage of the cycle that is green for the subject movement. This is called the *platoon ratio*, R_p.

$$R_p = PVG/PTG \tag{8.1}$$

where

R_p platoon ratio,
PVG percentage of vehicles arriving on GREEN, and

Table 8.2 Relationship between arrival type and platoon ratio (R_p)—1985 HCM

Arrival type	Platoon ratio, R_p
1	0.0–0.50
2	0.51–0.85
3	0.86–1.15
4	1.16–1.50
5	>1.50

Source "Highway Capacity Manual," *Special Report 209*, Transportation Research Board, Washington D.C., 1985 [1], Table 9-2, pp. 9–8, used with permission of the National Academy of Science, Courtesy of National Academies Press

PTG percentage of time that the signal is GREEN for the subject lane group (PTG = G/C * 100).

Logically, for the base condition, defined as Arrival Type 3 (random arrivals), the ratio would be 1.00. For good progression, the value would be higher than 1.00, and for poor progression, the value would be less than 1.00. Beyond this basic logic, there was no data in 1985 from which to calibrate the relationship between platoon ratio and arrival type, which is shown in Table 8.2.

One of the signalization conditions required is the minimum pedestrian green time. This was used as a check against the actual green time to determine whether or not pedestrians are safely accommodated by the signalization. This could be entered directly or the minimum green time could be estimated as:

$$G_p = 7.0 + \frac{L_p}{4} - Y \tag{8.2}$$

where

G_p minimum pedestrian green time (s),
L_p length of the crosswalk, taken as the distance from the curb to the midpoint of the farthest travel lane (ft),
Y sum of the yellow (y) plus all-red (ar) interval (s).

The "4" in the equation is the assumed walking speed of pedestrians (4.0 ft/s).

The equation as stated essentially assumes that pedestrians have 7 s to *begin* crossing the street. It also assumes that pedestrians are permitted to *complete* crossing the street during the yellow and all-red intervals. This is a matter of agency policy, and some agencies permit crossing only during GREEN and YELLOW intervals, or only during GREEN intervals. The equation would have to be modified for application in such cases.

As part of the Input Module, the 1985 HCM offers suggested default values for key variables when site-specific data is unavailable. Like all default values, they should be used with extreme caution. Default values given are shown in Table 8.3.

Table 8.3 Default values suggested in the 1985 HCM

Parameter	Default value
Conflicting pedestrian flow rate (peds/h)	Low ped flow—50 peds/h Moderate ped flow—200 peds/h High ped flow—400 peds/h
Percent heavy vehicles	2%
Peak hour factor	0.90
Grade	0%
Number of buses	0 buses/h
Number of parking maneuvers	20 maneuvers/h (if parking exists)
Arrival type	3

Source "Highway Capacity Manual," *Special Report 209*, Transportation Research Board, Washington D.C., 1985 [1], Table 9-3, pp. 9–8, used with permission of the National Academy of Science, Courtesy of National Academies Press

8.2.2 The 1994 Update to the 1985 HCM

Two new inputs were added in 1994 input module:

- Lost time per phase was added as explicit number. Text was added describing green time, effective green, and lost times. The input worksheet had the user enter lost time per phase, and the text defined this as the sum of the start-up lost time and the clearance lost time.
- An additional arrival type, Arrival Type 6, was added to describe urban streets with short blocks and almost perfect progression.

Many additional default values were also suggested, as shown in Table 8.4.

8.2.3 The 1997 Update to the 1985 HCM

There was one change to the inputs in the 1997 HCM. Rather than enter the total lost time per phase (the sum of the start-up lost time plus clearance lost time), start-up lost time and a new variable, extension of effective green, *e*, was entered.

Entering extension of effective green allowed variation in the clearance lost time, based upon the actual length of the *yellow* and *all red* intervals. This provided a more accurate measure of lost time per phase, t_L, and thus effective green time. Extension of effective green (e) is the time that is still used after the end of the actual green

Table 8.4 Default values suggested in the 1994 update

Parameter	Default value
Conflicting pedestrian flow rate (peds/h)	Low ped flow—50 peds/h Moderate ped flow—200 peds/h High ped flow—400 peds/h
Percent heavy vehicles	2%
Peak hour factor	0.90
Grade	0%
Number of buses	0 buses/h
Number of parking maneuvers	20 maneuvers/h (if parking exists)
Arrival type	3 for lane groups with through movements, If intersection is isolated 4 if coordinated 3 for lane groups without through movements
Signal type	Pretimed
Cycle length range	60–120 s
Lost time	3.0 s/phase
Yellow plus all red	4.0 s/phase
Area type	Non-CBD
Lane width	12 ft

Note Taken from "Highway Capacity Manual," *Special Report 209*, Transportation Research Board, Washington D.C., 1994 [4], Table 9-3, pp. 9–11, used with permission of the National Academy of Science, Courtesy of National Academies Press

time displayed, that is, it is the portion of the yellow change and all-red clearance intervals that vehicles use to enter the intersection.

The clearance lost time can then be calculated using Eq. (8.3).

$$\ell_2 = Y - e \tag{8.3}$$

where all terms are as previously defined.

The 1997 HCM recommended a default value of 2.0 s for e, where specific information is not available. Because clearance lost time is calculated rather than an input, this allowed for comparisons of analyses when the clearance time (yellow plus all red) is changed.

8.2.4 The 2000 HCM Input Module

There were no changes made to the Input Module in the 2000 HCM.

8.3 The Volume Adjustment Module

8.3.1 The 1985 Highway Capacity Manual

The Volume Adjustment Module includes analysis steps that concern the manipulation of demand volumes. This includes three key steps: (1) volumes are converted to flow rates, (2) demand flow rates are assigned to appropriate lane groups, and (3) an optional lane utilization factor may be applied. The second is a critical step, as the remainder of the methodology considers each lane group separately.

Converting volumes to flow rates simply involves division by the peak hour factor (PHF):

$$v_i = V_i / PHF \tag{8.4}$$

where all terms are as previously defined. The manual directs users to subtract any right turn on red volume from the number of right-turning vehicles before conversion to flow rate.

Demand flow rates must then be assigned to appropriate lane groups. In general, this is a simple task: Exclusive left-turn and/or right-turn lanes are established as separate lane groups. All through lanes and shared left- and/or right-turn and through lanes form a single lane group.

Where exclusive LT or RT lanes exist, the affected turning movements and through vehicles are physically separated by regulation and traffic markings, sometimes augmented by physical barriers. Figure 8.2 shows the combinations of lane groups that typically occur on intersection approaches.

There is one difficulty that may arise at an intersection approach which must be checked before finalizing lane group definitions: the existence of a defacto left-turn lane. In some cases, there are enough left-turners to fully occupy the left lane of a shared-lane approach. This, in effect, sets up an exclusive left-turn lane even where one is not established by regulation. Equations 8.5 and 8.6 provided a simple means of testing to see if this is likely to occur.

$$v_{LE} = v_L \left(\frac{1800}{1400 - v_o} \right) \tag{8.5}$$

where

v_{LE} equivalent left-turn flow rate in terms of through vehicles (veh/h),
v_L actual left-turn flow rate (veh/h), and
v_o opposing flow rate (veh/h).

The equivalence is based upon general guidelines in which the normal saturation flow rate for through passenger cars is 1800 pc/hg/ln. For left turns against an opposing flow, the saturation flow rate is taken as (approximately) $1400 - v_o$ pc/hg/ln.

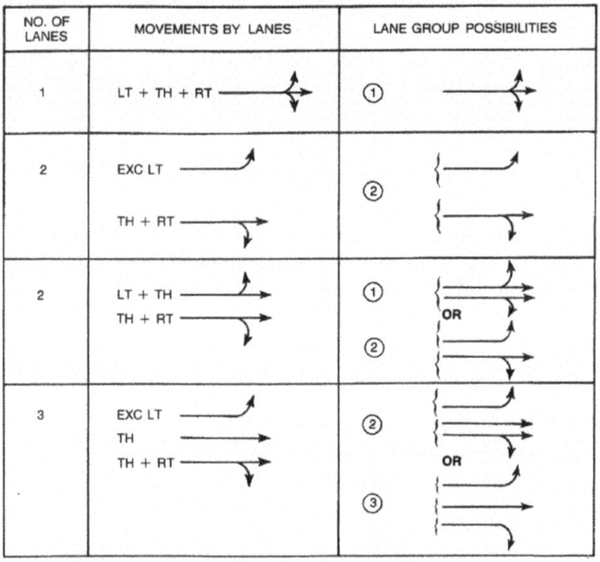

Fig. 8.2 Typical lane groups at signalized intersection approaches. *Source* "Highway Capacity Manual," *Special Report 209*, Transportation Research Board, Washington D.C., 1985 [1], Fig. 9-10, pp. 9–10, used with permission of the National Academy of Science, Courtesy of National Academies Press

A defacto left-turn lane will probably be formed if the equivalent left-turn flow rate exceeds the average flow rate in other lanes of the lane group, assuming that all left turns are in the left lane, that is, if:

$$v_{LE} \geq \frac{v - v_L}{N - 1} \tag{8.6}$$

where N is the number of lanes in the lane group, and all other terms are as previously defined.

If it appears that a defacto left-turn lane exists, it is treated as such and established as a separate exclusive LT lane group.

The last adjustment of the Volume Adjustment Module addresses lane distribution in multilane groups. Lane group flow rates may be adjusted to reflect unequal lane distribution. When this is done, the remainder of the analysis and its results reflect operating conditions in the worst lane of the lane group. It is also possible to ignore this adjustment, in which case the rest of the analysis and the results reflect average conditions within the lane group. The adjustment is applied as shown in Eq. 8.7. The adjustment factors are shown in Table 8.5.

$$v = v_g * U \tag{8.7}$$

Table 8.5 Default lane utilization adjustment factor—1985 HCM

No. of through lanes in group (excluding lanes used by LT vehicles)	Lane utilization factor (U)
1	1.00
2	1.05
≥3	1.10

Source "Highway Capacity Manual," *Special Report 209*, Transportation Research Board, Washington D.C., 1985 [1], Table 9-4, pp. 9–10, used with permission of the National Academy of Science, Courtesy of National Academies Press

where

v adjusted demand flow rate for lane group (veh/h),
v_g unadjusted demand flow rate for lane group (veh/h),
U lane utilization adjustment factor.

8.3.2 The 1994 Update to the 1985 HCM

In 1994, three changes to the volume adjustment module:

- The test for a defacto left-turn lane was removed from this module and placed in the saturation flow rate module, where the proportion of left turns in the shared lane, P_L, is computed. If P_L is greater than one, the method directs the user to assume that a defacto left-turn lane exists and to start over in the volume adjustment module with a separate exclusive left-turn lane group. Finding P_L will be discussed in the saturation flow rate module section.
- Default lane utilization factors were added for double exclusive left-turn lanes ($U = 1.03$) and for double exclusive right-turn lanes ($U = 1.13$). Lane utilization factors for through or shared lanes remained the same.
- Directions for estimating the number of right turns on red were added for one case. When the RTOR is made from an exclusive lane during the left-turn phase of the cross street, then the number of right turners that should be subtracted from the right-turn volume can be assumed to equal the left-turn per lane volume occurring during the protected phase. For example, during a northbound protected left-turn phase, if RTOR is allowed, the eastbound right-turn vehicles can easily make the right turn at the same time; the number of right turners using RTOR is assumed to equal the number of left turners using the protected phase.

8.3.3 The 1997 Update to the 1985 HCM

A major change was made in the 1997 Volume Adjustment Module. The lane adjustment factor, U, was removed from this module. A new factor to account for unequal lane utilization was added to the saturation flow rate module as the lane utilization factor, f_{LU}. The benefit of this change was that there was no longer an increase in volume for vehicles that were not really at the intersection. Instead, the saturation flow rate is decreased to account for the unequal usage of lanes in a lane group. For example, the lane utilization adjustment factor for two through lanes of 1.05 was removed, and a lane utilization factor of 0.95 was added as in the saturation flow module.

8.3.4 The 2000 Highway Capacity Manual

There were no changes in the 2000 HCM Volume Adjustment Module.

8.4 The Saturation Flow Rate Module

8.4.1 General Form of the Saturation Flow Module 1985—2000

In the Saturation Flow Module, the ideal saturation flow rate is adjusted to reflect characteristics that vary from the defined base, to estimate the total saturation flow rate under prevailing conditions in *each* lane group. Equation 8.8 shows the equation used in the 1985 and 1994 HCMs.

$$s = s_o N \prod f_i \tag{8.8}$$

where

s total saturation flow rate for the lane group (veh/hg),
s_o ideal saturation flow rate per lane,
N number of lanes in the lane group, and
f_i multiplicative adjustment factor for condition "i."

In the 1985 HCM, there were eight adjustment factors f_i, accounting for the following non-base or standard conditions:

f_w adjustment for lane width,
f_{HV} adjustment for heavy vehicles,
f_g adjustment for approach grade,

f_p adjustment for parking activity,
f_{bb} adjustment for local bus blockage,
f_a adjustment for area type,
f_{RT} adjustment for right turns, and
f_{LT} adjustment for left turns.

In the 1985 HCM, the ideal saturation flow rate was given as 1800 pc/hg/ln. In the 1994 HCM, the ideal saturation flow rate was increased to 1900 pc/hg/ln based on research done at Polytechnic University [9]. In the 2000 HCM, they stopped using the term "ideal saturation flow rate" and changed it to "base saturation flow rate."

In the 1997 HCM, a ninth adjustment factor was added for lane utilization, and thus the equation became:

$$s = s_o \; N f_w f_{HV} f_g f_p f_{bb} f_a f_{RT} f_{LT} f_{LU} \tag{8.9}$$

where

f_{LU} adjustment for lane utilization all other terms are as previously defined.

In the 2000 HCM, two additional factors were added to account for the effects of bicycles and pedestrians on saturation flow rate, as shown in Eq. 8.10.

$$s = s_o \; N f_w f_{HV} f_g f_p f_{bb} f_a f_{RT} f_{LT} f_{LU} f_{Rpb} f_{Lpb} \tag{8.10}$$

where

f_{Rpb} adjustment for the effect of pedestrians and bicycles on right-turn vehicles, and
f_{Lpb} adjustment for the effect of pedestrians and bicycles on left-turn vehicles.

Most of these adjustment factors are easily determined using simple equations or look-up tables. However some of the factors require their own additional worksheets to find. The left-turn adjustment factor is particularly complex and will be discussed later in a separate section.

8.4.2 The Simple Adjustment Factors: 1985 HCM

Table 8.6 summarizes all adjustment factors in the 1985 HCM *except* for the right- and left-turn adjustment factors. The right-turn adjustment factor is shown in Table 8.7.

There was a lot of disagreement in the Highway Capacity Committee over the inclusion of a factor for area type. The data base did show a 10% decrease in saturation flow rate for intersections in a CBD location compared to other parts of the urban area. The data, however, did not give a statistically significant result. The HCM does account for many of the factors that lower saturation flow rates in urban areas, such as high pedestrian volumes in crosswalks, buses stopping, and parking movements. What is not clear is whether the area type factor accounts for impacts not in the HCM model or whether it double counts these impacts.

Table 8.6 Saturation flow adjustment factors for signalized intersections 1985 HCM

Adjustment factor for lane width (f_w)

Lane width (ft)	8	9	10	11	12	13	14	15	≥16
f_w	0.87	0.90	0.93	0.97	1.00	1.03	1.07	1.10	Use 2 lanes

Adjustment factor for heavy vehicles (f_{HV})

%HV	0	2	4	6	8	10	15	20	25	30
f_{HV}	1.00	0.99	0.98	0.97	0.96	0.95	0.93	0.91	0.89	0.87

Adjustment factor for grade (f_g)

	Downhill			Level	Uphill		
% Grade	−6	−4	−2	0	+2	+4	+6
f_g	1.03	1.02	1.01	1.00	0.99	0.98	0.97

Adjustment factor for parking (f_p)

No. of lanes	No parking	Number of parking maneuvers/hour, N_m				
		0	10	20	30	40
1	1.00	0.90	0.85	0.80	0.75	0.70
2	1.00	0.95	0.92	0.89	0.87	0.85
3	1.00	0.97	0.95	0.93	0.91	0.89

Adjustment factor for local bus blockage (f_{bb})

No. of lanes	Number of buses stopping per hour, N_B				
	0	10	20	30	40
1	1.00	0.96	0.92	0.88	0.83
2	1.00	0.98	0.96	0.94	0.92
3	1.00	0.99	0.97	0.96	0.94

(continued)

Table 8.6 (continued)

Adjustment factor for area type (f_a)	
Type of area	Factor
CBD	0.90
All other areas	1.00

Source "Highway Capacity Manual," *Special Report 209*, Transportation Research Board, Washington D.C. 1985 [1], Tables 9-5–9-10, pp. 9–12, used with permission of the National Academy of Science, Courtesy of National Academies Press

Table 8.7 Right-turn adjustment factors (f_{RT}) at signalized intersections 1985 HCM

Type of lane group	Adjustment factor (f_{RT})
Exclusive RT lane; Protected RT phasing	0.85
Exclusive RT lane; Permitted RT phasing	$f_{RT} = 0.85 - (PEDS/2100) \quad PEDS \leq 1700$ $f_{RT} = 0.05 \quad PEDS > 1700$
Exclusive RT lane; Prot + Perm RT phasing	$f_{RT} = 0.85 - (1 - P_{RTA})(PEDS/2100)$ $f_{RT}(\text{min}) = 0.05$
Shared RT lane; Protected RT Phasing	$f_{RT} = 1.0 - 0.15P_{RT}$
Shared RT lane; Permitted phasing	$f_{RT} = 1.0 - P_{RT}[0.15 + (PEDS/2100)]$ $f_{RT}(\text{min}) = 0.05$
Shared RT lane; Prot + Perm RT phasing	$f_{RT} = 1.0 - P_{RT}[0.015 + (PEDS/2100)(1 - P_{RTA})]$ $f_{RT}(\text{min}) = 0.05$
Single-lane approach; Permitted RT phasing	$f_{RT} = 0.90 - P_{RT}[0.135 + (PEDS/2100)]$ $f_{RT}(\text{min}) = 0.05$
Double RT lane; Protected RT phasing	0.75

PEDS = Flow rate of pedestrians in the conflicting crosswalk, peds/h
P_{RTA} = Proportion (decimal) of right turns taking place in the protected portion of the phase
P_{RT} = Proportion (decimal) of right turns in the lane group
Source "Highway Capacity Manual," *Special Report 209*, Transportation Research Board, Washington D.C., 1985 [1], Table 9-11, pp. 9–13 and 9–14, *equations only*, used with permission of the National Academy of Science, Courtesy of National Academies Press

The right-turn adjustment factor depends upon the geometry and signalization established for right turns. Turns may be made from an exclusive right turn or a shared lane. The right-turn signalization may be permitted, protected, or protected plus permitted. In practical terms, the incidence of protected or protected plus permitted right turns is rare. Nevertheless, the methodology accommodates all of the potential configurations. Table 8.7 gives equations for the right-turn adjustment factor for all of these cases.

The adjustments incorporate a number of basic understandings. The base adjustment is 0.85. Simply because of the right-turn maneuver, which generally occurs on a short-radius, right-turn saturation flow rates are *at least* 15% less than the ideal. The factor is further reduced by pedestrian interference during permitted phases. At a conflicting crosswalk flow rate of 1700 peds/h, the minimum factor of 0.05 is reached. This reflects a situation in which all right turns are made as "sneakers" after pedestrians clear the crosswalk. All other equations reflect special conditions that somewhat modify the basic numbers.

Table 8.8 Equations for adjustment factors—1994 update

Adjustment factor for	Equation
Lane width, $W \geq 8$ ft	$f_w = \frac{W-12}{30}$
Heavy vehicles	$f_{HV} = \frac{100}{100+\%HV(E_T-1)}$
Grade, $-6 \leq \%G \leq +10$	$f_g = 1 - \frac{\%G}{200}$
Parking, $0 \leq N_M \leq 180$ maneuvers per hour, and $f_p \geq 0.05$	$f_p = \frac{N-0.10-\left(\frac{18 N_m}{3600}\right)}{N}$
Bus blockage, $0 \leq N_B \leq 250$ and $f_{bb} \geq 0.05$	$f_{bb} = 1.0 - \left(\frac{14.4 N_B}{3600}\right)$

8.4.3 *The Simple Adjustment Factors: 1994 Update*

In 1994, equations were added for calculating all of the various factors described in the previous section. This allowed their use when values outside the ranges given in the tables of the 1985 HCM occurred (with certain limitations). The equations are given in Table 8.8 [4].

The 1994 HCM heavy vehicle factor set E_T equal to 2.0 passenger cars per heavy vehicle. Although an equation was not given in the 1985 HCM, the factors were based on a truck equivalency factor of 1.5 passenger cars per heavy vehicle.

The right-turn adjustment factors did not change in 1994, with the *exception* of the right-turn factor for double exclusive right-turn lanes was removed. This was done because a lane utilization factor was added for dual exclusive right turn lanes which increased the right-turn volume.

8.4.4 *The Simple Adjustment Factors: 1997 Update*

An additional factor was added for lane utilization, f_{LU}, at the same time as the lane utilization adjustment was removed from the volume adjustment module. This factor adjusts the saturation flow rate to account for vehicles distributing themselves unequally in a lane group that has more than one lane. If the actual volume in the lane with the highest volume is known, the factor can be calculated with Eq. 8.11. Alternatively, the factor can be found in Table 8.9.

$$f_{LU} = \frac{v_g}{v_{g1} N} \tag{8.11}$$

where

f_{LU} lane utilization factor,
v_g demand flow rate for the lane group, veh/h,

Table 8.9 Default lane utilization adjustment factors—1997 update

Lane group movement(s)	Number of lanes in lane group	Lane utilization factor
Through or shared lane(s)	1	1.00
	2	0.95
	3	0.91
Exclusive left-turn lane(s)	1	1.00
	2	0.97
Exclusive right-turn lane(s)	1	1.00
	2	0.88

v_{g1} demand flow rate for the single lane within the lane group with the highest volume, veh/h/ln, and
N number of lanes in the lane group.

While the value for area type adjustment, f_a, remained the same, additional text was added to advise the user where and when to use this value. The 1997 HCM clarifies that this factor should be used where the geometry and traffic and/or pedestrians are such that headways are increased, thereby affecting capacity [5]. This essentially allows analysts to eliminate this factor (essentially setting it to 1.00) unless some unusual situation exists that would further reduce saturation flow rates, even after all other existing reductions and factors have been applied.

8.4.5 The Simple Adjustment Factors: 2000 HCM

There were two new factors added to the saturation flow rate equation to account for the effect of pedestrian and bicycle interference on right (f_{Rpb}) and left (f_{Lpb}), turning vehicles. These two factors were not simple lookup values. The final calculation for each factor is accomplished using Eqs. 8.12 and 8.13, but there are a number of intermediate computations as well.

$$f_{Rpb} = 1 - P_{RT}\left(1 - A_{pbt}\right)(1 - P_{RTA}) \tag{8.12}$$

$$f_{Lpb} = 1 - P_{LT}\left(1 - A_{pbt}\right)(1 - P_{LTA}) \tag{8.13}$$

where

f_{Rpb} the pedestrian/bicycle adjustment factor for right turns,
f_{Lpb} the pedestrian/bicycle adjustment factor for left turns,
P_{RT} the proportion of right turns in the lane group,
P_{LT} the proportion of left turns in the lane group,
A_{pbt} adjustment for the permitted portion of the phase,

P_{RTA} the proportion of the total right turn green time that is protected, and
P_{LTA} the proportion of the total left turn green time that is protected.

Figure 8.3 depicts the conflict zones where pedestrians and bicycles interfere with vehicle movements. It is the modeling of these conflict zones that defines these two adjustment factors.

Right-turning vehicles conflict with both pedestrians and bicycles immediately upon beginning their maneuver. Left-turning vehicles must first cross the opposing traffic before conflicting with pedestrians. Left turners are assumed not to conflict with bicycles, although clearly some conflicts could arise.

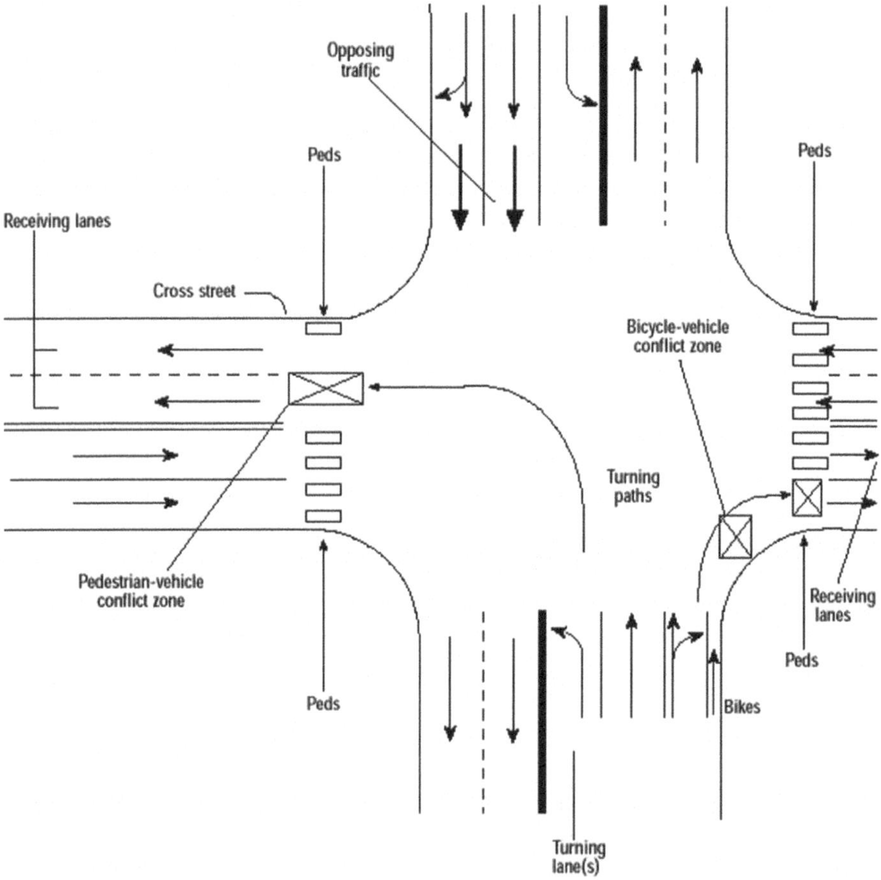

Fig. 8.3 Pedestrian and bicycle interference. *Source* "Highway Capacity Manual", Washington DC, 2000 [6], Exh D16-1, pp. 16–135, used with permission of the National Academy of Science, Courtesy of National Academies Press

8.4.5.1 The Right Turn Bike/Pedestrian Interference Factor

Since pedestrians and bicycles only move in the green phase, their volumes are first converted to a flow rate during the green phase, using Eqs. 8.14 and 8.15.

$$v_{pedg} = v_{ped}(C/g_p) \tag{8.14}$$

$$v_{bikg} = v_{bik}(C/g) \tag{8.15}$$

where

v_{pedg} pedestrian flow rate during the green phase, peds/hg,
v_{ped} pedestrian volume, peds/h,
v_{bikg} bicycle flow rate during the green phase, bikes/hg,
v_{bik} bicycle volume, bikes/h,
C cycle length, s,
g_p green phase for pedestrians (walk plus flashing don't walk), s, and
g effective green phase for vehicles, s.

Next, the proportion of green time during which pedestrians and bicycles occupy their respective conflict areas of the crosswalk, OCC_{pedg} and OC_{bikg}, are given by Eqs. 8.16 and 8.17, respectively.

$$\begin{aligned} OCC_{pedg} &= \tfrac{v_{pedg}}{2000} && for\; v_{pedg} \le 1000 \text{ peds/hg} \\ OCC_{pedg} &= 0.40 + \left(\tfrac{v_{pedg}}{10{,}000}\right) && for\; 1000 < v_{pedg} < 5000 \end{aligned} \tag{8.16}$$

$$\begin{aligned} OCC_{bikg} &= 0.02 + \left(\tfrac{v_{bikg}}{2700}\right) && for\; v_{bikg} \le 1900 \text{ bikes/hg} \\ OCC_{bikg} &= 0.72 && for\; v_{bikg} > 1900 \text{ bikes/hg} \end{aligned} \tag{8.17}$$

Equation 8.16 for pedestrian flow rates less than 1000 peds/hg assumes that each pedestrian blocks the crosswalk conflict area for 1.8 s (3600/1.8 = 2000). For higher flow rates and thus more congested conditions, Eq. 8.16 assumes that each additional pedestrian occupies the crosswalk conflict area for another 0.36 s (3600/0.36 = 10,000).

If there are no bicycles (or bicycles do not conflict with right-turning vehicles in the intersection), then the proportion of time that the relevant conflict zone is occupied is equal to the pedestrian occupancy:

$$OCC_r = OCC_{pedg} \tag{8.18}$$

where

OCC_r relevant occupancy for both left and right turning vehicles.

When right turns do conflict with both pedestrians and bicycles, the percentage of time that the conflict zone is occupied is higher. A portion of each mode's occupation time (OCC_{pedg} and OCC_{bikg}), however, overlaps. Thus the proportion of time that right turn vehicles cannot use the available green due to conflicting pedestrians and bicycles occupying the conflict areas is:

$$OCC_r = OCC_{pedg} + OCC_{bikg} - \left(OCC_{pedg} * OCC_{bikg}\right) \quad (8.19)$$

The next step is to estimate the adjustment for the permitted portion of the phase, A_{pbT}.

The adjustment factor, A_{pbT}, is dependent upon the ability of the driver to avoid a conflict by having a choice of turning into more than one lane. Where the number of receiving lanes, N_{rec}, equals the number of turning lanes, N_{turn}, the turning vehicle must turn into a designated lane, and turns may only be made during the time that the conflict zone is unoccupied. Where the number of receiving lanes, N_{rec}, is greater than the number of turning lanes, N_{turn}, the driver has a choice of lanes into which to turn, and thus has some flexibility to move around and avoid pedestrians or bicycles. Equation 8.20 shows the model for this adjustment.

$$\begin{aligned} A_{pbT} &= 1 - OCC_r \quad \quad if\ N_{rec} = N_{tum} \\ A_{pbT} &= 1 - 0.6OCC_r\ if\ N_{rec} > N_{tum} \end{aligned} \quad (8.20)$$

The factor for adjusting the saturation flow rate due to the interference of pedestrians and bicycles on right-turning vehicles can be calculated using Eq. 8.12.

8.4.5.2 The Left Turn Pedestrian Interference Factor

For left turns, it is assumed that only pedestrians will provide significant interference. Thus, bicycles are not included in the estimation of the adjustment factor.

Finding the relevant occupancy for permitted lefts turns is somewhat more complicated than for right turns. During the opposing queue clearance time, g_q, left-turning vehicles are blocked from moving. The occupancy rate, therefore, must consider the proportion of time that left-turning vehicles can depart and also conflict with pedestrians.

When the opposing queue clearance time is greater than the time of the pedestrian phase, then there is no conflict time and no pedestrian effect on saturation flow, i.e., $f_{Lpb} = 1.0$.

$$OCC_r = 0\,and\ f_{Lpb} = 1.0 \quad for\ g_q \geq g_p \quad (8.21)$$

When the opposing queue clears before the pedestrian phase ends, the relevant conflict time depends upon the amount of time available and the probability that a gap will occur for the left turning vehicle when the conflict area is occupied. This is found estimated using Eq. 8.22.

$$OCC_r = OCC_{pedg}[1 - 0.50(g_q/g_p)]\left[e^{-(5/3600)\,v_o}\right] \; for\, g_q < g_p \qquad (8.22)$$

where all terms have been previously defined. The calculation of queue clearance time (g_q) is discussed with the left-turn adjustment factor (f_{LT}).

Once the occupancy is known, the permitted phase adjustment factor, A*pbt*, is found using Eqs. 8.20 and 8.13.

When left turns are made from a one-way street, i.e., left turns behave as right turns, Eq. 8.21 should be used and $f_{Lpb}=f_{Rpb}$. Note that for left turns that only have a protected left-turn phase during which pedestrians are not allowed to cross, there would be no adjustment and f_{Lpb} would be 1.0.

8.4.5.3 Other Adjustment Factors

Because the 2000 HCM added specific adjustments for the pedestrian and bicycle interference to left- and right-turns, the right-turn adjustment factor (f_{RT}) had to be modified. Prior to 2000, f_{RT} included a component to account for pedestrian interference, which had to be removed to avoid double-counting this effect. The modification led to a much simpler f_{RT}, estimated using Eq. 8.23.

$$\begin{aligned} f_{RT} &= 0.85 && for\ exclusive\ RT\ lanes \\ f_{RT} &= 1.0 - 0.15\ P_{RT} && for\ shared\ RT/TH\ lanes \\ f_{RT} &= 1.0 - 0.135\ P_{RT} && for\ one-lane\ approaches \end{aligned} \qquad (8.23)$$

There was no similar modification made to the left-turn factor (f_{LT}), as the pre-2000 version did not include any component reflecting pedestrian or bicycle interference.

None of the other simple adjustment factors were modified in the 2000 HCM.

8.4.6 *The Left-Turn Adjustment Factor (f_{LT})*

Where left turns are made completely on protected left-turn phases (whether from shared or exclusive LT lanes), adjustment factors are relatively simple and straightforward. Where permitted left turns are involved, however, the complex nature of the interaction of left-turning vehicles against an opposing flow makes the process extremely complicated.

8.4.6.1 The 1985 HCM: "Simple" Cases

Table 8.10 shows the left-turn adjustment factors for fully protected LT movements from either an exclusive LT lane or a shared lane. It also shows the adjustment for

Table 8.10 Adjustment factors for left turns at a signalized intersection—1985 HCM

Type of lane group	Adjustment factor (f_{LT})
Exclusive LT lane; Protected phasing	0.95
Exclusive LT lane; Protected phasing	0.95
Shared LT lane; Protected phasing	$f_{LT} = 1/(1.0 + 0.05P_{LT})$
Shared LT lane; Prot + Perm phasing	$f_{LT} = \frac{1400-v_o}{(1400-v_o)+(235+0.435v_o)P_{LT}}$ $v_o \leq 1220$ veh/h $f_{LT} = \frac{1}{1+4.525P_{LT}}$ $v_o > 1220$ veh/h
Double LT lane; Protected phasing	0.92

P_{LT} = Proportion (decimal) of left turns in the lane group
V_o = Opposing flow rate (veh/h)
Source "Highway Capacity Manual," *Special Report 209*, Transportation Research Board, Washington D.C., 1985 [1], excerpts from Table 9-12, pp. 9–15, used with permission of the National Academy of Science, Courtesy of National Academies Press

the most complex case: protected plus permitted phasing from a shared lane. In the 1985 HCM, an equation was adopted that represents a gross estimation.

8.4.6.2 The Left-Turn Adjustment: Permitted Left Turns—1985 HCM

Permitted left-turns, whether from an exclusive LT lane or a shared lane, represent a complex set of interactions resulting from the conflict of left-turning vehicles with an opposing through flow. Figure 8.4 illustrates the basic operating characteristics at work in the case of a permitted left turn made from a shared lane [9].

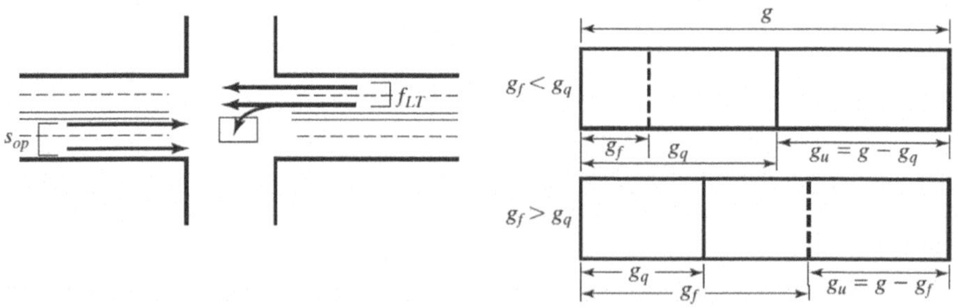

Fig. 8.4 Interactions between permitted left turns and the opposing flow at signalized intersections. *Source* Roess et al. [9], Fig. 22-7, p. 529. Reprinted with permission of Pearson Education Inc, NY, New York

When a left-turning vehicle arrives in the shared lane, it must wait for a gap in the opposing vehicle flow. While it does so, it effectively blocks the shared lane.

In terms of the use of green time, the GREEN phase must be segmented into three parts:

- At the beginning of the GREEN phase, the opposing queue of vehicles begins to depart. Until this queue clears the intersection, left turns are essentially blocked. The amount of time that it takes the opposing queue of vehicles to clear the intersection is designated "g_q"—often referred to as the queue clearance time (s).
- No blockage of the shared lane occurs, however, until the first left-turning vehicle arrives at the stop line. The time until the first left-turning vehicle arrives is designated "g_f" (s). Note that, as shown in Fig. 8.4, g_f can be larger than, equal to, or smaller than g_q. However in the 1985 HCM model, g_f is always less than g_q, as will be seen in the derivation of g_f discussed below.
- After the opposing queue of vehicles clears, left turns face an unsaturated opposing flow through which some left turns may be made. The time during which this occurs is designated "g_u" (s).

The importance of these three components of GREEN time cannot be understated. Effectively, left-turn operations (and therefore the left-turn adjustment factor) are vastly different during different portions of the GREEN time. Consider that in the shared left-turn/through lane:

- Before the arrival of the first left-turning vehicle, left turns have *no impact* on operations, i.e., the left-turn adjustment factor should be one 1.00 during g_f.
- After the first left arrives, there is the remaining time it takes for the opposing queue to clear. There is a time ($g_q - g_f$) during which a left-turning vehicle blocks the shared lane, and no one moves. Thus, during ($g_q - g_f$), the left-turn adjustment factor should be 0.00.
- After the opposing queue clears, left-turning vehicles can filter through an unsaturated opposing flow at a rate that is dependent upon the opposing flow rate. Because this occurs only after *both* the opposing queue has cleared *and* the first left-turning vehicle arrives, the length of this period, g_u, is the remaining time left in the effective green.

The 1985 HCM is the first manual to deal with the difference between actual and effective green time. This issue is discussed during the presentation of the Capacity Analysis Module.

Because these sub-portions of the GREEN time were so critical, models for their prediction had to be developed. In 1985, the models were all theoretical, and involved some basic assumptions. In subsequent years, field studies produced calibrated equations for these estimates. Even today, however, the models include both calibrated and theoretical components.

The refinement of dividing the GREEN time into its component sub-portions was added late in the process of developing 1985 HCM procedure. Indeed, the model was originally developed over dinner on the back of a paper placemat. Without any data for calibration, the conundrum for modeling development was that to estimate the saturation flow rate in one direction, the saturation flow rate in the opposing direction had to be known. Thus the computations begin with a rough estimate of the saturation flow rate in the opposing lane group.

Step 1: Estimate the Opposing Saturation Flow Rate

$$s_{op} = \frac{1800\, N_o}{1 + P_{LTo}(400 + v_M)/(1400 - v_M)} \tag{8.24}$$

where

s_{op} opposing saturation flow rate (vphg),
N_o number of opposing lanes (*not* including exclusive LT or RT lanes,
P_{LTo} proportion of left turns in the opposing lane group (decimal),
v_M subject (mainline) through flow (*not* including LTs or RTs from exclusive lanes) (vph).

The denominator of the equation represents a weighted average through-vehicle equivalent that the opposing lefts must cross through. Note that the opposing saturation flow rate is only being adjusted for left turns and none of the other factors that adjust the ideal saturation flow rate in Eq. 8.5. Thus if there were no left turners in the opposing flow, the opposing saturation flow rate would be the ideal 1800 pc/h/ln times the number of lanes.

Step 2: Compute the Ratio of Opposing Flow to Opposing Saturation Flow Rate

$$Y_o = v_o / s_{op} \tag{8.25}$$

where Y_o is referred to as the "opposing flow ratio" and all other variables are as previously defined.

Step 3: Estimate Unsaturated Green Time, g_u.

The flow ratio for the opposing flow, Y_o, represents the decimal proportion of real time that must be available as GREEN to process the opposing flow. If the opposing flow is 500 pc/h and the opposing saturation flow rate is 1500 pc/hg, then it is logical that the GREEN time must be at least 500/1500 = 0.333 of real time in order to process the demand. The *g*/*C* ratio represents the decimal proportion of the cycle length that is effectively green, thus unless $g/C > Y_o$, there can be no g_u. The value of g_u is estimated as:

$$g_u = \frac{C\left(\frac{g}{C} - Y_o\right)}{1 - Y_o} = \frac{g - CY_o}{1 - Y_o} \tag{8.26}$$

where all terms are as previously defined.

The unsaturated portion of the green phase, "g_u" is derived by estimating the number of vehicles queued during each cycle, assuming random arrivals. This is established as the number of vehicles that arrive (Cv_o) minus the vehicles that arrive during unsaturated green ($g_u v_o$). Vehicles queued are then set equal to the time it would take to clear them at saturation flow rate s_o, giving equation $Cv_o - g_u v_o = (g - g_u)s_o$. Solving for g_u gives Eq. 8.26 above.

Step 4: Estimate P_L, the Proportion of Left Turns in the Shared Left Lane

The proportion of left turns in the shared left-turn lane is needed in order to calculate g_f, the time that the first left turner arrives at the front of the queue. The number of left turning vehicles is known, so it is easy to calculate the proportion of left turns in the entire lane *group* (P_{LT}). The model then attempts to establish a left-turn adjustment factor for the shared left lane *only*. The proportion of left turns in the shared left lane is estimated as:

$$P_L = P_{LT}\left[1 + \left(\frac{(N-1)g}{f_s g_u + 4.5}\right)\right] \tag{8.27}$$

where

$$f_s = \frac{875 - 0.625\, v_o}{1000}$$

P_L proportion of left-turning vehicles in the shared left lane (decimal),
P_{LT} proportion of left-turning vehicles in the lane group (decimal),
N number of lanes in the subject lane group,
g effective green time for the lane group (s),
f_s left-turn saturation factor, or the proportion of the unsaturated green time available to be used by left turners,
g_u unsaturated portion of the GREEN, after clearance of the opposing queue (s), and
v_o opposing rate of flow, veh/h.

The proportion of left-turning vehicles in the shared lane was derived by splitting the volume in each lane in the same proportion as the usable green time in each lane. The 4.5 value in the denominator is the time used by two sneakers at the rate of 1600 veh/h.

Step 5: Estimate the Duration of g_f

The duration of the initial portion of the GREEN interval, g_f, is estimated using the probabilistic model:

$$\begin{aligned} g_f &= \tfrac{2P_T}{P_L}\left(1 - P_T^{0.5 g_q}\right) \\ g_q &= g - g_u \end{aligned} \tag{8.28}$$

where P_T is the decimal proportion of through vehicles in the lane group and all other variables are as previously defined. Where the lane group is an exclusive LT lane group, $g_f = 0.0$ s (it is assumed that there is a left-turning vehicle waiting for service).

The probabilistic g_f model is derived from the expected time to the arrival of the first left-turning vehicle, $2P_T/P_L$, times the probability of there not being all through vehicles in the subject queue during the time it takes for the opposing queue to clear, $1 - P_T^{0.5gq}$. This effectively limits the time until the first left-turning vehicle arrives to a value within the opposing queue's clearance time. The 1985 HCM model does not consider the case of a left turning vehicle not arriving during g_q.

Step 6: Estimate the Left-Turn Equivalent (E_L) During the Filtration Portion of the GREEN interval (g_u)

The passenger car equivalent for left turns made during g_u, when they filter through an unsaturated opposing flow is computed as:

$$E_L = \frac{1800}{1400 - v_o} \tag{8.29}$$

where all variables are as previously defined.

Step 7: Compute the Left-Turn Adjustment Factor for the Shared Left-Turn Lane

At this point, a left-turn adjustment factor for the shared left-turn lane is estimated for the various green periods. Fundamentally, before the first left-turning vehicle arrives (g_f), the effective left-turn adjustment factor is 1.0. During the filtration period (g_u), the effective left-turn adjustment factor is given by $1/[1 + P_L\ (E_L - 1)]$.

Another component affecting left turns is the number of "sneakers" that complete their turn after the opposing flow clears, generally on the YELLOW indication. It is assumed that each "sneaker" per cycle adds 2 s of effective green time. Up to two "sneakers" per cycle can be accommodated, but this is affected by the likelihood that two left-turning vehicles follow each other when the light turns YELLOW. For modeling purposes, it is assumed that one sneaker makes the turn and that the possibility of a second sneaker depends upon the percent left turns in the lane, thus the number of "sneakers" per cycle is taken to be $(1 + P_L)$.

The overall left-turn adjustment factor may now be computed as:

$$f_m = \frac{g_f}{g} + \frac{g_u}{g}\left[\frac{1}{1 + P_L(1 + E_L)}\right] + \frac{2}{g}(1 + P_L) \tag{8.30}$$

where f_m is the left-turn adjustment factor, as applied *only* to the lane from which left turns occur, and all other variables are as previously defined.

Note that if the permitted left turn takes place from an exclusive lane, the basic model does not change, but g_f would always be 0.0

Step 8: Compute the Left-Turn Adjustment Factor for the Lane Group

The final step in this process is to compute a left-turn factor for the lane group as a whole. This is done by taking f_m which applies to the shared through and left-turn lane, and assuming that the effective factor for other lanes of the lane group is 1.00. Then:

$$f_{LT} = \frac{f_m + (N - 1)}{N} \tag{8.31}$$

where all terms are as previously defined.

Because the steps of this methodology are more easily seen by examining the worksheet, it is shown here in Fig. 8.5.

8.4.6.3 The Left-Turn Adjustment: Permitted Left Turns—1994 Update

The left-turn adjustment factor had major changes in two of the left turn cases. (1) The special procedure for permitted left turns changed based on research done by Polytechnic University [10, 11], with a separate methodology added for when the opposing direction is a single lane. (2) The method of analyzing protected-permitted phasing from an exclusive left-turn was changed.

Additionally, the factor of 0.92 for double exclusive left-turn lanes with protected phasing was removed. This was done for the same reason that the factor for double exclusive right-turn lanes was removed. The lane utilization factor increases the left-turn volume by a factor a 1.03, which has the same effect of reducing the saturation flow rate by a factor of 0.03 over the single exclusive lane factor of 0.95.

The left-turn factor for all other simple cases remained the same.

Lastly, direction was added for the case of left turns that are not opposed by through vehicles at any time, but do conflict with pedestrians. In this case, the 1994 update directs the user to apply the right-turn factor as the left-turn factor.

The permitted left-turn adjustment model in the 1994 update has two separate methodologies, and therefore separate worksheets, for finding the left-turn factor, f_{LT}. One method is for approaches opposed by a multilane approach and another for approaches opposed by a single-lane approach. Although the general form of the two models is the same, there are some unique differences were introduced for the one-lane opposing approach.

The basic model for permitted left turns, which divided the green time into three distinct portions (g_f, g_q, and g_u) remained the same in the 1994 update. The changes occurred in how the values of the different portions were determined.

The changes were based upon Polytechnic University research [10, 11] and recommendations by the HCQSC. The Polytechnic research was based on data collected at 25 intersections located in four cities across the country. Regression models were developed for both g_f (the time until the first left turning vehicle arrives) and g_q (the time for the opposing queue to clear) separately for single-lane approaches and multilane approaches.

SUPPLEMENTAL WORKSHEET FOR LEFT-TURN ADJUSTMENT FACTOR, f_{LT}				
INPUT VARIABLES	EB	WB	NB	SB
Cycle Length, C (sec)				
Effective Green, g (sec)				
Number of Lanes, N				
Total Approach Flow Rate, v_a (vph)				
Mainline Flow Rate, v_M (vph)				
Left-Turn Flow Rate, v_{LT} (vph)				
Proportion of LT, P_{LT}				
Opposing Lanes, N_o				
Opposing Flow Rate, v_o (vph)				
Prop. of LT in Opp. Vol., P_{LTO}				
COMPUTATIONS	EB	WB	NB	SB
$S_{op} = \dfrac{1800\,N_o}{1 + P_{LTO}\left[\dfrac{400 + v_M}{1400 - v_M}\right]}$				
$Y_o = v_o / S_{op}$				
$g_u = (g - CY_o) / (1 - Y_o)$				
$f_s = (875 - 0.625\,v_o) / 1000$				
$P_L = P_{LT}\left[1 + \dfrac{(N-1)\,g}{f_s g_u + 4.5}\right]$				
$g_q = g - g_u$				
$P_T = 1 - P_L$				
$g_f = 2\,\dfrac{P_T}{P_L}\left[1 - P_T^{\,0.5\,g_q}\right]$				
$E_L = 1800 / (1400 - v_o)$				
$f_m = \dfrac{g_f}{g} + \dfrac{g_u}{g}\left[\dfrac{1}{1 + P_L\,(E_L - 1)}\right] + \dfrac{2}{g}(1 + P_L)$				
$f_{LT} = (f_m + N - 1) / N$				

Fig. 8.5 Worksheet for permitted left-turn adjustment factor—1985 HCM. *Source* "Highway Capacity Manual," Special Report 209, Transportation Research Board, Washington D.C., 1985 [1], pp. 9–78, used with permission of the National Academy of Science, Courtesy of National Academies Press

Case 1: Subject Approach Opposed by a Multilane Approach

The Polytechnic research created algorithms based on regression analysis for both g_f and g_q. The multilane regression equations were:

$$g_f = Ge^{-0.882\,LTC^{0.717}} - t_L \tag{8.32}$$

$$g_q = 9.532\,v_{OLC}^{0.560}\,qr_o^{0.819} - t_L \tag{8.33}$$

where

G	Actual green time for the permitted phase, s,
LTC	Left turns per cycle or v_{LT} * C/3600,
v_{LT}	Adjusted left-turn flow rate, veh/h,
C	Cycle length,
t_L	lost time per phase, s,
v_{OLC}	Opposing flow rate per lane per cycle or $v_oC/3600N_o$,
v_o	Adjusted opposing flow rate, veh/h,
N_o	Number of opposing lanes,
qr_o	Opposing queue ratio or $1 - R_{po}(g_o/C)$,
R_{po}	Platoon ratio for the opposing flow, based on arrival type, and
g_o	Effective green for opposing flow, s.

Only the regression equation for g_f was accepted by the HCQSC. The Committee felt that the g_q model did not make sense intuitively because it should logically be sensitive to the opposing flow rate. In an unpublished memo to the HCQSC [2], a different model was recommended and adopted into the 94 update, as follows:

$$g_q = \frac{v_{OLC}qr_o}{0.5 - \left[v_{OLC}\left(\frac{1-qr_o}{g_o}\right)\right]} - t_L \tag{8.34}$$

where the value in brackets being subtracted from 0.5 must be less than or equal to 0.49 and $0.0 \le g_q \le g$.

Equation (8.34) represents the peak queue that builds during the red portion of the phase divided by the net departure rate. It is derived by setting the vehicles that depart during g_q equal to the vehicles that arrive during the red plus vehicles that join the queue during the queue clearance time (assuming random arrivals). The 0.5 in Eq. 8.34 assumes vehicles depart at 2 s/veh (or a saturation flow rate of 1800 pc/hg/ln, as in the 1985 HCM).

Notice that in all of the above equations, each green time then subtracts t_L, the total lost time for the phase. In the 1994 update, all of the lost time experienced by a phase was assigned to the beginning of a phase. This was done both for simplicity (the user would not need to know exact values of start-up lost time and clearance lost time) and because, in most cases, more of the lost time occurs at the beginning of the phase.

The calculation of g_u in the 1994 update was found by subtracting the larger of g_q or g_f from the effective green time. Thus the 1994 update allowed for the case (contrary to the 85HCM) of the first left-turning vehicle arriving after the opposing queue cleared.

$$\begin{aligned} g_u &= g - g_q \; for \; g_q \geq g_f \\ g_u &= g - g_f \; for \; g_q < g_f \end{aligned} \tag{8.35}$$

The next change to the procedure was to the equation for calculating f_m, the left-turn factor that applies only to the lane from which left turns occur. It was changed in two ways.

1. Sneakers were removed from the equation. The argument over whether sneakers should be added as in the 1985 HCM or set as a practical minimum as in the 1994 update goes back to the 1965 HCM, which has passages that suggest both approaches. It was observed in the Polytechnic research [9] that the number of sneakers that actually completed the turn during the clearance interval was much smaller than that being predicted in the 1985 HCM, that is, $(2/g)(1 + P_L)$. For the 1994 update, this equation for sneakers was instead set as a minimum value for f_m. Justification for this was that saturation flow rate is observed by counting vehicles as they cross the stop line. Sneakers most often wait in the intersection so adding them into the equation would be double counting the sneakers.
2. The left-turn equivalent was changed. The simple equation of the 1985 HCM was replaced by a table from a paper by Messer and Fambro [2].

The left-turn equivalency value was renamed E_{L1}, and replaces E_L from the 1985 HCM. The values for E_{L1} are found in Table 8.11.

The equation for f_m was changed in two ways: (1) the term for "sneakers" was eliminated, and (2) the value of the left-turn equivalent changed to those shown in Table 8.11.

$$f_m = \frac{g_f}{g} + \frac{g_u}{g}\left[\frac{1}{1 + P_L(E_{L1} - 1)}\right] \tag{8.36}$$

where all terms are as previously defined.

In the 1994 update, the identification of a defacto left-turn lane, should one exist, is done at this point in the analysis: If $P_L \geq 1.0$, then it is assumed that a defacto left-turn exists. The user must start over in the volume adjustment module and create an exclusive left-turn lane. All left turn volume is then assigned to the exclusive lane and calculations are redone.

The last change to the permissive left-turn adjustment model was to the equation for the lane group f_{LT}. In the 85HCM, it was assumed that left turns only affected lane from which left turns were made. It was assumed that the adjustment applied to all other lanes in the lane group was 1.00. Based upon the Polytechnic research (9), it was found that left turns also affected adjacent lanes. In the 1994 update, a constant adjustment of 0.91 was applied to adjacent lanes. This was a simplification, as the

Table 8.11 Left-turn equivalents, E_{L1}, for permitted left turns during unsaturated green, g_u—1994 update

Total no. of signal phases	Type of left turn lane	No. of opposing lanes	Opposing flow, v_o						
			0	200	400	600	800	1000	≥1200
2	Shared	1	1.05	2.0	3.3	6.5	16*	16*	16*
		2	1.05	1.9	2.6	3.6	6.0	16*	16*
		≥3	1.05	1.8	2.5	3.4	4.5	6.0	16*
	Exclusive	1	1.05	1.7	2.6	4.7	10.4*	10.4*	10.4*
		2	1.05	1.6	2.2	2.9	4.1	6.2	10.4*
		≥3	1.05	1.6	2.1	2.8	3.6	4.8	10.4*
>2	Shared	1	1.05	2.2	4.5	11*	11*	11*	11*
		2	1.05	2.0	3.1	4.7	11*	11*	11*
		≥3	1.05	2.0	2.9	4.2	6.0	11*	11*
	Exclusive	1	1.05	1.8	3.3	8.2*	8.2*	8.2*	8.2*
		2	1.05	1.7	2.4	3.6	5.9	8.2*	8.2*
		≥3	1.05	1.7	2.4	3.3	4.6	6.8	8.2*

*Generally indicates turning capacity only available at end of phase-"sneakers" only

Source "Highway Capacity Manual," *Special Report 209*, Transportation Research Board, Washington D.C., 1994 [4], Table 9-7, pp. 9–19, used with permission of the National Academy of Science, Courtesy of National Academies Press

impact on adjacent lanes would be expected to be related to the impact on the lane from which left turns were made (f_m). The study data base was, however, insufficient to calibrate a relationship, and the HCQSC decided that applying a constant factor to adjacent lanes was preferable to ignoring any impact on them. Note that the adjustment to adjacent lanes only applies when the *subject* lane group has 2 or more lanes.

$$\begin{aligned} f_{LT} &= \frac{f_m + 0.91(N-1)}{N} \quad for\ N > 1 \\ f_{LT} &= f_m \quad\quad\quad\quad\quad\ \ for\ N = 1 \end{aligned} \tag{8.37}$$

where N is the number of lanes in the subject lane group.

The worksheet for the multilane approach methodology is shown in Fig. 8.6.

Case 2: Subject Approach Opposed by a Single Lane Approach

The first difference of the single-lane approach are the equations used for calculating g_f and g_q. Both regression equations from the Polytechnic University research [9] are used directly. Equations 8.32 and 8.33 are repeated here for convenience:

$$\begin{aligned} g_f &= Ge^{-0.882\,LTC^{0.717}} - t_L \\ g_q &= 9.532\,v_{OLC}^{0.560}\,qr_o^{0.819} - t_L \end{aligned}$$

where all variables are as previously defined.

The second change was that the 1994 HCM introduced a new factor to the opposing queue clearance time when the opposing approach is a single lane. The theoretical structure of the HCM permissive left-turn model (shown in Fig. 8.4) does not allow for left turns being made during g_q, the time during which the opposing queue clears through the intersection. However, when the opposing approach is a single lane, opposing left-turning vehicles create gaps that may be used by the subject approach's left-turning vehicles. Of course, this is only true after the first left-turning vehicle arrives on the subject approach, and that first left turner must arrive before the opposing queue clears, that is g_f must be less than g_q ($g_f < g_q$).

When left-turning vehicles are opposed by a single lane approach, during the critical interval between the arrival of the first left on the subject approach and the clearance of the opposing queue, $g_q - g_f$, left turns may in fact occur. The left turn equivalency for these left turns would naturally be different than for left turns made during the unsaturated green time, g_u. The model created for this case is a probabilistic model documented in Ref. [1]. This second left-turn equivalency value was named E_{L2}.

The derivation of the model was as follows: At the time g_f, by definition, a left turn vehicle is at the stop line waiting to make the turn. That first left turn vehicle will wait for a gap in the opposing queue created by an opposing left-turn vehicle. The expected waiting time for that first opposing left-turn vehicle can be found by considering the probability of the first, second, third, etc., opposing vehicle being a left turner. The E_{L2} equivalency is shown in Eq. 8.38.

SUPPLEMENTAL WORKSHEET FOR PERMITTED LEFT TURNS *** For Use Where the Subject Approach is Opposed by a Multilane Approach ***				
APPROACH	EB	WB	NB	SB
Enter Cycle Length, C				
Enter Actual Green Time For Lane Group, G				
Enter Effective Green Time For Lane Group, g				
Enter Opposing Effective Green Time, g_o				
Enter Number of Lanes in Lane Group, N				
Enter Number of Opposing Lanes, N_o				
Enter Adjusted Left-Turn Flow Rate, v_{LT}				
Enter Proportion of Left Turns in Lane Group, P_{LT}				
Enter Adjusted Opposing Flow Rate, v_o				
Enter Lost Time per Phase, t_L				
Compute Left Turns per Cycle: $LTC = v_{LT}\ C/3600$				
Compute Opposing Flow per Lane, Per Cycle: $v_{olc} = v_o\ C/(3600\ N_o)$				
Determine Opposing Platoon Ratio, R_{po} (Table 9-2 or Eq [9-7])				
Compute g_f** $= G \exp(-0.882\ LTC^{0.717}) - t_L$, $g_f \le g$				
Compute Opposing Queue Ratio: $qr_o = 1 - R_{po}\ (g_o/C)$				
Compute g_q using equation 9-17, $g_q \le g$				
Compute g_u: $g_u = g - g_q$ if $g_q \ge g_f$; $g_u = g - g_f$ if $g_q < g_f$				
Compute $f_s = (875 - 0.625\ v_o)/1000$, $f_s \ge 0$				
Compute P_L† $= P_{LT}\ [1 + \{(N-1)g/(f_s g_u + 4.5)\}]$				
Determine E_{L1} (Figure 9-7)				
Compute $f_{min} = 2(1 + P_L)/g$				
Compute f_m: $f_m = [g_f/g] + [g_u/g][1/\{1 + P_L(E_{L1}-1)\}]$; min $= f_{min}$; max $= 1.00$				
Compute $f_{LT} = [f_m + 0.91\ (N-1)]/N$‡				

** For special case of a single-lane approach opposed by a multilane approach, see text.
† For $P_L \ge 1$ and N>1, assume a defacto left-turn lane and re-do calculations.
‡ For multiple exclusive left-turn lanes, $f_{LT} = f_m$

Fig. 8.6 Supplemental worksheet for f_{LT} for opposing multilane approaches 1994 update. *Source* "Highway Capacity Manual," Special Report 209, Transportation Research Board, Washington D.C., 1994 [4], pp. 9–113, used with permission of the National Academy of Science, Courtesy of National Academies Press

$$E_{L2} = \frac{1 - P_{THO}^{n}}{P_{LTO}} \tag{8.38}$$

where

P_{THO} proportion of through and right turning vehicles in the opposing single-lane approach, $1 - P_{LTO}$,

P_{LTO} proportion of left-turning vehicles in the opposing single lane approach, and

n maximum number of opposing vehicles that could arrive during the time period $g_q - g_f$, where $n = (g_q - g_f)/2$.

The time $g_q - g_f$ was called g_{diff} for simplicity. The model for the left turn factor for approaches opposed by a single lane approach is shown in Eq. 8.41.

$$f_{LT} = f_m = \frac{g_f}{g} + \frac{g_{diff}}{g}\left[\frac{1}{1 + P_L(1 + E_{L2})}\right] + \frac{g_u}{g}\left[\frac{1}{1 + P_L(1 + E_{L1})}\right] \tag{8.39}$$

where all terms are as previously defined. The worksheet for opposing single-lane approaches is shown in Fig. 8.7.

Special Cases: Additional Notes

If the subject approach has multiple lanes and the opposing approach is a single lane, then the time g_f would be calculated with the multilane equation for g_f. Additionally, the f_{LT} of Eq. 8.39 would have to be adjusted for the effect on other lanes in the subject lane group. Thus f_{LT} would become:

$$f_{LT} = \frac{f_m + 0.91(N - 1)}{N}$$

where all terms are as previously defined.

If the subject approach is a single lane approach that is opposed by a multilane approach, gaps would not be opened during the opposing queue clearance time. Thus the multilane method would be used except that g_f would be calculated using the single lane equation for g_f, Eq. 8.32, and f_{LT} would be equal to f_m.

8.4.6.4 The Left-Turn Adjustment: Compound Left Turns from an Exclusive Lane—1994 Update

The model for compound phases from exclusive left-turn lanes provided in the 1994 HCM was much more detailed than in the 1985 HCM. The 1994 HCM analyzed the protected left-turn and permitted left-turn phases separately. Essentially, starting in the saturation flow rate module, the compound phase is split into two lane groups: one for the protected portion of the phase and another for the permitted portion. Separate values for the saturation flow rate in each portion of the compound phase are used. This division into two lane groups continues into the capacity analysis module.

SUPPLEMENTAL WORKSHEET FOR PERMITTED LEFT TURNS *** For Use Where the Subject Approach is Opposed by a Single lane Approach ***				
APPROACH	**EB**	**WB**	**NB**	**SB**
Enter Cycle Length, C				
Enter Actual Green Time for Lane Group, G				
Enter Effective Green Time for Lane Group, g				
Enter Opposing Effective Green Time, g_o				
Enter Number of Lanes in Group, N				
Enter Adjusted Left-Turn Flow Rate, v_{LT}				
Enter Proportion of Left Turns in Lane Group, P_{LT}				
Enter Proportion of Left Turns in Opposing Flow, P_{LTo}				
Enter Adjusted Opposing Flow Rate, v_o				
Enter Lost Time Per Phase, t_L				
Compute Left Turns per Cycle: $LTC = v_{LT}\ C/3600$				
Compute Opposing Flow per Lane, per Cycle: $v_{olc} = v_o\ C/3600$				
Determine Opposing Platoon Ratio, R_{po} (Table 9-2 or Eq [9-7])				
Compute $g_f{}^{**} = G \exp(-0.860\ LTC^{0.629}) - t_L \quad g_f \le g$				
Compute Opposing Queue Ratio: $qr_o = 1 - R_{po}(g_o/C)$				
Compute $g_q = 4.943\ v_{olc}{}^{0.762}\ qr_o{}^{1.061} - t_L \quad g_q \le g$				
Compute g_u: $g_u = g - g_q$ if $g_q \ge g_f$ $g_u = g - g_f$ if $g_q < g_f$				
Compute $n = (g_q - g_f)/2,\ n \ge 0$				
Compute $P_{THo} = 1 - P_{LTo}$				
Determine E_{L1} (Figure 9-7)				
Compute $E_{L2} = (1 - P_{THo}{}^n)/P_{LTo}$				
Determine $f_{min} = 2(1 + P_{LT})/g$				
Compute $f_{LT}{}^{**} = f_m = [g_f/g] + [(g_q - g_f/g][1/\{1 + P_{LT}(E_{L2} - 1)\}]$ $+ [g_u/g][1/(1 + P_{LT}(E_{L1} - 1)]$ $\min = f_{min};\ \max = 1.00$				

** For special case of a multilane approach opposed by a single-lane approach, or when $g_f > g_q$, see text.

Fig. 8.7 Supplemental worksheet for f_{LT} for opposing single-lane approaches 1994 update. *Source* "Highway Capacity Manual," Special Report 209, Transportation Research Board, Washington D.C., 1994 [4], pp. 9–114, used with permission of the National Academy of Science, Courtesy of National Academies Press

For the protected portion of the phase, the saturation flow rate uses a constant left-turn factor of 0.95 (see Table 8.10). For the permitted portion of the phase, the saturation flow rate uses a left-turn factor calculated using the special procedures for permitted left turns, with some notable modifications that account for offsets between portions of the green phase for opposing approaches.

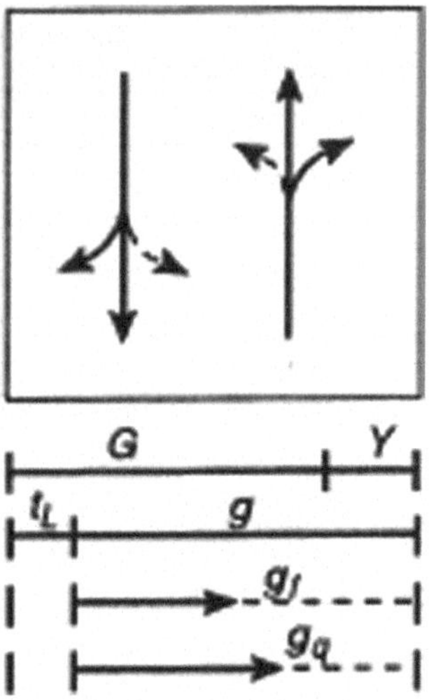

Fig. 8.8 Permitted left turns in the 1994 update. *Source* "Highway capacity manual," Special Report 209, Transportation Research Board, Washington D.C., 1994 [4], Fig. 9-8, pp. 9–22, used with permission of the National Academy of Science, Courtesy of National Academies Press

As for other permitted left-turn phases, finding the adjustment factor for the permitted portion of the compound phase requires the calculation of g_f, g_q and g_u. In cases of compound phasing, these values will have to be adjusted for when they occur in the phase, since the equations for these variables are referenced to the beginning of effective green. With compound phasing, the subject movement may not be allowed to move at the beginning of effective green. Additionally, because all lost time is assigned to the beginning of a phase, one portion of a compound phase will not be assigned any lost time.

Figure 8.8 shows how the 1994 update standard case for a fully-permitted left-turn phase is handled. All cases of compound phasing are a modification of this case.

Because the 1994 update placed *all* lost time (both start-up and clearance) at the *beginning* of the GREEN indication, the equations for g_f and g_q are (described in previous sections) all indexed to a time t_L seconds after the beginning of the actual green time, G. Equations 8.32 and 8.34 are used to compute g_f and g_q when the opposing approach has multiple lanes, and Eqs. 8.32 and 8.33 respectively when the opposing approach is a single lane.

Where a leading and/or lagging green phase exists, the values of g_f and g_q are initially computed as if the phase were fully permitted. The results, however, must then be adjusted to account for the presence of either the leading or lagging green phase(s). Figure 8.9 illustrates the impact.

Where both a leading and lagging green phase exist, there are three sub-portions of the phase: *1. Leading Portion, 2. Permitted Portion, and 3. Lagging Portion.* If only a leading phase exists, then Portion 3 is eliminated. If only a lagging phase exists, then Portion 1 is eliminated.

Given that we are looking for the impact of g_f and g_q on the *permitted* portion of the phase (Portion 2), and that the standard computations for these are indexed to the start of the Green for that movement, adjustments must be made. For example,

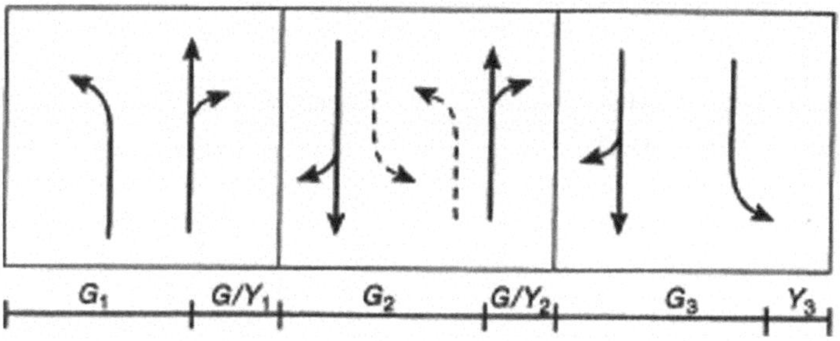

Fig. 8.9 Leading and/or lagging green phases in the 1994 update. *Source* "Highway Capacity Manual," Special Report 209, Transportation Research Board, Washington D.C., 1994 [4], Fig. 9-8, pp. 9–23, used with permission of the National Academy of Science, Courtesy of National Academies Press

the green for the leading left turn starts in Portion 1. For the model, however, we only want to use that part of g_f and g_q that encroach on Portion 2 of the phase. Also, note that during the interval G/Y_1, the leading left-turn movement sees yellow plus all red (Y), while the leading through movement sees green. Similarly, during the interval G/Y_2, the lagging left-turn and through movements sees green, while the leading through movement sees yellow plus all red.

Table 8.12 shows the equations for modifications to the g_f and g_q computations for leading/or and lagging green phases.

Additional detail on these derivations and their interpretation is found in the 1994 update, as well as a discussion of the case of a standard NEMA phasing with an exclusive LT phase followed by a leading LT phase in the peak LT direction. These modifications were retained through the 1997 update, and in the 2000 HCM.

8.4.6.5 The Left-Turn Adjustment—1997 Update

The handling of the left-turn adjustment in the 1997 update was, for the most part, the same as in the 1994 update. Three specific changes were introduced, however:

- A change in the computation of the opposing volume, v_o,
- Simplified left-turn equivalents for permitted left turns, E_{L1}, and
- An improved formula for estimating the proportion of left-turning vehicles in a shared left-turn/through lane, P_L.

The opposing flow rate, v_o, used in estimating g_q was modified to account for unequal use of opposing lanes. This was made necessary because the adjustment for lane distribution was moved to the saturation flow module.

$$v_o = \frac{V_o}{f_{LU}} \tag{8.40}$$

Table 8.12 Equations for modified green times (g_f and g_q) for leading and/or lagging green phases—1994 update

Case	Mvts	Equations
Lead only	Lead	$G* = G_1 + G/Y_1 + G_2$ $g* = G_2 + Y_2$ $g_f* = g_f - G_1 - G/Y_1 + t_L$ $g_q* = g_q + t_L$ $NOTE: G_3 = Y_3 = 0$
	Perm only	$G^* = G_2$ $g* = G_2 + Y_2 - t_L = g$ $g_f* = g_f$ $g_q* = g_q - G_1 - G/Y_1$
Lag only	Lag	$G* = G_2 + G/Y_2 + G_3$ $g^* = G_2 + G/Y_2 - t_L$ $g_f* = g_f$ $g_q* = g_q$ $NOTE: G_1 = G/Y_1 = 0$
	Perm only	$G* = G_2$ $g* = G_2 + G/Y_2 - t_L$ $g_f* = g_f$ $g_q* = g_q$
Lead and lag	Lead	$G* = G_1 + G/Y_1 + G_2$ $g* = G_2 + G/Y_2$ $g_f* = g_f - G_1 - G/Y_1 + t_L$ $g_q* = g_q - t_L$
	Perm	$G^* = G_2 + G/Y_2 + G_3$ $g* = G_2 + G/Y_2 - t_L$ $g_f* = g_f$ $g_q* = g_q - G_1 - G/Y_2$

Note In all cases, g_f and g_q are computed as if no leading or lagging phase existed, and in all cases, g_f* and g_q* are limited to values between 0 and g*

where V_o is the opposing flow rate per lane, and all other variables are as previously defined.

The left-turn equivalent, E_{L1}, applied to permitted left turns made during the unsaturated portion of green time, g_u, was simplified, as shown in Table 8.13.

The effect of the modification was to reduce the impact of permitted left turns on the saturation flow rate. For opposing flow rates greater than 1200 veh/h, Eqs. 8.41 and 8.42 are used.

For exclusive LT lanes:

Table 8.13 Equivalents for permitted left turns made during g_u 1997 update

Type of LT lane	Effective opposing flow rate (veh/h) $v_o = V_o/f_{LU}$						
	1	200	400	600	800	1000	1200
Shared	1.4	1.7	2.1	2.5	3.1	3.7	4.5
Exclusive	1.3	1.6	1.9	2.3	2.8	3.3	4.0

Source "Highway Capacity Manual," Special Report 209, Transportation Research Board, Washington D.C., 1997 [5], pp. 9–20

$$E_{L1} = \frac{s_{TH}}{s_{LT}} \tag{8.41}$$

For shared LT/TH lanes:

$$E_{L1} = \frac{s_{TH}}{s_{LT}} - 1 \tag{8.42}$$

where

$$s_{LT} = \frac{v_o e^{-v_o t_c/3600}}{1 - e^{-v_o t_f/3600}} \tag{8.43}$$

and:

E_{L1} equivalent for permitted left turns during g_u,
s_{TH} saturation flow rate of through traffic = 1900 pc/hg/ln,
s_{LT} saturation flow rate of filtering left turns during g_u, pc/hg/ln,
t_c critical gap = 4.5 s, and
t_f follow-up headway = 2.5 s (exclusive LT lanes); 4.5 s (shared LT/TH lanes).

The values of E_{L1} used in the 1997 update essentially resulted from gap acceptance modeling, and are based on theory, not data.

The last change to the left-turn adjustment in the 1997 update resulted from an unpublished memo to the signalized intersection subcommittee (of the HCQSC), which recommended an improvement in the model for determining the proportion of left turns in shared LT/TH lanes [3].

In the 1985 HCM and the 1994 update, E_L was computed using Eq. 8.27, repeated here for convenience:

$$P_L = P_{LT}\left[1 + \left(c\frac{(N-1)g}{f_s g_u + 4.5}\right)\right]$$

where

$$f_s = \frac{875 - 0.625\, v_o}{1000}$$

In the 1997 update, the adjustment to g_u was replaced by E_{L1} and the time g_f was used in the denominator, as this is the time being used by through vehicles in the shared lane before the arrival of the first left-turning vehicle.

Introducing these produced the following equation for P_L.

$$P_L = P_{LT}\left[1 + \left(\frac{(N-1)g}{g_f + \frac{g_u}{E_{L1}+4.24}}\right)\right] \tag{8.44}$$

where all variables are as previously defined. The 4.24 is the time for two sneakers at the default saturation flow rate of 1700 veh/hg/ln.

8.4.6.6 The Left-Turn Adjustment—2000 HCM

Only one change to the left-turn adjustment was introduced in the 2000 HCM.

For the case of compound phasing from a shared lane, the single equation used in previous versions (see Table 8.6) was changed to be consistent with the methodology for compound phasing from an exclusive LT lane. Thus, separate saturation flow rates for the protected and permitted portions of the phase were calculated, and the two portions were effectively treated as separate phases in the analysis.

8.5 The Capacity Analysis Module

In the capacity analysis module, the results of volume adjustment computations and saturation flow rate computations are combined to produce a view of how the intersection is operating. Three key variables are estimated in this module:

- The capacity of each lane group (c) (veh/h),
- The v/c ratio for each lane group (decimal), and
- The critical v/c ratio for the intersection as a whole (decimal).

This is also the module in which critical lane groups are identified, and in which a most critical variable is determined: the sum of the critical lane group v/s ratios: $\sum (v/s)_{ci}$.

Note that, like the saturation flow module, the capacity analysis module uses effective green time (g) as opposed to actual green time (G).

8.5.1 The 1985 HCM Capacity Analysis Module

Most aspects of the Capacity Analysis Module are relatively straightforward, although some introduce moderate degrees of complexity.

The capacity of each lane group is simply computed from the saturation flow rates resulting from the Saturation Flow Rate Module.

$$c_i = s_i(g_i/c) \tag{8.45}$$

where

c_i capacity of lane group i (veh/h),
s_i saturation flow rate for lane group i (veh/hg),
g_i effective green time for lane group i (s), and
C signal cycle length (s).

The v/c ratio for a given lane group is given the symbol "X" for simplicity, and for consistency with signal literature in which the variable is often referred to as the *degree of saturation*. Then:

$$X_i = v_i/c_i \tag{8.46}$$

where all terms are as previously defined.

The values of X_i are critical. Any lane group with a value of $X_i > 1.00$ will fail, i.e., saturated queues will develop, and some vehicles will have to wait for more than one cycle to get through the intersection.

A final parameter is the critical v/c ratio (X_c) for the intersection as a whole. This value is essentially the sum of the critical lane volumes (v_{ci}) divided by the sum of the capacities of the critical lane groups (c_{ci}). This concept can be manipulated into the form:

$$X_c = \frac{\sum v_{ci}}{\sum c_{ci}} = \sum (v/s)_{ci} \left[\frac{C}{C - L}\right] \tag{8.47}$$

where

X_c critical v/c ratio,
$(v/s)_{ci}$ the ratio of demand flow rate to saturation flow rate for critical movement i,
C cycle length (s), and
L total lost time per cycle (s).

Determining the critical v/c ratio requires that the critical lane groups be identified. This is done by comparing the flow ratios (v/s) for each lane group and signal phase. Where phases do not overlap, the task is quite simple:

- For each signal phase, the critical lane group will be the lane group with the highest v/s ratio for that phase.
- The critical lane group v/s ratios are then added to find $\sum_i (v/s)_{ci}$.
- For each signal phase, there will be *one* and *only one* critical lane group.

Where overlapping phases do exist, the task becomes a bit more difficult. The critical path through the signal will involve the combination of lane groups and phases that produces the highest sum of critical lane v/s ratios. For a more comprehensive discussion of how this determined, consult the textbook by Roess et al. [8].

When the critical v/c ratio, X_c, is greater than 1.00, the sum of the critical lane demand flows is greater than the sum of the capacities of critical lane groups to accommodate them. This situation represents a failed intersection. Moreover, it cannot be fixed without changes in the geometry and/or signalization that lead to increased capacities.

Where X_c and all lane group X_i values are less than 1.00. The intersection may be expected to operate without unstable queues (i.e., all queues forming during RED intervals are fully dissipated during the next GREEN interval).

The most interesting case occurs when $X_c < 1.00$, but some lane group X_i values are greater than 1.00. In this case, the basic geometry and signalization can be made to work by reallocating GREEN time from lane groups with $X_i < 1.00$ to lane groups with $X_i > 1.00$. No changes to signal phasing, cycle length, nor intersection geometry are necessary to produce a situation in which *all* X_i and X_c are less than 1.00. Again, consult Ref. 8 for a fuller description of how the reallocation of green time can be accomplished.

Where compound phasing exists, a specific problem arises: how is the demand volume allocated to the protected and permitted portions of the compound phase(s)?

In the 1985 HCM, essentially, only the protected portion of the phase is considered, and all left-turn volume is initially assigned to the protected portion of the phase. If after calculating the capacity and v/c ratio for the protected phase, it is found that the v/c ratio is higher than acceptable, volume can be assigned to the permitted phase and subtracted from the protected phase. The volume moved from the protected phase may range from 2 vehicles per cycle up to the maximum capacity of the permitted phase, estimated from Eq. 8.48.

$$c_{LT} = (1400 - v_o) * (g/C)_{PLT} \tag{8.48}$$

where

c_{LT} the capacity of the left-turn permitted phase
$(g/C)_{PLT}$ the g/C ratio for the permitted phase

and all other terms are as previously defined.

The volume assigned to the permitted phase is then subtracted from the protected phase, and the calculations for capacity and v/c ratio are redone for the protected phase. This type of phasing may, therefore, involve iterative computations. The same is true if initial signal timings are not known. After estimating an initial timing, the calculations might have to be redone if capacity deficiencies are identified.

8.5.2 The 1994 Update

The capacity analysis module in the 1994 was the same as that in 1985 HCM, except for the case of compound phasing from an exclusive left-turn lane.

As described in saturation flow rate module for the 1994 update, the protected left- and permitted left-turn phases are handled separately (as separate lane groups) with different saturation flow rates in each portion of the phase.

For compound phasing, the first step is to calculate capacity of each portion of the phase using $c = s * g/C$. Next, the left-turn volume must be apportioned to the two portions of the phase. To do this, the assumption is made that all of the left-turn volume will occur in whichever portion of the phase is first, up to the capacity of that phase ($v/c = 1.0$), and remaining volume, if any, is then assigned to the second portion of the phase.

Often the left-turn volume is less than the capacity of the first phase, particularly when the first phase is the protected phase. In that case, zero volume is assigned to the second phase.

In order to calculate the capacity of the two portions of the compound left-turn phases, the effective green times of each portion must be determined separately. The total lost time for an individual phase is assigned to the beginning of that phase. The effective green times for the protected and permitted portions of the phase will thus depend on which phase is displayed first, because only the first phase will be assigned any lost time.

Using this approach, if the compound phase is *protected plus permitted,* i.e., the protected portion comes first, then the effective green times are:

$$\begin{aligned} g_{prot} &= G_{prot} + Y_{prot} - t_L \\ g_{perm} &= G_{perm} + Y_{perm} \end{aligned} \tag{8.49}$$

If, however, the compound phase is *permitted plus protected*, i.e., the permitted portion comes first, then the effective green times are found as:

$$\begin{aligned} g_{perm} &= G_{perm} + Y_{perm} - t_L \\ g_{prot} &= G_{prot} + Y_{prot} \end{aligned} \tag{8.50}$$

where

- g_{prot} effective green time for the protected portion of the phase, s,
- G_{prot} actual green time shown for the protected portion of the phase, s,
- Y_{prot} yellow plus all red shown after the protected green phase, before the permitted phase begins, s,
- Y_{perm} yellow plus all red shown after the permitted green Phase, before the protected phase begins, s, and
- t_L total lost time for the phase, s.

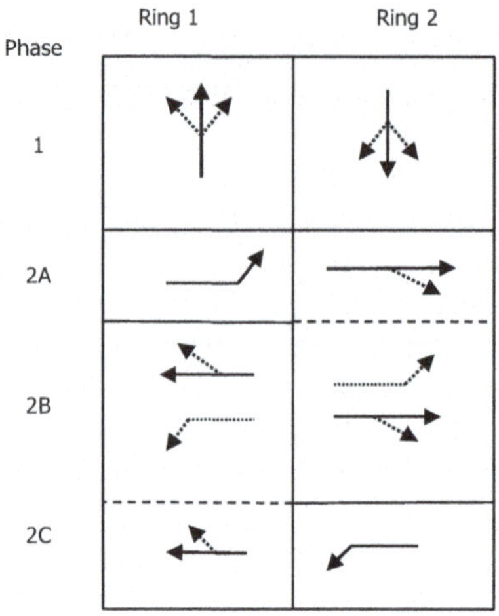

v/s Ratios			
1985 HCM		1994 Update	
NB 0.28	SB 0.24	NB 0.28	SB 0.24
EBLT 0.25	EBTR 0.36*	EBLT 0.15	EBTR 0.36*
WBTR 0.40** EBLT 0.00	WBLT 0.00 EBTR -----	WBTR 0.40* EBLT 0.09	WBLT 0.10 EBTR ------
WBTR -----	WBLT 0.34	WBTR -----	WBLT 0.20

*Includes Phases 2A and 2B
**Includes Phases 2B and 2C

Fig. 8.10 Determining critical movements—1985 HCM and 1994 update

In the 1985 HCM and its successors, the determination of critical lane groups was based upon finding a path through the signal ring diagram that resulted in the highest sum of critical lane group v/s ratios. Where phases do not overlap (no compound phasing), this is a simple task, as each phase in the signal plan is unique, with one or two lane groups moving simultaneously in each. For each phase, the highest v/s ratio for the lane groups moving identifies the critical lane group for that phase.

Where compound phasing exists, the situation becomes a bit more difficult. Consider the signal ring diagram shown in Fig. 8.10. It shows a single phase for NB and SB movements, and a lead-lag compound phasing for EB and WB movements.

In the 1985 HCM, the situation was simplified by assuming that *all* of the left-turning traffic occurred during the protected portion of the compound phase. This left only two choices for the sum of critical movements in the compound phase. In 1994, however, demand volume was split between the protected and permitted portions of a compound phase, making two additional choices possible.

Figure 8.10 illustrates a common compound phasing with leading and lagging greens. Phase 1 handles all NB and SB movements, so the selection of critical movements is simple: Either the NB or SB movements are critical. The highest v/s ratio for these movements designates the critical path. For both the 1985 and 1994 methodology, the critical v/s ratio for Phase 1 is NB—0.28.

It is in Phase 2 that the situation gets more complicated. There are three sub-phases, and many movements overlap, i.e., they move in more than one sub-phase:

- EB movements go in Phases 2A and 2B. The EBLT is protected in Phase 2A and permitted in Phase 2B.

- WB movements go in Phases 2B and 2C. The WB LT is permitted in Phase 2B and protected in Phase 2C.

The 1985 HCM assumes that all left turns are made during the protected portion of a compound phase. Therefore, there are only two possible critical paths through Phase 2:

- EBLT + WBTR (0.25 + 0.40 = 0.65)
- EBTR + WBLT (prot) (0.36 + 0.34 = 0.70).

The maximum sum of v/s ratios determines the critical path, which is, in this example, the EBTR + WBLT = 0.70.

In 1994, however, the left-turn demand is split between the protected and permitted portion of the phase. This, in effect, reduces the v/s ratio during the protected portion of the phase, but adds another v/s ratio during the permitted portion of the phase. The original two choices of the 1985 HCM remain:

- EBLT(prot) + WBTR (0.15 + 0.40 = 0.55)
- EBTR + WBLT (prot) (0.36 + 0.20 = 0.56).

However, there are two additional choices that present themselves:

- EBLT(prot) + EBLT (perm) + WBLT(prot) = (0.15 + 0.09 + 0.20 = 0.44)
- EBLT(prot) + WBLT (perm) + WBLT(prot) = (0.15 + 0.10 + 0.20 = 0.45).

In this case, the highest sum of critical v/s ratios is 0.56, and the critical movements are the EBTR and the WBLT(prot).

As the critical v/s ratio for Phase 1 is 0.28, the sum of critical v/s ratios for the signal shown is 0.28 + 0.70 = 0.98 for the 1985 HCM and 0.28 + 0.56 = 0.84 for the 1994 update. The reduction is a real reflection of the impact of the compound phasing, which is more or less ignored in the 1985 HCM.

Another critical aspect of determining the critical path is the number of lost times that are encountered on it. Remember, that for both the 1985 and 1994 manuals, all lost time is accrued at the *beginning* of the phase.

In this particular example, there is no change in this situation from 1985 to 1994. In both cases, one movement begins in Phase 1 (NB). For the 1985 solution, for Phase 2, the EBTR movement starts in Phase 2A, and the WBLT movement starts in Phase 2C. Thus, three (3) lost times exist along the critical path. There is, however, NO movement on the critical path that *begins* in Phase 2B, so no lost time is experienced there.

The critical path, however, changes for the 1994 solution. The NB movement still begins in Phase 1. In Phase 2, however, the EBTR movement begins in Phase 2A and continues in Phase 2B. Only one lost time is experienced. The WBLT (Prot) begins in Phase 2C, and involves another lost time. Again, there are three (3) lost times along the critical path. In this case, there is no critical movement that *begins* in Phase 2B, so there is no lost time assessed there.

There are more complex signalizations that could exist using the 1994 approach. Some may involve a left-turn movement that moves on a protected basis in one phase, and on a permitted basis in another, where the two are not consecutive. If such a movement were on the critical path, the one movement would include *two* lost times, because it literally begins *twice* within one signal cycle. Suffice it to say that correct assessment of the total lost time in a signal cycle is important in all cases. In general, total lost time in a phase will vary from $L = 2t_L$ to $L = 4t_L$, depending upon the phase design and critical movements. The 1994 update contains additional discussion of this issue, with examples.

8.5.3 The 1997 and 2000 Capacity Analysis Modules

There were no changes made to the Capacity Analysis Module in the 1997 update or the 2000 HCM.

8.6 The Level of Service Module

Since 1985, the Level of Service Module has been the place where delay measures have been estimated, and levels of service assigned. Over the years the specific delay measure used has changed somewhat, as have the algorithms and models used to predict it. Level of service is defined first for each individual lane group, and can be aggregated for each approach that contains more than one lane group.

The aggregation of delay measures and levels of service for the intersection as a whole has been controversial, and has generally been discouraged, although the computation of a value is not difficult numerically. The principal argument against aggregation for the whole intersection is that such an aggregation would represent an average over the various lane groups. Its use, without looking at individual lane group values, could easily mask serious problems, even a failure, of one or more lane groups, even when the average for the whole looks like a stable value.

8.6.1 The 1985 Highway Capacity Manual

The 1985 HCM was the first to use delay as the service measure determining level of service. Specifically, the parameter used was *average individual stopped-time delay*, which is the average time a vehicle spends stopped while getting through the intersection. Stopped-time delay excludes time lost due to acceleration and deceleration, as well as time spent moving while within the queue. It was chosen because it was a relatively easy parameter to measure in the field.

The level of service criteria for signalized intersections in the 1985 HCM is shown in Table 8.14.

Table 8.14 Level of service criteria for signalized intersections—1985 HCM

Level of service	Stopped delay per vehicle (s)
A	≤5.0
B	5.1–15.0
C	15.1–25.0
D	15.1–40.0
E	40.1–60.0
F	>60.0

Source "Highway Capacity Manual," *Special Report 209*, Transportation Research Board, Washington D.C., 1985 [1], Table 9-1, pp. 9–4, used with permission of the National Academy of Science, Courtesy of National Academies Press

The move to delay as a service measure was not without controversy. Early use of the 1965 HCM demonstrated that load factor was not a particularly good measure to describe overall service quality. Delay was recommended by many, and was generally accepted as a reasonable measure to classify the quality of operations in any given lane group, approach, or intersection.

It, however, also introduced a new problem in interpretation. Because delay was not directly related to the v/c ratio, the important boundary between LOS E and F became difficult to interpret. It is possible, for example, for an intersection approach to operate at a v/c ratio that is marginally larger than 1.00, and yet still have a delay that is less than 60 s/veh. In such cases, the LOS would *not* be F. The traditional understanding that when demand exceeded capacity, failure (LOS F) occurred was no longer valid. The reverse was also true—delays higher than 60.0 s/veh could be experienced when v/c ratios are still less than 1.00. The traditional understanding that LOF F occurred when (and only when) demand exceeded capacity was no longer true either.

Because of this, the 1985 HCM clearly indicates that to fully understand the operating conditions on a signalized intersection lane group or approach, *both* the level of service *and* the v/c ratio must be considered. Unfortunately, in practice, the v/c ratio, no longer directly linked to level of service, is often ignored.

With average individual stopped-time delay as the service measure, the methodology had to provide a model for estimating this critical value. Delay for a lane group is estimated as:

$$d = d_i * PF$$

$$d_i = 0.38\, C \frac{(1 - g/C)^2}{1 - (g/C)X} + 173\, X^2 \left[(X - 1) + \sqrt{(X - 1)^2 + (16X/c)} \right] \quad (8.51)$$

where

d average individual stopped-time delay (s/veh) for a lane group,

d_i average individual stopped-time delay (s/veh) for a lane group, assuming that random arrivals are occurring,
C signal cycle length (s),
g effective green time for the lane group (s),
X degree of saturation, or v/c ratio, for the lane group, and
c capacity of the lane group (veh/h), and
PF progression adjustment factor.

The first term of the delay equation in Eq. 8.51, is essentially Webster's Uniform Delay Formula [4], adjusted to reflect stopped delay instead of approach delay. Approach delay includes time lost during deceleration and acceleration while stopped does not. The second term represents additional delay due to the fact that arrivals *are not* uniform, as well as additional delay due to overflow situations in which some vehicles must wait for two or more signal cycles to complete their desired maneuver. The second term is a form of Akcelik's Equation for overflow delay [5]. The random delay equation, d_i, is only valid for X values up to 1.2.

The term for uniform delay makes some simplifying assumptions. The equation is derived by finding the area inside the triangle made between cumulative arrivals and departures during a signal cycle. The simplifying assumption is made that there is only one arrival rate. This will be discussed in more detail in Chap. 9, which covers the 2010 and 2016 methodologies, when this assumption was removed.

The base delay estimation assumes that arrivals at the signal are random. This is, of course, rarely true. When progression is good, the delay will be quite a bit *less* than it would under random arrivals. When the progression is bad, however, the delay would be quite a bit *more* than it would be under random arrivals. The equation models this by multiplying the random delay value by the progression factor (PF). Values of the progression factor are given in Table 8.14, and are based upon the five defined arrival types (AT) described in the Input Module discussion earlier in this chapter.

The 1985 HCM also allows delays to be averaged to estimate an average delay for an approach which includes more than one lane group, and for the intersection as a whole.

$$d_A = \frac{\sum_i d_i v_i}{\sum_i v_i}$$
$$d_I = \frac{\sum_A d_A v_A}{\sum_A v_A} \tag{8.52}$$

where

d_i average delay for lane group i (s/veh),
d_A average delay for approach A (s/veh),
d_I average delay for the overall intersection (s/veh),
v_i demand flow rate for lane group i (veh/h), and
v_A demand flow rate for approach A (veh/h) (Table 8.15).

Table 8.15 Progression factor for estimating lane group delay at a signalized intersection—1985 HCM

Type of signal	Lane group type	v/c Ratio (X)	Arrival type				
			1	2	3	4	5
Pretimed	TH, RT	≤0.6	1.85	1.35	1.00	0.72	0.53
		0.8	1.50	1.22	1.00	0.82	0.67
		1.0	1.40	1.18	1.00	0.90	0.82
Actuated	TH, RT	≤0.6	1.54	1.08	0.85	0.62	0.40
		0.8	1.25	0.98	0.85	0.71	0.50
		1.0	1.16	0.94	0.85	0.78	0.61
Semi-actuated	Main St TH, RT[a]	≤0.6	1.85	1.35	1.00	0.72	0.42
		0.8	1.50	1.22	1.00	0.82	0.53
		1.0	1.40	1.18	1.00	0.90	0.65
Semi-actuated	Side St TH, RT[a]	≤0.6	1.48	1.18	1.00	0.86	0.70
		0.8	1.20	1.07	1.00	0.98	0.89
		1.0	1.12	1.04	1.00	1.00	1.00
	All LT[b]	All	1.00	1.00	1.00	1.00	1.00

[a]Semi-actuated signals are typically timed to give all extra green time to the main street. This effect should be taken into account in the allocation of green times

[b]This category refers to exclusive LT lane groups with protected phasing only. Where LTs are included in a lane group encompassing an entire approach, use the factor for the overall lane group type. Where heavy LTs are intentionally coordinated, apply factors for the appropriate through movement

Source "Highway Capacity Manual," *Special Report 209*, Transportation Research Board, Washington D.C., 1985 [1], Table 9-12, pp. 9–20, used with permission of the National Academy of Science, Courtesy of National Academies Press

8.6.2 *The 1994 Update to the 1985 HCM*

There were three changes made in the level of service module in the 1994 HCM. The first was a change to the calculation of delay, d, the second was to the adjustment for progression effects, and the third was an additional change to the calculation of uniform delay, d_i, for compound phases made from an exclusive lane. Other than these three changes, the remainder of the level of service module remained the same.

In the 1985 HCM, the two parts of the delay equation, the uniform delay and the incremental delay, were not calculated separately and the progression factor was applied to the entire delay (see Eq. 8.51). In the 1994 HCM, the two parts are always calculated separately, as seen in Eqs. 8.53–8.55, and a new factor for progression is applied only to d_1, the uniform delay.

$$d_1 = 0.38\,C\frac{(1-g/C)^2}{1-[g/C][\min(X,1)]} \tag{8.53}$$

$$d_2 = 173\,X^2\left[(X-1)+\sqrt{(X-1)^2+\frac{mX}{c}}\right] \tag{8.54}$$

$$d + d_1 DF + d_2 \tag{8.55}$$

The equation for uniform delay, d_1, uses values of v/c ratio up to 1.0, or where volume equals capacity. The 1985 HCM did not recognize that all delay for v/c ratios (X) greater than 1.00 is included in the second term of the equation. The new progression factor (DF) is only applied to uniform delay because the effect of progression on overflow delays was thought to be minimal.

The equation for overflow delay, d_2, calculates delay due to random arrivals as well as delay caused by cycle failures occurring during the fifteen minutes of analysis. In Eq. 8.54, the variable *m* replaced the value of 16 in the 1985 HCM. This variable was added to account for platooning due to progression type. The value of 16 in the 1985 HCM represented the value for random arrivals, because the second term was also multiplied by the PF to account for progression. The values are smaller when there is some level of coordination present, because demand varies less when it is organized into platoons by upstream signals.

The adjustment to delay in the 1994 update was improved, based on the research in Ref. [6]. Also, the 1994 update added a sixth arrival type (AT-6). Arrival Type 6 better described locations with near perfect progression. It is defined as a location with short blocks that can form very dense platoons that stay together because of the short distances between intersections, as well as low side-street volume. The definition of the other five arrival types did not change. The relationship of arrival types to the platoon ratio (R_p), however, was changed to accommodate the addition of AT-6 (Table 8.16).

The factor for delay adjustment, DF, which was now only applied to uniform delay, is determined by one of two factors.

Table 8.16 Relationship between arrival types and the platoon ratio 1994 update

Arrival type	Range of platoon Ratio, R_p	Default value of R_p	Progression quality
1	$\leq$0.50	0.333	Very poor
2	>0.50 $\leq$ 0.85	0.667	Unfavorable
3	>0.85 $\leq$ 1.15	1.000	Random arrivals
4	>1.15 $\leq$ 1.50	1.333	Favorable
5	>1.50 $\leq$ 2.00	1.667	Very favorable
6	>2.00	2.000	Exceptional

Source "Highway Capacity Manual," *Special Report 209*, Transportation Research Board, Washington D.C., 1994 [4], pp. 9–10, used with permission of the National Academy of Science, Courtesy of National Academies Press

- Where signals are uncoordinated, there is no progression impact. However, the type of controller (actuated, pre-timed) can affect delay, and the delay factor (DF) is equal to the control-type factor (CF).
- Where signals are coordinated, the progression impact is measured by the progression factor (PF), and DF = PF.

The Progression Factor (PF) is computed as:

$$PF = \frac{(1 - P)f_p}{1 - g/C} \tag{8.56}$$

where

P proportion of vehicles arriving on green,
g/C green to cycle length ratio, or the proportion of green time available to the movement,
f_p factor to account for when the platoon arrives.

The P in Eq. 8.56 reflects delay being affected by the proportion of vehicles that arrive during the green. A greater proportion of vehicles arriving during the green means decreased delay, while more vehicles arriving during the red means increased delay.

The f_p adjustment factor accounts for delay also being impacted by when the beginning of the platoon arrives at the stop line. Arrival types 5 and 6 assume the first vehicle arrives just as the light turns green. Referenced to this, arrival type 4 assumes late arrivals, meaning the first vehicle in the platoon arrives during the green, but after initiation of the green. The f_p factor for arrival type 4 is 1.15. For arrival type 1, the worst progression, it is assumed that the first vehicle in the platoon arrives at the initiation of the red. Relative to this, arrival type 2 assumes early arrivals, meaning the platoon arrives during the red (after red initiation, but before the initiation of green). The f_p factor for arrival type 2 is 0.93.

Table 8.17 shows control type factors (CF), progression factors (PF) for default values of P, the proportion of vehicles arriving on green, and the incremental delay factor, m. As noted, the delay factor (DF) is determined for either the type of control or for the effects of progression, but not both. If the intersection is isolated, then the delay adjustment factor is based on the factor for controller type, *CF*. If the intersection is part of an arterial, i.e., coordinated, or has lane groups that are not actuated, then the delay adjustment factor is based on the progression adjustment factor, *PF*.

A new worksheet was added in the 1994 update to calculate uniform delay for compound phases. The previous models for delay allow only for one arrival rate and one discharge rate. Where there is compound phasing, however, the 1994 saturation flow module calculates different discharge rates for the protected phase and the permitted phase. This new worksheet accounts for this difference in departure rate. Five cases are defined for which varying equations are given to calculate d_1. The five cases are grouped by the type of phasing (leading or lagging left) as well as by the

Table 8.17 Uniform delay adjustment factor (DF)—1994 update

Controller-type adjustment factor (CF)		
Controller type	Non-coordinated intersections	Coordinated intersections
Pretimed	1.00	PF (see below)
Semi-actuated		
Traffic actuated lane groups	0.85	1.0
Non-actuated lane groups	0.85	PF (see below)
Fully actuated	0.85	Treat as semi-actuated

Progression factor (PF) $PF = (1 - P)f_p/(1 - g/C)$

Green ratio (g/C)	Arrival type (AT)					
	AT-1	AT-2	AT-3	AT-4	AT-5	AT-6
0.20	1.167	1.007	1.000	1.000	0.833	0.750
0.30	1.286	1.063	1.000	0.986	0.714	0.571
0.40	1.445	1.136	1.000	0.895	0.555	0.333
0.50	1.667	1.240	1.000	0.767	0.333	0.000
0.60	2.001	1.395	1.000	0.576	0.000	0.000
0.70	2.556	1.653	1.000	0.256	0.000	0.000
Default f_p used	1.00	0.93	1.00	1.15	1.00	1.00
Default R_p used	0.333	0.667	1.000	1.333	1.667	2.000
m[a]	8	12	16	12	8	4

[a]For uncoordinated signals, use m = 16

Notes

DF = CF for uncoordinated signals; DF = PF for coordinated signals

Tabulation is based upon default values of f_p and R_p

$P = R_p$ (g/C), maximum value = 1.00

PF may not exceed 1.00 for AT-3 through AT-6

Source "Highway Capacity Manual," *Special Report 209*, Transportation Research Board, Washington D.C., 1994 [4], pp. 9–28, used with permission of the National Academy of Science, Courtesy of National Academies Press

queue that may remain at the end of either the protected or permitted portion of the phase, that is, if X_{perm} or X_{prot} is greater than 1.0.

Figure 8.11 shows queue accumulation polygons for the five cases. For each of the cases, the queue at the time the protected left phase begins and at the time the permitted left phase begins is calculated. Figure 8.12 shows all the calculations.

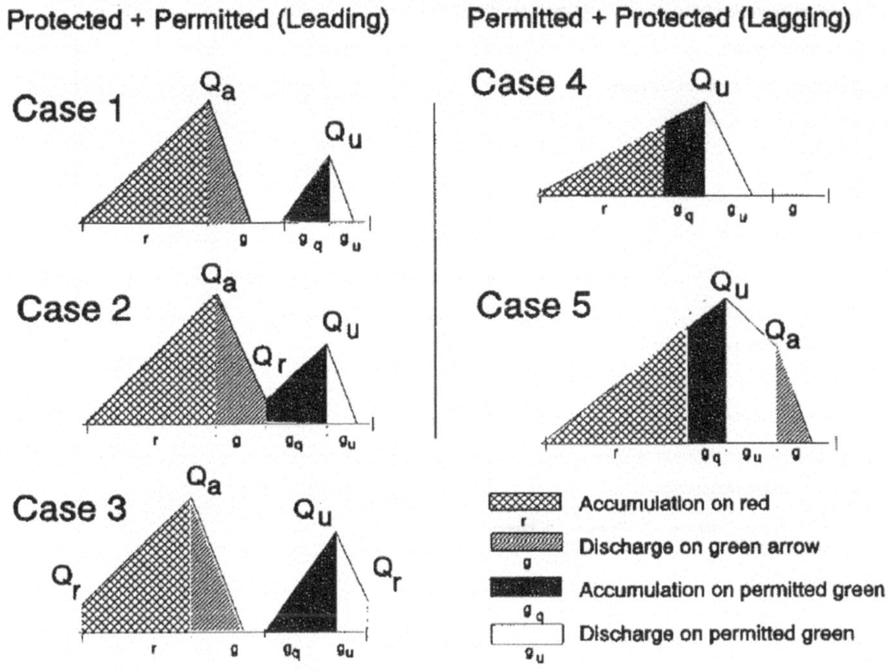

Fig. 8.11 Queue accumulation polygons—compound phasing 1994 update. *Source* "Highway Capacity Manual," *Special Report 209*, Transportation Research Board, Washington, D.C., 1994 [4], Fig. 9-12, pp. 9–30, used with permission of the National Academy of Science, Courtesy of National Academies Press

8.6.3 The 1997 Update to the 1985 HCM

The major change in the 1997 update was in the definition of level of service. This produced changes in its calculation, and its threshold values. Whereas previous editions defined level of service in terms of average stopped delay per vehicle, the 1997 HCM defined LOS in terms of average control delay per vehicle (also called signal delay). Control delay includes all delay added to a trip by the placement of the signal. Included along with stopped delay is the deceleration delay, queue move up time, and acceleration delay.

A factor of 1.3 was used to convert stopped delay to the larger control delay. The level of service thresholds were then also increased by a factor of 1.3. Changes to the delay model are documented in Refs. [17, 18].

The change to control delay made the delay used in the signalized intersection chapter consistent with the definition of delay used in the other interrupted flow chapters in the manual (unsignalized intersections and arterials), thus making it easier for comparisons among these to be made.

The average control delay represents the delay that is experienced by all vehicles that have arrived during the subject analysis period expressed as a per vehicle delay

SUPPLEMENTAL UNIFORM DELAY WORKSHEET FOR LEFT TURNS WITH PRIMARY AND SECONDARY PHASES

------------ INPUT DATA: ----------------		EBLT	WBLT	NBLT	SBLT
Adj. LT vol from Vol. Adjustment Worksheet,	v				
v/c ratio from Capacity Worksheet,	X				
Signal timing intervals:					
Primary phase effective green	g				
Secondary phase green intervals	g_q				
(From Supplemental Permitted LT Worksheet)	g_u				
Cycle length, C ______ Red ($C-g-g_q-g_u$)	r				
Arrival and Departure rates (veh/sec)					
Arrivals: $v/(3600(\max(X,1.0)))$	q_a				
Primary ph. departures: $s/3600$	s_p				
Secondary ph. depatrures: $s(g_q+g_u)/(g_u \times 3600)$	s_s				

---------------- COMPUTATIONS -------------

	Protected + Permitted (Leading Lefts)	Permitted + Protected (Lagging Lefts)	EBLT	WBLT	NBLT	SBLT
X_{perm}	$q_a(g_q+g_u)/s_sg_u$	$q_a(r+g_q+g_u)/s_sg_u$				
X_{prot}	$q_a(r+g)/s_pg$	N/A				

CASE

		EBLT	WBLT	NBLT	SBLT
1. $X_{perm} <= 1.0$ & $X_{prot} <= 1.0$ 2. $X_{perm} <= 1.0$ & $X_{prot} > 1.0$ 3. $X_{perm} > 1.0$ & $X_{prot} <= 1.0$	4. $X_{perm} <= 1.0$ 5. $X_{perm} > 1.0$				

Uniform queue sizes at transition points

		EBLT	WBLT	NBLT	SBLT
Queue at beginning of green arrow	Q_a				
Queue at beginning of unsaturated green	Q_u				
Residual queue	Q_r				
Uniform delay from formulas below	d_1				

CASE	Q_a	Q_u	Q_r	DELAY FORMULAS
1.	q_ar	q_ag_q	0	$d_1 = [.38/(q_aC)]\,[rQ_a + Q_a^2/(s_p-q_a) + g_qQ_u + Q_u^2/(s_s-q_a)]$
2.	q_ar	$Q_r + q_ag_q$	$Q_a - g(s_p-q_a)$	$d_1 = [.38/(q_aC)]\,[rQ_a + g(Q_a+Q_r) + g_q(Q_r+Q_u) + Q_u^2/(s_s-q_a)]$
3.	$Q_r + q_ar$	q_ag_q	$Q_u - g_u(s_s-q_a)$	$d_1 = [.38/(q_aC)]\,[g_qQ_u + g_u(Q_u+Q_r) + r(Q_r+Q_a) + Q_a^2/(s_p-q_a)]$
4.	0	$q_a(r+g_q)$	0	$d_1 = [.38/(q_aC)]\,[(r+g_q)Q_u + Q_u^2/(s_s-q_a)]$
5.	$Q_u - g_u(s_s-q_a)$	$q_a(r+g_q)$	0	$d_1 = [.38/(q_aC)]\,[(r+g_q)Q_u + g_u(Q_u+Q_a) + Q_a^2/(s_p-q_a)]$

Fig. 8.12 Worksheet for uniform delay computations—compound phasing—1994 update. *Source* "Highway Capacity Manual," *Special Report 209*, Transportation Research Board, Washington D.C., 1994 [4], Fig. 9-21, pp. 9–46, used with permission of the National Academy of Science, Courtesy of National Academies Press

time. An additional term, d_3, was added to account for vehicles that arrived in a previous time period but were not processed in that previous time period. Delay is estimated as:

$$d = d_1 PF + d_2 + d_3 \tag{8.57}$$

where

d average control delay (s/veh),
d_1 uniform control delay (s/veh),
d_2 additional control delay for overflow conditions and randomness (s/veh),
d_3 additional delay due to pre-existing queue at the start of the analysis period (s/veh), and
PF progression adjustment factor.

Uniform delay, d_1 is calculated using Eq. (8.58).

$$d_1 = \frac{0.50\,C(1 - g/C)^2}{1 - [\min(X, 1)][g/C]} \tag{8.58}$$

where all terms are as previously defined.

Equation 8.58 is exactly the same equation as the d_1 equation of the 1994 HCM, except that the factor of 0.38 in Eq. 8.56 has been increased by a factor of 1.3 (0.38 * 1.3 = 0.494), rounded up to 0.5 in Eq. (8.58) to account for the change to control delay.

The progression factors used for coordinated pretimed signals did not change, but the adjustment for controller type was removed. Instead, a factor for actuated controller settings was added to the d_2 equation, shown below. Incremental delay, that is delay caused by non-uniform arrivals and overflow queues, is calculated using Eq. 8.59.

$$d_2 = 900T\left[(X-1) + \sqrt{(X-1)^2 + \left(\frac{8kIX}{cT}\right)}\right] \tag{8.59}$$

where

T the duration of the analysis period (h),
k an incremental delay factor that accounts for controller settings,
I an upstream filtering/metering adjustment factor,
c capacity of the lane group (veh/h).

All other variables as are previously defined.

There were quite a few changes in the formula for d_2. The new model allows longer analysis periods than the typical 15 min (T = 0.25 h), is valid for v/c ratios greater than 1.2, as well as being sensitive to actuated control settings and variable

Table 8.18 Delay adjustment for controller type (k) 1997 update

Unit extension U (s)	Degree of saturation—v/c Ratio (X)					
	≤0.50	0.60	0.70	0.80	0.90	1.00
≤2.0	0.04	0.13	0.22	0.32	0.41	0.50
2.5	0.08	0.16	0.25	0.33	0.42	0.50
3.0	0.11	0.19	0.27	0.34	0.42	0.50
3.5	0.13	0.21	0.28	0.35	0.43	0.50
4.0	0.15	0.22	0.29	0.36	0.43	0.50
4.5	0.19	0.25	0.31	0.38	0.44	0.50
5.0*	0.23	0.28	0.34	0.39	0.45	0.50
Pretimed or unactuated movement	0.50	0.50	0.50	0.50	0.50	0.50

*For U > 0.50, extrapolate to find k, with $k_{max} = 0.50$
Straight-line interpolation for X and U is permissible to find k to the nearest 0.00
Source "Highway Capacity Manual," *Special Report 209*, Transportation Research Board, Washington D.C., 1997 [6], Table 6-17, pp. 9–29, used with permission of the National Academy of Science, Courtesy of National Academies Press

demands. There is also a factor to account for the filtering and metering effects from upstream signals, *I*.

The *k* factor values are shown in Table 8.18 and are a function of the unit extension and the degree of saturation (for actuated lane groups). The table shows that smaller unit extensions result in lower values of *k* and therefore d_2, except for when the lane group is at or over capacity. Where the lane group is at or over capacity, the benefits of the controller are not realized because the actuated controller will behave, more-or-less, as a pretimed signal.

The *I* factor is an upstream filtering factor that accounts for the way demand is metered from an upstream signal. It is dependent on the degree of saturation upstream. As the degree of saturation upstream gets higher, there is less variability in the arrivals downstream and thus d_2 decreases. For an isolated intersection, the *I* value is 1.0. For platooned arrivals, *I* is calculated with Eq. 8.60.

$$I = 1.0 - 0.91[\min(X_u, 1)]^{2.68} \tag{8.60}$$

where X_u is the weighted average (by relative flows) of degrees of saturation (v/c ratios) of all contributing movements at the upstream intersection.

T values larger than 15 min ($T = 0.25$ h) should be used when demand is not able to be processed in 15 min, but volumes are unchanged for the entire *T* analysis time. The equation for d_2 in previous HCM editions (Eq. 8.54) only allowed for a 15-minute analysis, that is, $T = 0.25$. The 173 in Eq. 8.54 is increased by a factor of 1.3 (converting stopped delay to control delay), and 173 * 1.3 = 225. The 900 in Eq. 8.59, for T = 0.25, equals 900 * 0.25 = 225.

Residual delay or delay caused by vehicles that are not processed in the previous time period are calculated with the d_3 term. When there is no residual queue from the

Table 8.19 Values for calculating initial queue delay [24] 1997 update

Case	X	Q_b	t	U
Initial queue and all arriving demand met in time T	$\leq$1.0	>0	$\frac{Q_b}{c\,(1-X)}$	0
Decreasing unmet demand at end of T	<1.0	>0	T	$1 - \frac{cT}{Q_b}(1-X)$
Increasing unmet demand at end of T	>1.0	> 0	T	1

previous time period, $d_3 = 0$. When there is a residual queue, there are three cases to consider [16].

1. When both the residual queue and all arriving demand during T are processed within time T.
2. When there is demand at the end of time period T, but that demand is decreasing.
3. When demand at the end of time period T is increasing.

Calculation of d_3 is accomplished using Eq. 8.61.

$$d_3 = \frac{1800\, Q_b(1-u)t}{cT} \tag{8.61}$$

where

Q_b initial queue (veh),
t duration of T used to process the initial queue (h), and
u delay parameter.

Table 8.19 shows the three different cases and the values used in Eq. 8.61 for each case.

For all of these cases, the uniform delay, d_1, must be adjusted for the fact that the initial queue will operate at capacity, i.e., $X = 1.0$. During time t, the time it takes for the initial queue to be processed, uniform delay is calculated with $X = 1.0$ and no progression factor is applied. For the remainder of the analysis period $(T - t)$, uniform delay is calculated with the actual X of the lane group. Thus:

$$d_1 = d_s\left(\frac{t}{T}\right) + d_u PF\left(\frac{T-t}{T}\right) \tag{8.62}$$

where d_s is the uniform delay during the queue clearance time (or the saturated portion of the time), and d_u is the uniform delay during the time after the queue has cleared (or the unsaturated portion of the time).

Determining the initial queue for calculating d_3 is not an easy or typical value measured from field studies. The 1997 update introduces a method for analyzing multiple periods, where the unmet demand from a previous analysis period is used as the input of initial queue, Q_b, for the following analysis period. An analysis should therefore begin with an unsaturated time period (i.e., $Q_b = 0$). The unmet demand at the end of each time period is calculated using Eq. 8.63.

Table 8.20 Level of service criteria for signalized intersections 1997 update

Level of service	Control delay (s/veh)
A	≤10
B	>10 ≤ 20
C	>20 ≤ 35
D	>35 ≤ 55
E	>55 ≤ 80
F	>80

Source "Highway Capacity Manual," 4th, Transportation Research Board, Washington D.C., 2000 [6], Exhibit 16-2, pp. 16–2, used with permission of the National Academy of Science, Courtesy of National Academies Press

$$Q_{b,1+i} = \max\{0, [Q_{bi}\_cT(X-1)]\} \quad (8.63)$$

This value of $Q_{b,1+i}$ is then used as Q_b for the next analysis period.

One additional change in the 1997 update was the addition of a method for calculating the average green times to be used in an actuated controller analysis. The full method was not documented in the 1997 update, but a summary of the method, some simple examples, and a detailed discussion of the inputs needed was added as an appendix. Inputs for this method included maximum and minimum green times, minimum gap, gap reduction rate, and others. The full method is documented in Ref. [9].

With this new research, for the first time, the 1997 update was made sensitive to the parameters of an actuated controller. Because the entire methodology was not documented in the HCM, most software developers took years to include it in their software, and thus it was not used until the 2000 HCM was published.

Finally, because the service measure for level of service was changed to reflect control delay, the criteria were also changed, and are shown in Table 8.20.

8.6.4 The 2000 Highway Capacity Manual

There were no changes to the delay calculations in the 2000 HCM, or in the level of service criteria. However, a new model for estimating queues was added as an additional measure of effectiveness.

During the 1997 Summer Meeting of the HCQSC, it was agreed that a model should be included for estimating the back of queue, that is, the furthest position to which the queue would extend. Vehicles that are in queue are defined as fully stopped vehicles. This is measured from the stop line to the last vehicle that begins to move forward after the start of green. The measure will be affected by the amount of red time, demand, capacity, and the arrival type. The measures calculated include the average back of queue, various percentile back of queues, and the queue storage ratio.

The queue storage ratio is especially useful for identifying spillback occurrence, that is, does the queue extend into the upstream intersection, or does an exclusive lane bay extend longer than the bay length, which would in turn affect the performance of the adjacent lane group.

Figure 8.13 shows the worksheet used to calculate back of queue. The methodology used was developed from Refs. [20, 21]. The major parameters and equations on the worksheet will be discussed, but the equations will not be repeated in this text.

The "back of queue" consists of two parts:

$$Q = Q_1 + Q_2 \tag{8.64}$$

Q_1 is the queue caused by the signal normally cycling through its phases, that is, the vehicles that arrive on the red signal and those that join the queue during the green. It assumes unsaturated conditions and uniform arrivals, which are then adjusted for arrival type by multiplying by a progression factor.

Q_2 is the additional queue caused by random fluctuations in demand per cycle, as well as the effects due to oversaturation. Random fluctuations in demand can cause occasional cycle failures (cycles when the arriving queue does not clear the intersection). Cycle failures can occur even in unsaturated conditions. Q_2 also accounts for conditions where demand exceeds capacity for the entire analysis period. Lastly, Q_2 accounts for an initial queue at the start of the analysis period. In the equation for Q_2 (see Fig. 8.13), the term Q_{bL} is the initial queue at the beginning of the time period. This is found during a multi-period analysis calculation of d_3.

If there is an initial queue remaining from the previous time period, it is important to start the analysis with an unsaturated period, and let the methodology calculate any remaining queue that will be an input to the following analysis period.

Because the back of queue models used are for individual lane analysis, all of the input variables shown at the top of Fig. 8.13 must be entered as per lane values. Values used in previous worksheets must be converted from lane group parameters to per lane parameters by dividing by the number of lanes in the lane group. All equations for calculating Q_1 and Q_2 are given in Fig. 8.13.

It should be noted that small delay time does not necessarily mean short queues or vice versa. For example, at an approach with extremely bad progression and a very small proportion of red time in the cycle, many vehicles will have to queue during the red, causing a large back of queue. Since the red signal is very short, however, none of those queued vehicles are delayed too long. Vice versa, at an approach with a large proportion of red time and a small arrival volume, most vehicles will have to queue and be delayed a long time, but since there are only a few arrivals per cycle, the back of queue will be small.

Percentile back of queue calculations are also shown in the worksheet of Fig. 8.13. The percentile back of queue gives the queue length that will be surpassed in (1-Q%) of the cycles. For example, if the average back of queue is calculated as Q = 10 vehicles, at a pretimed signal the 95th percentile queue would be 17 vehicles. This means that the queue would be greater than 17 vehicles in only 5% of cycles. For the same case, the 70th percentile queue would be 12 vehicles, and the queue would

BACK-OF-QUEUE WORKSHEET

General Information

Project Description ______________________________

Average Back of Queue

	EB			WB			NB			SB		
	LT	TH	RT	LT	TH	RT	LT	TH	RT	LT	TH	RT
Lane group												
Initial queue per lane at the start of analysis period, Q_{bL}												
Flow rate per lane, v_L (veh/h)												
Saturation flow rate per lane, s_L (veh/h)												
Capacity per lane, c_L (veh/h)												
Flow ratio, v_L/s_L												
v/c ratio, $X_L = v_L/c_L$												
Effective green time, g (s)												
Green ratio, g/C												
Upstream filtering factor, I												
Proportion of vehicles arriving on green, P												
Platoon ratio, R_p $R_p = \left(\frac{P}{g/C}\right)$												
Effects of progression adjustment factor, PF_2 $PF_2 = \frac{\left(1 - R_p \frac{g}{C}\right)\left(1 - \frac{v_L}{s_L}\right)}{\left(1 - \frac{g}{C}\right)\left[1 - R_p\left(\frac{v_L}{s_L}\right)\right]}$												
First-term queued vehicles, Q_1 (veh) $Q_1 = PF_2 \frac{\frac{v_L C}{3600}\left(1 - \frac{g}{C}\right)}{\left[1 - \min(1.0, X_L)\left(\frac{g}{C}\right)\right]}$												
Second-term adjustment factor, k_B $k_B = 0.12\, I\left(\frac{s_L g}{3600}\right)^{0.7}$ (pretimed signals) $k_B = 0.10\, I\left(\frac{s_L g}{3600}\right)^{0.6}$ (actuated signals)												
Second-term queued vehicles, Q_2 $Q_2 = 0.25 c_L T\left[(X_L - 1) + \sqrt{(X_L - 1)^2 + \frac{8 k_B X_L}{c_L T} + \frac{16 k_B Q_{bL}}{(c_L T)^2}}\right]$												
Average number of queued vehicles, Q $Q = Q_1 + Q_2$												
Percentile Back of Queue	☐ 70%			☐ 85%			☐ 90%			☐ 95%		☐ 98%
Percentile back-of-queue factor,[1] $f_{B\%}$												
Percentile back-of-queue, $Q_{\%}$ (veh), $Q_{\%} = Q f_{B\%}$												
Queue Storage Ratio												
Average queue spacing, L_h (ft)												
Available queue storage, L_a (ft)												
Average queue storage ratio, $R_Q = \frac{L_h Q}{L_a}$												
Percentile queue storage ratio, $R_{Q\%} = \frac{L_h Q_{\%}}{L_a}$												

Notes

1. $f_{B\%} = p_1 + p_2 e^{\left(\frac{-Q}{p_3}\right)}$, where p_1, p_2, and p_3 are obtained from Exhibit G16-5.

Fig. 8.13 Back-of-queue worksheet—2000 HCM. *Source* "Highway Capacity Manual", Transportation Research Board, Washington D.C., 2000 [6], used with permission of the National Academy of Science, Courtesy of National Academies Press

exceed 12 vehicles in only 30% of cycles. For the usual 15-minute analysis period, and a typical range of cycle lengths, the 95th percentile queue would be exceeded in less than one cycle.

8.7 Conclusions

The signalized intersection methodology remains the most complex point analysis methodology in the HCM, and it contains many sub-models that invite almost continuous updating and improvement.

The 1985 HCM brought a major change to the signalized intersection procedure. It was the first time that level of service based on delay was introduced, and the structure of modern critical movement analysis was introduced to the manual.

The 1994 update of the signalized intersection chapter did not change the basic methodology or the format of the methodology. There were many changes however. The most significant changes were an update to the left-turn special procedure model for permitted lefts turns and the handling of compound phases (protected plus permitted).

The 1997 update included many significant changes, the most notable being the change from stopped delay to control delay as the measure of effectiveness for defining level of service.

Whereas the entire 2000 HCM was considered a major update to the manual, and included a new organization and changes to many of the methodologies, the signalized intersection methodology had only a few changes. Most important were the addition of a model for queue length and the addition of two factors to the saturation flow rate module accounting for the effect of pedestrians and bicycles on left and right turning vehicles.

Through all of these versions of the HCM (1985–2000), the format of worksheets detailing the methodologies did not change. The next chapter will discuss the 2010 and 2016 HCM models for signalized intersections, which did away with worksheets, and radically altered the approach to analysis. Even with all of the changes, however, the 2010/2016 HCM remains a "critical movement analysis" methodology. The most radical change was that the base case should be an actuated signal. While this reflected the movement toward coordinated actuated signal systems, it, made treatment of pretimed signals more difficult, and masked the fundamental fact that critical movement analysis models are themselves based upon pretimed signalization.

References

1. Highway capacity manual. Special Report 209, Transportation Research Board, Washington, DC, 1985
2. Messer CJ, Fambro DB (1977) Critical lane analysis for intersection design. Transportation research record 644. Transportation Research Board, Washington, DC
3. Reilly et al (1982) Signalized intersection capacity study. Final report. NCHRP project 3, vol 28, no 2, JHK & Associates
4. Highway Capacity Manual (1994) Special report 209. Transportation Research Board, Washington, DC
5. Highway Capacity Manual (1997) Special report 209. Transportation Research Board, Washington, DC
6. Highway Capacity Manual (2000) Special report 209. Transportation Research Board, Washington, DC
7. Highway Capacity Manual (2010) 5th edn. Transportation Research Board, Washington DC
8. The Highway Capacity Manual (2016) A guide for multimodal mobility analysis, 6th edn. Transportation Research Board, Washington, DC
9. Roess RP, Prassas ES, McShane WR (2019) Traffic engineering, 5th edn. Pearson-Prentice Hall, Upper Saddle River, NJ
10. Roess et al. (1989) Levels of service in shared-permissive left-turn lane groups, final report. Transportation Training and Research and Center
11. Prassas E, Roess R (1993) The left turn adjustment for permitted turns from shared lane groups: another look. Transportation research record 1398. Washington DC
12. Courage K (1992) Unpublished memo to the Signalized Intersection Subcommittee of the HCQSC
13. Rouphail N (1992) Unpublished memo to the signalized Intersection subcommittee
14. Webster F, Cobbe B (1966) Traffic signals. Her Majesty's Stationary Office, London
15. Akcelik R (1981) Traffic signals: capacity and timing analysis, Australian Road Research Board report 123. Australian Road Research Board, Kew, Victoria, Australia
16. Fambro D, Chang E, Messer C (1989) Effects of the quality of traffic signal progression on delay. NCHRP project 3-28C
17. Fambro DB, Rouphail NM, Sloup PR, Daniel JR, Li J, Anwar M, Engelbrecht RJ (1996) Highway capacity revisions for chapters 9 and 11. Report FHWA-RD-96-088. FHWA, U.S. Department of Transportation
18. Fambro D, Rouphail N Generalized delay model for signalized intersections and arterial streets
19. Transportation Research Record 1572 (1996). Transportation Research Board, Washington, DC
20. Courage K et al (1996) Capacity analysis of traffic-actuated intersections. Final report NCHRP project 3-48
21. Akcelik R (1980) Time dependent expressions for delay, stop rate, and queue length at traffic signals. Internal report AIR 367-1. Australian Road Research Board
22. Akcelik R A queue model for HCM2000. Technical note. ARRB Transport Research Ltd., Vermont South, Australia
23. Roess RP, Prassas ES, McShane WR (2011) *Traffic Engineering*, 4th edn. Pearson-Prentice Hall, Upper Saddle River, NJ

Chapter 9
The 2010 and 2016 Highway Capacity Manual Signalized Intersection Methodology

The 2010 HCM [1] *signalized intersection methodology completely changed the organization and emphasis of the chapter on isolated signalized intersections. Previous versions of the signalized intersection chapter presented a pretimed methodology, with an appendix added in 1997 containing an outline for determining an average cycle length and phase times for an actuated controller. The 2010 HCM and the 2016 HCM* [2] *are organized to present the methodology for an actuated signal, with pretimed signal phasing handled as a subset of the actuated controller methodology by setting all the recall settings to maximum.*

Beginning with the 2010 HCM, calculations can no longer be accomplished by hand due to multiple iterative layers in the methodology. Thus in 2010 worksheets were done away with, and much of the documentation of the methodology is in software code only. The underlying framework of the methodology, however, does remain the same, with the outputs continuing to be capacity, v/c ratio, delay, level of service, and queue storage ratio.

There were relatively few changes between the 2010 and 2016 versions of the methodology which are presented towards the end of this chapter.

An additional change introduced in the 2010 HCM signalized intersection chapter was that it became multimodal. The pedestrian and bicycle methodologies for the intersection, previously in a separate chapter, were added to this chapter, but are not discussed in this book.

E. S. Prassas and R. P. Roess, *The Highway Capacity Manual: A Conceptual and Research History Volume 2*, Springer Tracts on Transportation and Traffic 12,
https://doi.org/10.1007/978-3-030-34480-1_9

9.1 Key Changes to the Signalized Intersection Methodology in the 2010 HCM

The key changes to the automobile methodology for signalized intersections in 2010 were:

- The definition of a lane group changed. The analysis of a shared lane (left/through or through/right lane) is done as a separate lane group instead of being combined with through-only lanes.
- A more robust method was added for the calculation of uniform delay, d_1,
- Total lost time is no longer all assigned to the beginning of the phase, but is placed where it actually occurs, that is, start up lost time at the beginning of the phase and clearance lost time at the end.

9.1.1 The Definition of Lane Groups

The 2000 HCM defined separate lane groups for exclusive-only turn lanes (left-only, right-only) with all other lanes combined into one lane group. In 2010, these groupings are now called *movement* groups.

The 2010 HCM then added another type of lane group. Exclusive left- and right-turn lanes are still separate lane groups, but now exclusive through-only lanes are also a separate lane group and any shared lane is a separate lane group. The analysis unit is still the lane group, but now an intersection approach may have more lane groups than in previous methodologies. The possible lane groups are now:

1. Left-only exclusive lane(s)
2. Right-only exclusive lane(s)
3. Through-only exclusive lane(s)
4. Shared left/through lane
5. Shared through/right lane
6. Shared left/right lane (T-intersections).

Figure 9.1 shows some common movement and lane groups combinations. Interestingly, the HCM has moved towards adopting other well-respected signalized intersection procedures, such as the Australian [3] and Canadian [4] methodologies, both of which perform a lane-by-lane analysis.

Because any shared lane is now a separate lane group, when there is another lane group that serves one of the movements in the shared lane, a new methodology was added to determine how the vehicles divide themselves among the lane groups. For example, if an approach consists of a left/thru lane and a thru/right lane, an estimate of how many thru vehicles are in each shared lane must be calculated. This is accomplished by equalizing the volume to saturation flow rate ratios (v/s) of the lane groups that share a movement. The methodology is described later in the chapter.

Number of Lanes	Movements by Lanes	Movement Groups (MG)	Lane Groups (LG)
1	Left, thru., & right:	MG 1:	LG 1:
2	Exclusive left:	MG 1:	LG 1:
	Thru. & right:	MG 2:	LG 2:
2	Left & thru.:	MG 1:	LG 1:
	Thru. & right:		LG 2:
3	Exclusive left:	MG 1:	LG 1:
	Through:	MG 2:	LG 2:
	Thru. & right:		LG 3:

Fig. 9.1 Common movement and lane groups on a signalized intersection approach. *Source* "Highway Capacity Manual", Transportation Research Board, 2010 [1], Exhibit 18–12, pp. 18–34, used with permission of the National Academy of Science, Courtesy of National Academies Press

9.1.2 Changes To Estimating Uniform Delay, D_1

All previous versions of the HCM since 1985 calculated uniform delay by assuming that there is just one constant arrival rate during the cycle and one constant departure rate (saturation flow rate) during effective green time. The queue accumulation polygon (QAP) for such cases is always a triangle, as shown in Fig. 9.2, and uniform delay is found by calculating the area of the triangle [5].

The QAP in Fig. 9.2 accurately represents only certain lane groups, such as exclusive through lanes or an exclusive turn lane with protected-only phasing. However, if, for instance, a left turn has permitted phasing, then the saturation flow rate (departure rate) changes during different portions of the effective green time. Previous versions of the HCM ignored these changes when calculating uniform delay. The 2010 HCM has removed this restriction so that uniform delay can be better approximated for lane groups that have non-constant arrival and/or departure rates or that have multiple green times within the cycle.

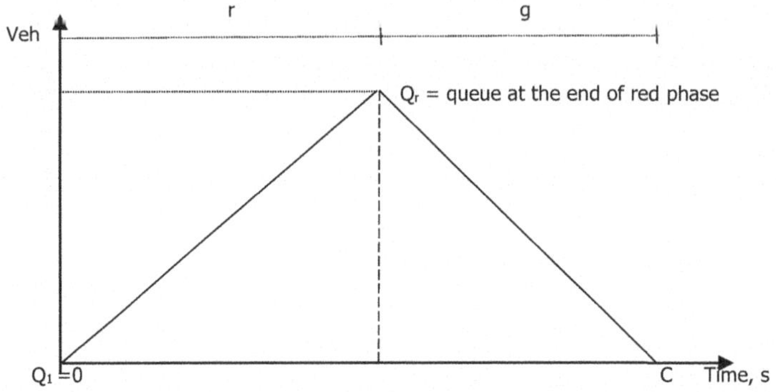

Fig. 9.2 Queue accumulation polygon for a cycle with one arrival rate and one departure rate

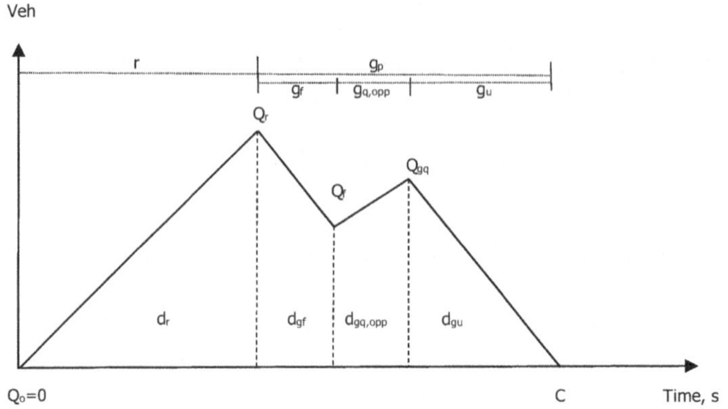

Fig. 9.3 Queue accumulation polygon for a shared left/through lane with permitted phasing

In the 2010 HCM, uniform delay is calculated with a more realistic queue accumulation polygon (QAP), using an incremental queue accumulation method (IQA) [6, 7]. The QAP created is divided into polygons for which the arrival and departure rate do not change. The area for each polygon is the calculated incremental delay. Total uniform delay for the cycle is then the sum of these incremental delays.

Figure 9.3 shows the queue accumulation diagram for a shared left/thru lane that has permitted phasing. The cycle length is divided into time intervals with constant arrival and departure rates. The triangle d_r represents the delay incurred during effective red time, r, where the saturation flow rate is zero (no departures) and the arriving vehicles accumulate to queue size Q_r. The permitted green phase time, g_p, is divided into three polygons. The polygon d_{gf} is the delay incurred during the interval of time before the first left-turning vehicle arrives at the stop line, g_f. During g_f the through vehicles move into the intersection and the queue decreases to Q_f. The polygon $d_{gq,opp}$ is the delay incurred after the first left arrives at the stop line and until the time the opposing queue clears, $g_{q,opp}$. During $g_{q,opp}$ the subject queue must wait, the saturation flow rate is zero, and thus the queue increases to Q_{gq}. The last polygon is triangle d_{gu}, which is the delay incurred during the unsaturated green time (the remaining green time after the opposing queue clears), g_u. Total uniform delay is calculated by summing the incremental delays, $d_r + d_{gf} + d_{gq,opp} + d_{gu}$. Calculating these values of the QAP polygon will be discussed in detail later in the chapter.

9.1.3 Calculating Effective Green Time

In the 2000 HCM, total lost time (start-up lost time plus clearance lost time, $l_1 + l_2$) was all assigned to the beginning of the phase. This had no effect on effective green when there was only a single phase per approach or when there were no overlapping phases, but when an approach has multiple phases, such as protected-permitted left-turn phasing, for instance, the phase that was displayed first was assumed to have an effective green time that subtracted both start-up and clearance lost time, and the phase that was displayed second received no lost time at all.

The 2010 HCM removed this simplification. For example, in a protected-permitted left-turn phase, the first phase displayed experiences start-up lost time and no clearance lost time, while the second phase displayed experiences no start-up lost time but does experience clearance lost time. This is true as long as there is no red time between the protected and permitted phases. If there is a red period between the protected and permitted phases, both the protected and permitted phases would experience start-up and clearance lost time.

9.1.4 *Peak Hour Factor,* PHF

In the 2000 HCM, *PHF* could be entered for each movement. However, entering *PHFs* for each movement analyzed the intersection for volumes that did not occur in the same fifteen minutes.

The 2010 HCM advises using actual counts for an operational analysis with the PHF set to 1.0. For other cases, the PHF is calculated for the entire intersection, using Eq. 9.1:

$$PHF = \frac{V_{60}}{4V_{15}} \tag{9.1}$$

where:

V_{60} is the vehicle count for one hour of all vehicles entering the intersection, vehs, and

V_{15} is the vehicle count of all vehicles entering the intersection for the peak 15-min within that hour.

9.1.5 *Interim Model Adjustments*

In the HCQSC summer meeting of 2012, a group of adjustments were made to the original 2010 model when a permitted left-turn movement is opposed by a multilane approach containing a shared left-turn lane. These were then published as errata to the 2010 HCM. In later sections, the original models and the interim changes are described. The change was motivated by reports from two software developers of unusually long delays when a permitted left-turn movement was opposed by a multilane approach with a shared permitted left-turn lane.

9.2 Analysis of a Pretimed Signalized Intersection

The analysis of a pretimed signal is similar to that in the 2000 HCM except for the changes described previously in Sect. 9.1. The following steps describe the 2010 HCM methodology for pretimed signalized intersections.

Step 1. Define Input Data

Required input data for a pretimed signalized intersection did not change from the 2000 HCM.

Step 2. Movement Groups and Lane Groups are Established

See Sect. 9.1.1 and Fig. 9.1 to determine movement groups and lane groups.

Step 3. Determine the Flow Rate in Each Movement and Lane Group

Step 3.1 Flow Rates for Movements Served by Only One Lane Group.

For movements that are served by only one lane group, the movement group flow rate is equal to the lane group flow rate and is found from the input volumes.

Figure 9.4 shows two approaches with the same volumes but different lane configurations. Tables 9.1 and 9.2 show the lane groups and flow rates in each lane group for Fig. 9.4a, b, respectively.

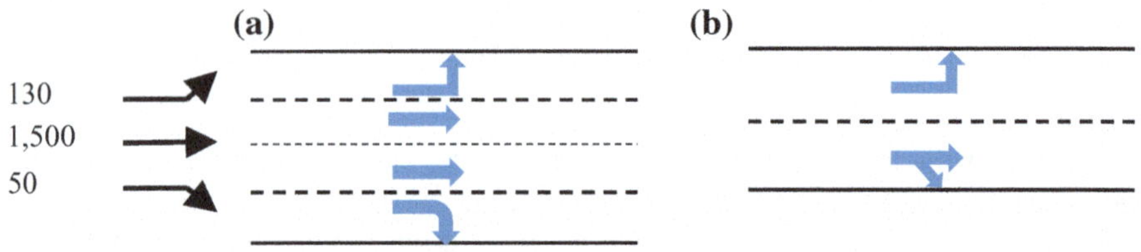

Fig. 9.4 Two example approaches where movement and lane group flow rates are the same (PHF = 1.00)

Table 9.1 Lane groups for Fig. 9.4a

Lane group	Number of lanes	Flow rate, vph
L	1	130
T	2	1500
R	1	50

Table 9.2 Lane groups for Fig. 9.4b

Lane group	Number of lanes	Flow rate, vph
L	1	130
TR	1	1550

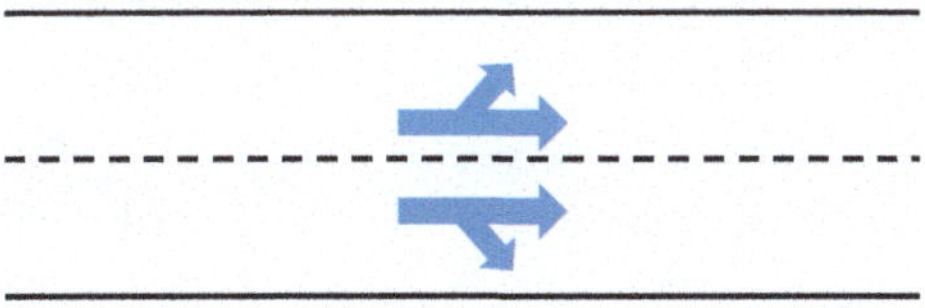

Fig. 9.5 Example approach with through movement served in two lane groups

Step 3.2 Flow Rates for Movements Served by More Than One Lane Group.

Figure 9.5 shows an example of an approach with a movement served by two lane groups. In this case, a through vehicle can choose to enter the left/through lane or the through/right lane. When one movement is serviced by two lane groups, that volume must be divided by equalizing the v/s ratios of the two lane groups [8]. This will predict the flow rate that will approximately give each lane group the same minimum service time. The methodology is an iterative procedure that will be described later in Sect. 9.2.1.

Step 4 Determine Saturation Flow Rate for Each Lane Group.

Calculation of saturation flow rate uses the same equation and factors as in the 2000 HCM, with a few changes. No new factors were added in 2010. Equation 9.2 is used to calculate saturation flow rate for through-only lanes. For right-turn lanes, the ped-bike and turn factors are handled directly in the saturation flow equation, using Eqs. 9.2 and 9.3. Note that for a right-turn lane with protected-only phasing, $f_{Rpb} = 1.00$. For a shared left-turn lane with protected-only phasing, saturation flow rate, s_{sl}, is found using Eqs. 9.2 and 9.4.

$$s_{TH} = s_o f_w f_{HV} f_g f_p f_{bb} f_a f_{LU} \tag{9.2}$$

$$s_r = \frac{s_{TH}}{1 + P_R * \left(\frac{E_R}{f_{Rpb}} - 1\right)} \tag{9.3}$$

$$s_{sl} = \frac{s_{TH}}{1 + P_L * \left(\frac{E_L}{f_{Lpb}} - 1\right)} \tag{9.4}$$

where all terms are as previously defined.

The changes in the 2010 HCM included:

- The exclusive and shared lane right-turn factors were made more general by including a default right-turn equivalency factor, $E_R = 1.18$
- The protected left-turn factors were made more general by including a default left-turn equivalency value for protected left turns, $E_{LT} = 1.05$
- The equation for the lane width adjustment factor, f_w, was replaced by a table with three width-based groups
- The right-turn adjustment factor for a single-lane approach was removed
- Base saturation flow rate for left-turn lane groups with permitted phasing is found directly using Eq. 9.5.

$$s_p = \frac{v_o e^{-v_o t_c/3600}}{1 - e^{-v_o t_f/3600}} \tag{9.5}$$

This saturation flow rate value is then adjusted by the other relevant factors (excluding f_{LT}).

- The last step of the ped-bike adjustment factor for shared lanes with permitted phasing was removed. This step adjusted the ped/bike factor for the percentage of turns in the lane. This is now handled in Eqs. 9.3 and 9.4.
- Some of the default values recommended for the various saturation flow rate factors have changed, and are listed later in Sect. 9.4.

Table 9.3 shows the adjustment factors in the 2010 HCM, with an asterisk next to those that are changed from the 2000 HCM (Table 9.4).

where:

P_{HV} percent heavy vehicles
P_g percent grade
N number of lanes in lane group
N_m number of parking movements/h
N_b number of buses/h stopping within 250 ft upstream or downstream of the intersection
v_g demand flow rate for lane group
v_{g1} demand flow rate for lane with highest demand flow rate in the lane group
E_R right-turn equivalent
E_{LT} left-turn equivalent
t_c critical headway, 4.5 s
t_f follow-up headway, 2.5 s
v_o opposing flow rate.

Step 5 Calculate Lane Group Capacities and v/c Ratios

Except for lane groups with permitted or compound phasing, capacity is found in the same manner as in the 2000 HCM, using Eqs. 9.6 and 9.7.

$$c_i = s_i * g/C \tag{9.6}$$

$$X_i = \frac{v_i}{c_i} \tag{9.7}$$

where:

c_i capacity of lane group i

Table 9.3 Saturation flow rate adjustment factors

Adjustment factor	Value
*Lane width, f_w	$f_w = 0$. For lane widths < 10.0 ft $f_w = 1.0$ for lane widths ≥ 10.0–12.9 ft $f_w = 1$. For lane widths > 12.9 ft
Heavy vehicle, f_{HV}	$f_{HV} = \frac{100}{100+P_{HV}(E_{HV}-1)}$
Grade, f_g	$f_g = 1 - \frac{P_g}{200}$
Parking, f_p	$f_p = \frac{N-0.10-\left(\frac{18N_m}{3600}\right)}{N}$
Bus blockage, f_{bb}	$f_{bb} = \frac{N-\left(\frac{14.4*N_B}{3600}\right)}{N}$
Area type, f_a	$f_a = 0.90$ for CBD $f_a = 1.0$ for all other areas
Lane utilization, f_{LU}	$f_{LU} = \frac{v_g}{(v_{g1}N)}$ or Table 7.4
*Right turns, f_{RT}	$f_{RT} = \frac{1}{1+P_R\left(\frac{E_R}{f_{Rpb}}-1\right)}$ $E_R = 1.18$
*Pedestrian and bicycle effect on right turners, f_{Rpb}	Calculate A_{pbT} from Chap. 8 $f_{Rpb} = 1.00$ *for protected rights* $f_{Rpb} = A_{pbT}$ *otherwise*
Pedestrian and bicycle effect on left turners, f_{Lpb}	Calculate A_{pbT} from Chap. 8 $f_{Lpb} = 1.00$ *for protected lefts* $f_{Lpb} = A_{pbT}$ *otherwise*
*Protected left turns, f_{LT}	$f_{LT} = \frac{1}{1+P_{LT}\left(\frac{E_{LT}}{f_{Lpb}}-1\right)}$ $E_{LT} = 1.05$
*Permitted left turn saturation flow rate, s_p	$s_p = \frac{v_o e^{-v_o t_c/3,600}}{1-e^{-v_o t_f/3,600}}$

s_i adjusted saturation flow rate of lane group i, vph
g effective green time for lane group i
C cycle length
v_i demand volume in lane group i
X_i volume to capacity ratio for lane group i.

The capacity of a lane group with permitted or compound phasing is found from the QAP. Figure 9.6 is the QAP of a shared left-turn lane opposing a multilane approach. The intervals of the permitted phase, g_p, are g_f, $g_{q,opp}$ and g_u as defined previously. Each of these intervals has a different saturation flow rate. Thus the capacity of each interval is found using the interval saturation flow rate and interval effective green time. The capacity of the lane group is the sum of the capacities of the intervals.

For convenience, the effective green times, saturation flow rates, and capacities of the intervals can be found during the QAP methodology for finding uniform delay.

Table 9.4 Default lane utilization adjustment factors

Lane group movements	Number of lanes in lane group (In)	Traffic in most heavily traveled lane (%)	Lane utilization adjustment factor f_{LU}
Exclusive through	1	100.0	1.000
	2	52.5	0.952
	3[a]	36.7	0.908
Exclusive left turn	1	100.0	1.000
	2[a]	51.5	0.971
Exclusive right turn	1	100.0	1.000
	2[a]	56.5	0.885

Note [a]If a lane group has more lanes than shown in this exhibit, it is recommended that field surveys be conducted or the smallest f_{LU} shown for that type of lane group be used
Source "Highway Capacity Manual", Transportation Research Board, Washington DC, 2010 [1], Exhibit 18–30, used with permission of the National Academy of Science, Courtesy of National Academies Press

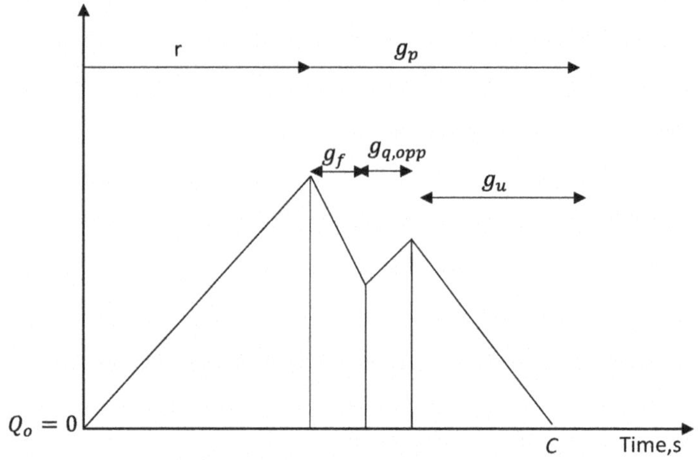

Fig. 9.6 QAP of shared left-turn lane with permitted phasing

Step 6 Calculate Uniform Delay, d_1

Step 6.1 Calculate Uniform Delay for Exclusive Lane Groups with Protected Phasing

Through-only lanes and exclusive lanes with protected only phasing have only one arrival rate and one departure rate (a simple triangle), as was shown previously in Fig. 9.2. The area of the triangle may be calculated with the equation for uniform delay in the 2000 HCM, shown in Eq. 9.8 with all terms as previously defined. This is still Webster's uniform delay [5], adjusted to reflect control delay.

$$d_1 = \frac{0.50\,C(1 - g/C)^2}{1 - [min\,(X, 1)]\,[g/C]} \tag{9.8}$$

Step 6.2 Calculate Uniform delay for All Other Lane Groups

Lane groups that have varying arrival and departure rates must use the QAP to calculate uniform delay. The methodology is also called the Incremental queue accumulation method (IQA). Using the IQA methodology for a simple triangle will give the same answer as Eq. 9.8.

The steps for performing an IQA analysis are as follows:

IQA Step 1.

Define the time intervals in the cycle where there is a change in either arrival flow rate or departure flow rate. Arrival flow rate can change due to the platooning conditions. The departure flow rate (saturation flow rate) can change, for example, when effective red changes to effective green, a protected phase changes to a permitted phase, left turns waiting for opposing queue to clear changing to the time after the opposing queue clears.

IQA Step 2.

Determine the duration of each interval, i.e., the effective green time.

IQA Step 3.

Find the saturation flow rate of each interval and the capacity of each interval.

IQA Step 4.

Find the capacity of the lane group, c_{LG}. The capacity of each interval is $c = s * g/C$, where g is found in Step 2 and s is found in Step 3. The summation of the intervals is the capacity of the lane group.

IQA Step 5.

Find arrival flow rate for each interval.

Arrival flow rates during the effective red times and effective green times are found using Eqs. 9.9 and 9.10, respectively. These values are dependent upon the percentage of vehicles that arrive during the green, P. This can be estimated from field data or by using the general arrival type categories of the 2000 HCM, as shown in Table 9.5 and Eq. 9.11.

$$v_r = \frac{(1 - P)VC}{r} \tag{9.9}$$

Table 9.5 Platoon ratios by arrival type

R_p	Arrival type
0.33	1
0.67	2
1.00	3
1.33	4
1.67	5
2.00	6

$$v_g = \frac{PVC}{g} \tag{9.10}$$

where:

v_r arrival flow rate during effective red, veh/h
v_g arrival flow rate during effective green, veh/h
P proportion of vehicles arriving on green
V average arrival flow rate, veh/h
C cycle length, s
R effective red, s
G effective green, s.

$$P = R_p * g/C \tag{9.11}$$

where:

P proportion of vehicles arriving on green
R_p ratio of the proportion of vehicles arriving on green to the proportion of the cycle that is effectively green.

IQA Step 6.

Check for oversaturated conditions. Calculate volume to capacity ratio, *X*, for the lane group by dividing the average arrival flow rate, *v*, by the lane group capacity.

$$X = \frac{V}{c_{LG}} \tag{9.12}$$

If *X* is greater than 1.0, the arrival volumes used in each interval must be adjusted because uniform delay assumes undersaturated conditions. In the 2000 HCM, the equation for uniform delay achieved this by limiting *X* to a maximum of 1.0 (as shown in Eq. 9.8). To accomplish the same thing in the IQA methodology, when *X* > 1.0, each interval's arrival volume must be divided by *X*.

$$v_{i,adj} = \frac{v_i}{X} \tag{9.13}$$

where:

$v_{i,adj}$ volume for interval i adjusted for oversaturated conditions.

Note: In the remaining steps, the variable $v_{i,adj}$ is used in place of v_i

IQA Step 7.

Find the queue at the beginning and end of each interval. For most phase types, the first interval of the cycle is assumed to have an initial queue of zero.

To find the queue at the end of the interval Eq. 9.14 is used.

$$q_{2,i} = q_{1,i} + \frac{v_i - s_i}{3600} * \Delta t_i \tag{9.14}$$

where:

$q_{2,i}$ queue at the end of interval i
$q_{1,i}$ queue at start of interval i, *where* $q_{1,i} = q_{2,i-1}$
v_i arrival flow rate during interval i (veh/h)
s_i saturation flow rate of interval i (veh/hg)
Δt_i duration of interval i.

IQA Step 8.

Calculate the incremental delay for the interval. Equation 9.15 calculates the incremental delay during the subject interval i.

$$D_i = \Delta t_i * \frac{q_{1,i} + q_{2,i}}{2} \tag{9.15}$$

where:

D_i incremental delay during interval i (veh-s).

IQA Step 9.

Steps 3 and 4 are repeated for all intervals, with the final queue of the previous interval becoming the beginning queue of the following interval.

IQA Step 10.

If when calculating the end queue of an interval, $q_{2,i}$, the value is a negative number, then the queue has cleared before the end of the green time. It is then necessary to find the point in time during the subject interval when the queue goes to zero because delay is only incurred while there is still a queue. After the queue goes to zero, vehicles arrive and immediately depart. Thus the interval is divided into two subintervals as follows:

- Subinterval 1: Time until queue dissipates during interval *i*
- Subinterval 2: the time in interval *i* after the queue clears. The queue at the beginning and end of subinterval 2 is zero. Vehicles arriving during subinterval 2 do not experience any delay.

To calculate the duration of subinterval 1, $\Delta t_{i,sub1}$, when the queue goes to zero, the number of vehicles arriving is set equal to the number of vehicles departing, as shown in Eq. 9.16. The number of arriving vehicles includes vehicles in queue at the start of the interval and those that join the queue. Equation 9.16 is then solved for $\Delta t_{i,sub1}$ as shown in Eq. 9.17.

$$q_{i,1} + \frac{V_g}{3600} * \Delta t_{i,sub1} = \frac{s_i}{3600} * \Delta t_{i,sub1} \tag{9.16}$$

$$\Delta t_{i,sub1} = \frac{3600 q_{i,1}}{s_i - V_g} \tag{9.17}$$

where:

$\Delta t_{i,sub1}$ duration of subinterval 1 in interval *i*.

IQA Step 11.

If the final queue in the cycle does not equal the initially assumed queue, then IQA Steps 6–9 must be iterated, with the initial queue at the start of next iteration equal to the final queue found in the previous iteration. The iteration continues until the end queue equals the start queue.

IQA Step 12.

When the final queue of the cycle equals the initial queue at the start of the cycle, the average uniform delay per vehicle can be found by summing the calculated incremental delays of each interval and dividing by the total number of arrivals during the cycle, using Eqs. 9.18 and 9.19.

$$d_1 = \frac{\sum_i^n D_i}{n_a} \tag{9.18}$$

$$n_a = \frac{\sum_i^n v_i}{3600} * C \tag{9.19}$$

where:

d_1 average uniform delay per vehicle, s/veh
D_i incremental delay of interval *i*, veh-s
N number of intervals
n_a arrivals in cycle, vehs.

Step 7 Effective Green Times and Saturation Flow Rates for Permitted Left Turns

The changes from the 2000 HCM in calculating the effective green time for permitted left-turn intervals include:

- A maximum value for the time before the first left-turn vehicle arrives at the stop line, $g_{f,\max}$, was added, as shown in Eq. 9.20:

$$g_{f,\max} = \frac{(1 - P_L)}{0.5P_L}\left(1 - [1 - P_L]^{0.5g_p}\right) - l_{1,p} \tag{9.20}$$

- The calculation of queue clearance time, g_q, for a single-lane approach does not use the regression equation used in the 2000 HCM, but rather uses the same equation for multilane approaches, which was found by setting the arrivals equal to the departures and solving for g_q, as shown in Eq. 9.21.

$$g_q = \frac{vC(1 - P)}{s - vC\left(\frac{P}{g}\right)} \tag{9.21}$$

where:

g_q queue clearance time, s
v arrival flow rate, vph
C cycle length, s
P proportion of vehicles arriving during the green time
S adjusted saturation flow rate.

g_q is found for all of the opposing lanes (except for an opposing shared left-turn permitted lane), and the largest $g_{q,opp}$ is used. Note: In the original 2010 HCM, all opposing lanes were evaluated. Added in an errata to the 2010 was the caveat that the opposing shared left-turn permitted lane is not considered in the calculation of $g_{q,opp}$.

- The unsaturated portion of the effective green time for the permitted phase, g_u, takes into account the start-up lost time of the opposing approach's queue clearance time:

$$g_u = G - l_{1,opp} - g_{q,opp} - l_{1,sub} + e_{sub} \tag{9.22}$$

To find the saturation flow rate of the various intervals of a permitted left- turn lane, the adjusted through-only saturation flow rate is found first, without considering the effect of turning vehicles:

$$s_{TH} = s_o f_w f_{HV} f_g f_p f_{bb} f_a \tag{9.23}$$

Table 9.6 shows the effective times for the intervals of a permitted left-turn phase. Also included in the table are the saturation flow rates of each interval.

Table 9.6 Effective times for lane groups with permitted left-turn phasing

	Effective time period	Effective green time, g (s)	Saturation flow rate, subject approach (veh/hg/ln)
1	$g_{p,sub}$ Effective green time of subject permissive phase	$G_p - l_{1,sub} + e_{sub}$	$s_p = \frac{v_{opp} e^{-v_o t_c/3,600}}{1-e^{-v_{opp} t_f/3,600}} * \prod f_i$ Where $\prod f_i$ is the product of factors affecting the lane group
2	g_f, Multi-Lane Time until first left-turn vehicle arrives at stop line on a multilane approach	$G_p e^{-0.882 LTC^{0.717}} - l_1 \geq 0$ $(\leq g_{f,\max})$	s_{TH} (Equation 9.20)
3	g_f, Single Lane Time until first left-turn vehicle arrives at stop line on a single-lane approach	$G_p e^{-0.860 LTC^{0.629}} - l_1 \geq 0$ $(\leq g_{f,\max})$	s_{TH} (Equation 9.20)
4	$^*g_{f,\max}$ Maximum value of g_f	$\frac{1-P_L}{0.5 P_L} * \left(1 - [1 - P_L]^{0.5 g_p}\right) - l_1 \geq 0$	s_{TH} (Equation 9.20)
5	$g_{q,opp}$ Queue clearance time of opposing approach when subject approach departure rate is zero waiting for opposing queue to clear	$\frac{v_{opp} C(1-P)}{s_{opp} - v_{opp} P\left(\frac{C}{g}\right)}$	0 Saturation flow rate of subject left turn while the opposing queue clears on an opposing multilane approach
6	$g_{u,sub}$ Unsaturated green time of the subject permitted left-turn phase, shared Lane	$G_{perm} - l_{1,opp} - g_{q,opp} - l_{1,sub} + e_{sub}$	$\frac{s_{TH}}{1+P_L\left(\frac{E_{L1}}{f_{Lpb}} - 1\right)}$ $E_{L1} = \frac{s_{TH}}{s_p}$
7	$g_{u,sub}$ Unsaturated green time of the permitted phase, exclusive lane	$G_{perm} - l_{1,sub} + e_{sub} - l_{1,opp} - g_{q,opp}$	s_p

(continued)

Table 9.6 (continued)

	Effective time period	Effective green time, g (s)	Saturation flow rate, subject approach (veh/hg/ln)
8	g_{diff} If the opposing approach is a shared single lane, g_{diff} is the time after the first left arrives at the stop line until the opposing queue clears when opposing left turns open gaps for each other	$g_{q,opp} - g_f \geq 0$	$\frac{s_{TH}}{1+P_L\left(\frac{E_{L2}}{f_{Lpb}}-1\right)}$ $E_{L2} = \frac{1-(1-P_{lto})^{n_q}}{P_{lto}} \geq 1.0$ $n_q = 0.5(g_{diff})$
9	r Effective red time	$C - g_{p,sub}$	0

where:

g effective green time
sub subject approach
G actual green time
l_1 start-up lost time
$l_{1,sub}$ start-up lost time of subject direction
$l_{1,opp}$ start-up lost time of opposing direction
e_{sub} extension of effective green time in subject direction
t_c critical headway, 4.5 s
t_f follow-up headway = 2.5 s. Note: In the original 2010, this value was 4.5 s when the subject left turn was served from a shared lane. In the errata to 2010, the follow-up headway is defaulted to 2.5 s for all cases.
$\prod f_i$ product of all adjustment factors to base saturation flow rate for a through-only lane group
P proportion of vehicles arriving during green
v_{opp} opposing volume per lane.

In the errata to the 2010 HCM, text was added giving the user the option to not include opposing right-turn volume made from an exclusive right-turn lane. Not including right-turn volume from an exclusive lane should be based on the engineer's knowledge of the local driver behavior and the geometry of the intersection. Not including the right-turn volume implies that the right-turn lane does not influence the gap acceptance of the permitted left turn. In this case, if the opposing approach consists of two lanes (one of which is the right-turn only lane), the opposing approach is considered to be a one-lane approach and therefore turns may occur during the period g_{diff}.

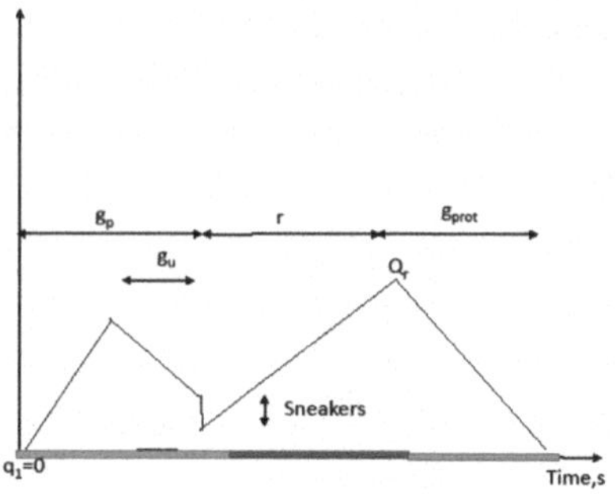

Fig. 9.7 QAP for an exclusive left turn lane with Lead protected-permitted phasing

Step 8 Finding Control Delay

For each lane group, a queue accumulation polygon (QAP) is created that divides the cycle length into time intervals that have a constant arrival rate and constant departure rate, as discussed earlier in Sect. 9.1.2. Each interval will have the shape of either a triangle or trapezoid and the incremental delay of each interval is found by calculating its area.

After the cycle is divided into intervals with constant arrival and departure rates, the queue at the beginning and end of the interval is found. For most cases, the first interval analyzed is the effective red time, where the queue is assumed to start at zero and grows during effective red. However, for a left-turn lane (either exclusive or shared) that has a leading protected phase and then permitted phasing, there will usually be a queue of vehicles when the leading protected phase begins, which clears during the protected phase. Thus the queue is assumed to be zero at the beginning of the permitted phase. For these cases then, the QAP starts at the beginning of the permitted left turn phase, as in Fig. 9.7, for an exclusive left-turn lane with leading protected-permitted phasing.

Control delay is found using Eq. 9.24:

$$d = d_1 + d_2 + d_3 \tag{9.24}$$

where:

d average control delay, s/veh,
d_1 uniform delay found from IQA method, s/veh,
d_2 oversaturated and random delay,
d_3 residual delay due to an initial queue.

Uniform delay is the only change in the delay calculations from the 2000 HCM. Level of service breakpoints have not changed from the 2000 HCM and are shown again here in Table 9.7:

Table 9.7 Level of service criteria for vehicles

Control delay (s/veh)	LOS (when v/c $\leq$ 1.0)*
$\leq$10	A
>10–20	B
>20–35	C
>35–55	D
>55–80	E
>80	F

*When v/c > 1.0, LOS = F

9.2.1 Movements Served by More Than One Lane Group

In Step 2.2 (determining movement and lane group flow rate), when a movement group consists of more than one lane group, at least one of the lane groups will be a shared lane. The volume that is served in the shared lane and another lane of the movement group must be divided between those lane groups. This is done by equalizing the v/s ratios of the subject lane groups.

For example, a two-lane approach with a left/thru lane and a thru/right lane; or a three-lane approach with an exclusive left-turn lane, a through-only lane, and a shared through/right lane would need to use this methodology to divide the through vehicles between the two lane groups that serve through vehicles.

To begin the methodology, an assumption of lane volumes and turning volume in the shared lane(s) must be made.

The initial flow rate assumed in the shared lane (either a left/thru or thru/right lane or both) is the total approach flow rate equally divided among all the lanes in the movement group. Additionally, it is initially assumed that the flow rate of turning vehicles in the shared lane is zero (see Eqs. 9.25–9.28). These initial assumptions seem counter-intuitive since the shared lanes include the average of all approach vehicles even when, for instance, a shared left/through lane could not possibly serve right-turning vehicles. These initial values, however, will change quickly in the following iterations, as vehicles are assigned to their proper lanes.

$$v_{sl} = (v_l + v_t + v_r)/(N_{sl} + N_t + N_{sr}) \tag{9.25}$$

$$v_{sr} = (v_l + v_t + v_r)/(N_{sl} + N_t + N_{sr}) \tag{9.26}$$

$$v_{sl,lt} = 0 \tag{9.27}$$

$$v_{sr,rt} = 0 \tag{9.28}$$

where:

v_{sl} flow rate in the shared-left lane

v_{sr} flow rate in the shared-right lane
N_{sl} number of shared-left lane (= 1 or 0)
N_{sr} number of shared-right lane (= 1 or 0)
N_t number of thru only lanes
$v_{sl,lt}$ left-turn volume in the shared-left lane
$v_{sr,rt}$ right-turn volume in the shared-right lane.

From these initial assumptions, the following steps are iterated until the *v/s* ratios are equal.

Equalize v/s Step 1. Estimate the flow rate in any exclusive lane(s) in the movement group, using Eqs. 9.29, 9.30, and 9.31.

Equation 9.29 calculates the left turn volume in the exclusive left-only lane(s), if one exists. Any left-turn volume assigned to the shared lane is subtracted from the total left-turn volume. The remaining left-turn volume is assigned to the exclusive lane(s), divided equally if more than one exclusive left-turn lane exists.

$$v_L = \frac{v_{lt} - v_{sl,LT}}{N_L} \geq 0 \tag{9.29}$$

v_L flow rate in exclusive left-turn lane(s), veh/h/ln
v_{lt} total left-turn flow rate, veh/h
$v_{sl,lt}$ left-turn flow rate in the shared left lane, veh/h
N_L Number of exclusive left-turn lanes.

In the first iteration, all left-turn volume is assigned to the exclusive left-turn lane(s).

Equation 9.30 does a similar calculation for the exclusive right-turn lane(s):

$$v_R = \frac{v_{rt} - v_{sr,rt}}{N_R} \geq 0 \tag{9.30}$$

v_R flow rate per lane in exclusive right-turn lane(s), veh/h/ln
v_{rt} total right-turn flow rate, veh/h
$v_{sr,rt}$ right-turn flow rate in the shared right lane, veh/h
N_R number of exclusive right-turn lanes.

Equation 9.31 estimates the thru volume in any exclusive through lanes:

$$v_T = \frac{V_{TH} - \left(v_{sl} - v_{sl,lt}\right) - \left(v_{sr} - v_{sr,rt}\right)}{N_T} \tag{9.31}$$

v_T flow rate per lane in exclusive thru lane(s), vphpl
v_{sl} total flow rate in shared left/thru lane, vph
$v_{sl.lt}$ left-turn flow rate in shared left/thru lane, vph
v_{sr} total flow rate in shared thru/right lane, vph

$v_{sr,rt}$ right-turn flow rate in shared thru/right lane, vph
v_{TH} total input thru flow rate, vph
N_T Number of exclusive thru lanes
$(v_{sl} - v_{sl,lt})$ volume in shared/left lane minus left-turn volume assigned to that shared lane = thru volume in shared lane
$(v_{sr} - v_{sr,rt})$ volume in shared/right lane minus right-turn volume in that shared lane = thru volume in shared lane.

Equalize v/s Step 2. Compute the proportion of turning vehicles in the shared lane(s)

Equations 9.32 and 9.33 compute the proportion of turning vehicles in the shared lane:

$$P_L = \frac{v_{sl,lt}}{v_{sl}} \leq 1.0 \tag{9.32}$$

$$P_R = \frac{v_{sr,rt}}{v_{sr}} \leq 1.0 \tag{9.33}$$

P_L proportion of lefts in shared left/thru lane
P_R proportion of rights in shared thru/right lane.

The proportion of turns in the shared lane will be used in the next step when finding saturation flow rate for each lane group.

Equalize v/s Step 3. Determine the saturation flow rate for the lane groups.

As part of the iteration process to equalize v/s ratios, the saturation flow rate must be calculated. Saturation flow rate uses the left-turn thru-car equivalences, E_{L1} and E_{L2}. However, as part of this methodology to equalize the v/s ratios, there are adjustments made to these equivalency values, which are used only within this methodology.

These adjustments are made to account for the probability of a driver making a lane change based on the flow rate in each lane group. The adjustment accounts for two factors: (1) the probability of a lane change being low when flow ratios are high due to the driver having less opportunity to find a sufficient gap, and (2) the probability of changing lanes being low at low flow ratios because drivers have less motivation to make a change.

The left-turn equivalencies, E_{L1} and E_{L2}, found in Table 9.6 are thus modified using Eqs. 9.34–9.37.

$$E_{L1,m} = \left(\frac{E_{L1}}{f_{Lpb}} - 1\right) P_{lc} + 1 \tag{9.34}$$

$$E_{L2,m} = \left(\frac{E_{L2}}{f_{Lpb}} - 1\right) P_{lc} + 1 \tag{9.35}$$

with:

$$P_{lc} = 1 - \left(\left[2\frac{v_{app}}{s_{lc}}\right] - 1\right)^2 \geq 0.0 \tag{9.36}$$

$$v_{app} = \frac{v_{lt} + v_{th} + v_{rt}}{N_{sl} + N_t + N_{sr}} \tag{9.37}$$

P_{lc} probability of a lane change
v_{app} total approach volume per lane, *veh/h/ln*, upstream of any exclusive turn bays
s_{lc} maximum saturation flow rate at which a lane change can be made, veh/h/ln, where $s_{lc} = 3600/\mathrm{t_{lc}}$
t_{lc} minimum gap required for a lane change is defaulted to 3.7 s
v_{lt} left-turn volume
v_{th} through volume
v_{rt} right-turn volume.

Notice that the numerator of v_{app} consists of the total input left, through, and right volume independent of what lane group they will be in. The denominator, on the other hand, consists of shared lanes and exclusive thru lanes only. This is because lane changes occur upstream before vehicles enter any exclusive left- or right-turn bay.

The saturation flow rate for a lane group with a permitted left-turn phase is found using Eq. 9.38. The portions of the green time used in Eq. 9.38 are those found in Table 9.6.

$$s_{sl} = \frac{s_{th}}{g_p}\left(g_f + \frac{g_{diff}}{1 + P_L(E_{L2,m} - 1)} + \frac{min(g_p - g_f, g_u)}{1 + P_L(E_{L1,m} - 1)} + \frac{3,600n_s^*}{s_{th}}\right) \tag{9.38}$$

with:

$$n_s^* = \frac{P_L}{1 - P_L}\left(1 - P_L^{n_s}\right)$$

$n_s^* = n_s$ for exclusive left-turn lane
$E_{L2,m}$ modified thru-car equivalent for a left-turn vehicle opposed by a single-lane approach, during g_{diff}
$E_{L1,m}$ modified thru-car equivalent during unsaturated green time (after opposing queue clears) and after the first left-turner arrives at the stop line
n_s^* expected number of sneakers per cycle in a shared lane
n_s number of sneakers per cycle in an exclusive lane (defaulted to 2).

Note about Eq. 9.38: In the original 2010 HCM, the third term for sneakers was not included. Instead there was a minimum value of s_{sl} set based on the sneakers, where $s_{sl,\min} = (1 + P_L) * (3600/g_p)$. In the errata to the 2010 HCM, the sneaker term was added to the equation.

When the lane group has protected left-turn phasing, the modified thru-car equivalent for the protected left-turn is found using Eq. 9.39.

$$E_{L,m} = (E_L - 1)P_{lc} + 1 \tag{9.39}$$

$E_{L,m}$ = modified thru-car equivalent for a protected left-turn vehicle, which will replace E_L in Table 9.6 to calculate saturation flow rate during the protected phase.

The modified thru-car equivalent for right-turning volume, $E_{R,m}$, found using Eq. 9.40, is used to replace E_R when calculating saturation flow rate for lane groups containing right-turn vehicles in Table 9.3.

$$E_{R,m} = \left(\frac{E_R}{f_{Rpb}} - 1\right)P_{lc} + 1 \tag{9.40}$$

Equalize v/s Step 4. Calculate flow ratio

The combined flow ratio is calculated using Eq. 9.41.

$$y^* = \frac{v_l N_l + v_{sl} N_{sl} + v_t N_t + v_{sr} N_{sr} + v_r N_r}{s_l N_l + s_{sl} N_{sl} + s_t N_t + s_{sr} N_{sr} + s_r N_r} \tag{9.41}$$

y^* combined flow ratio.

Equation 9.41 is a generalized form of the equation. Lane groups that are not part of the movement group for which the *v/s* ratios are being equalized will not be part of the equation. For example, if an approach has a shared left/thru lane, a thru-only lane, and an exclusive right-only lane, then the numerator will not include the volume in the exclusive right lane, $v_r N_r$, and the denominator will not include saturation flow rate of the exclusive right lane, $s_r N_r$. This is because the v/s ratios to be equalized are only those of the left/thru shared lane and the thru-only lane.

This equation will also change with the iterations if a defacto left- or right-turn lane is determined to exist ($P_L > 1, P_R > 1$).

Equalize v/s Step 5. Compute revised flow rates

For each lane group included in the calculation of step 4, the saturation flow rate of the lane group is multiplied by the combined flow ratio, y*, to get the volume that will equalize the v/s ratios of those lane groups, as shown in Eq. 9.42.

$$v_i = s_i y^* \tag{9.42}$$

The turn-movement flow rate in the shared lane(s) is computed by subtracting volume in the exclusive lane(s) from the total turn volume using Eqs. 9.43 and 9.44.

$$v_{sl,lt} = v_{LT} - v_L \geq 0.0 \tag{9.43}$$

$$v_{sr,rt} = v_{RT} - v_R \geq 0.0 \tag{9.44}$$

v_{LT} total left-turn flow rate

v_L left-turn flow rate in exclusive left-turn lane
v_{RT} total right-turn flow rate
v_R right-turn flow rate in exclusive right-turn lane.

Equalize v/s Step 6. Compare start and end flow rates of current iteration

The calculated total flow rate(s) in the shared lane(s) and calculated turning flow rate in that shared lane is compared with the flow rates used in step 1. If the difference is greater than 0.1, then another iteration is performed. This comparison can be made using Eq. 9.45.

$$Error = \sqrt[2]{(v_{sl,lt} - v_{sl.lt.old})^2 + (v_{sl} - v_{sl.old})^2 + (v_{sr,rt} - v_{sr.rt.old})^2 + (v_{sr} - v_{sr.old})^2} \quad (9.45)$$

If Error > 0.1, another iteration is performed, with the final values computed in Step 5 used as the inputs into step 1. The iteration process continues until Error found using Eq. 9.45 is less than 0.1.

When the iteration is complete, the flow rates that equalized the v/s ratios of the subject lane groups are used in the remainder of the HCM methodology. Continue the methodology at Step 3, finding saturation flow rate, using E_L and/or E_R instead of the modified equivalency values used herein.

9.3 Methodology for Actuated Signals

The steps in the signalized intersection methodology were illustrated using the "simplified" case of a pretimed signal, for which signal parameters are known. The base case for the 2010 and 2016 HCMs, however, is an actuated signal. This section discusses the additional layers of complexity that actuated signals introduce.

When a signalized intersection is fully actuated, the phase durations change in response to the random variation in the flow rate approaching the intersection. The methodology in the HCM finds average phase durations and cycle length, given the volumes that are being served in the time period being analyzed. The process is iterative, using the settings on the actuated controller (including minimum and maximum green times and passage time, among others) and the inputs needed for a pretimed analysis. The methodology begins using the input maximum green times as the starting green-phase durations and calculates the average phase durations based on the input volumes. The calculated average phase durations are then used as the initial times in the next iteration. This iteration process continues until the computed green times equal the initial green times.

An actuated signal phase consists of four intervals: the start-up lost time (an input value), the queue clearance time, the amount of time the phase is extended by randomly arriving vehicles, and the clearance time (yellow plus allred, also an input value). The phase duration can be represented by Eq. 9.46.

$$D_p = l_1 + g_q + g_e + Y \quad (9.46)$$

where:

D_p an actuated signal phase duration
l_1 start-up lost time
g_q time to clear queue
g_e time the green is extended after the queue clears
Y clearance time (yellow plus all-red times).

In order to estimate the average phase duration, the queue clearance time is calculated using the same equation as in the pretimed methodology, and shown again here in Eq. 9.47. Then the probability of a call to extend the green is determined along with how long the phase is extended. The green extension time depends on whether there is a gap in the traffic (called gapping out) or until the maximum time is reached (called maxing out).

$$g_q = \frac{vC(1 - P)}{\frac{s}{3600} - vC\left(\frac{P}{g}\right)} \quad (9.47)$$

The methodology is iterative because many of the intermediate values, such as the queue clearance time, are based on knowing the green and red times for that phase.

Figure 9.8 shows a dual ring phase diagram of an actuated controller that has eight possible phases. These phase numbers will be used later in the chapter.

The inputs needed for a fully actuated signalized intersection include all the inputs needed for a pretimed methodology, plus the following inputs specific to each phase of the actuated controller.

(a) Maximum Green Time

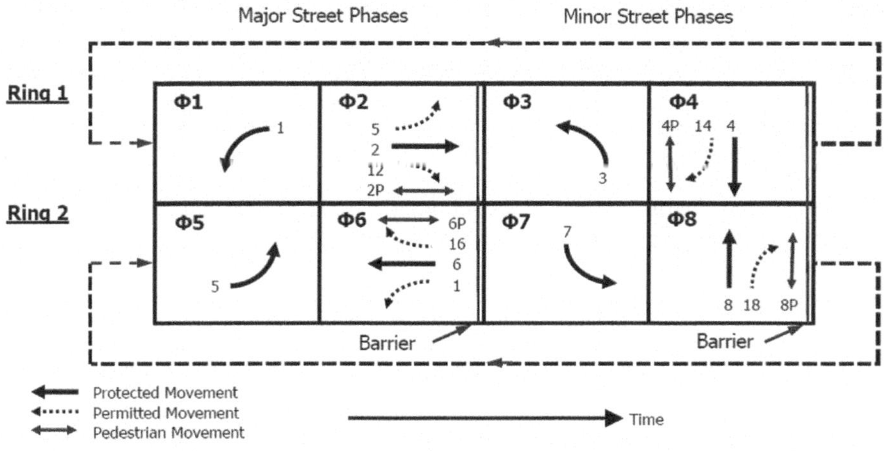

Fig. 9.8 Dual Ring Diagram. *Source* "Highway Capacity Manual: A Guide for Multimodal Mobility Analysis", Transportation Research Board, [2], Exhibit 19–2, pp. 19–9, 2016, used with permission of the National Academy of Science, Courtesy of National Academies Press

(b) Minimum Green Time
(c) Left-turn mode (protected, permitted, protected/permitted, lead or lag)
(d) Yellow and Al-Red Times
(e) Passage Time
(f) Recall Setting (no, min, max)
(g) Dual entry (yes, no)
(h) Enable Simultaneous Gap Out for Phase Group 1,2,5,6 (yes, no) and/or Phase Group 3,4,7,8 (yes, no)
(i) Is there a complementary protected right-turn green arrow with the left-turn phase? (yes, no).

The steps involved in finding the average phase times of a fully actuated signal are described below in steps 1 through 16. After the average phase times are computed, a complete pretimed analysis is done with the final computed average phase times.

Actuated Phase Times Step 1.

Starting Green Phase Durations. In the first iteration, the maximum green time inputs are used to calculate initial cycle length and estimate average phase durations. The calculated average phase durations are then used in the following iteration as the initial timings until the initial phase times and estimated phase times converge. Note that barrier phases must always end at same time. So phases 2 and 6 (as well as 4 and 8) must end together. If the yellow + all-red are not equal, then the larger of the two will be used.

Actuated Phase Times Step 2.

Perform Pretimed methodology steps up to but not including delay calculations (determine movement groups, lane groups, flow rates, saturation flow rates, and queue accumulation polygons). As part of the queue accumulation polygon, calculate beginning and ending queues using pretimed methodology IQA steps. When using the IQA methodology, as stated previously, delay does not need to be calculated, but the queue size at the beginning of each interval of the QAP is needed.

Actuated Phase Times Step 3.

Calculate queue clearance times. For each interval of the QAP, calculate the time to clear the queues calculated in actuated Step 3, using Eq. 9.45.

Actuated Phase Times Step 4.

Calculate the average rate of calls made from the detector to the controller to extend the green time. This call rate will depend on the arrival flow rate in all lane groups served by the phase, and is called the flow rate parameter for the phase. It is calculated using Eqs. 9.48, 9.49, and 9.50.

$$\lambda^* = \sum_{i=1}^{m} \lambda_i \tag{9.48}$$

$$\lambda = \frac{\varphi_i v_i}{1 - \Delta_i v_i} \tag{9.49}$$

$$\varphi_i = e^{-b_i \Delta_i v_i} \tag{9.50}$$

where:

λ^* call rate to extend phase, which is represented by the flow rate parameter for the subject actuated phase, veh/s
λ_I flow rate parameter for lane group i
M number of lane groups served by the subject actuated phase
v_i arrival flow rate for lane group in veh/s
Δ_I headway of a bunched vehicle stream in lane group i, ($\Delta = 1.5$ s for a single-lane lane group; $\Delta = 0.5$ s for multilane lane groups)
b_i bunching factor for lane group i ($b = 0.6$, 0.5, or 0.8 for lane groups with 1, 2, or ≥ 3 lanes, respectively).

Actuated Phase Times Step 5.

Calculate Maximum Allowable Headway (MAH). The MAH is the amount of time (or gap) that will retain the green indication. First, the MAH is calculated for each lane group served in the phase, then a combined MAH for the phase is computed when a phase serves more than one lane group. The methodology for determining MAH assumes that there is one detector per lane at the stop line, and that the inputs for detector length and mode for the through movements are the same as for any shared lane that contains a through movement.

For a pulse detector the MAH is equal to the passage time (PT).

For a presence detector, the computations for MAH are based on the relationship between the passage time (PT), the length of a vehicle (L_v), the detector length (L_{ds}), average vehicle speed (S_a) as shown in Fig. 9.9.

From Fig. 9.9, *MAH* can be calculated from Eq. 9.51.

$$MAH = PT + \frac{L_{ds} + L_v}{1.47 * S_a} \tag{9.51}$$

The length of the vehicle is calculated as the weighted average based on the percent heavy vehicles (P_{HV}) in the traffic stream, as follows:

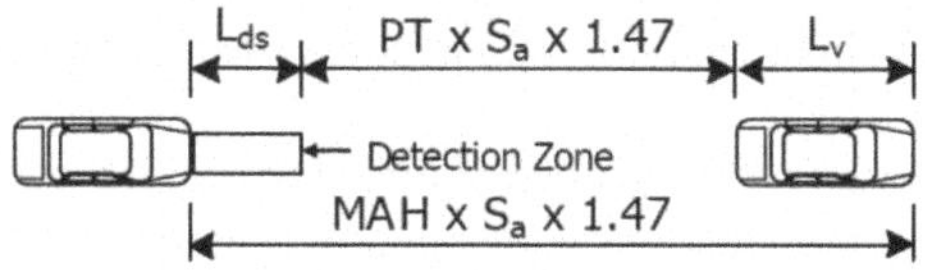

Fig. 9.9 Relationship among passage time, detector and vehicle length, vehicle speed, and maximum allowable gap. *Source* "Highway Capacity Manual: A Guide for Multimodal Mobility Analysis", Transportation Research Board, [2], Exhibit 31–2, pp. 31–8, 2016, used with permission of the National Academy of Science, Courtesy of National Academies Press

$$L_v = L_{pc}(1 - 0.01P_{HV}) + 0.01L_{HV}P_{HV} - D_{sv} \tag{9.52}$$

where:

L_v length of the vehicle detected
L_{pc} length of a passenger car (25 ft)
L_{HV} length of a heavy vehicle (45 ft)
D_{sv} distance between stored vehicles (8 ft).

The average speed on the intersection approach can be estimated from the posted speed limit (S_p) using Eq. 9.53.

$$S_a = 0.90 * \left(25.6 + 0.47S_p\right) \tag{9.53}$$

The general form of the equation shown in Eq. 9.51 is modified for different types of lane groups to account for things such as the longer headways of turning vehicles. The equations for all of the modifications for different lane group types are shown in Appendix of this chapter.

Actuated Phase Times Step 6.

Calculate the average number of extensions before the phase maxes out. This is done for each phase using Eq. 9.54.

$$n = v^*\left[G_{\max} - \left(g_q + l_1\right)\right] \geq 0 \tag{9.54}$$

where:

$v*$ arrival flow rate for the phase, which is the sum of the arrival flow rates of all lane groups served by the phase, veh/s.

Actuated Phase Times Step 7.

Calculate the probability of the green indication being extended by vehicles arriving after the queue clears.

$$p = 1 - \varphi^* e^{-\lambda^*(MAH^* - \Delta^*)} \tag{9.55}$$

where:

p probability of a vehicle arriving at a headway less than the MAH, which will extend the green
$\phi*$ combined proportion of unbunched vehicles in the phase $= e^{-\sum_{i=1}^{m} b_i \Delta_i v_i}$
m number of lane groups served by the subject actuated phase
v_i arrival flow rate for lane group in veh/s
b_i bunching factor for lane group i ($b = 0.6$, 0.5, or 0.8 for lane groups with 1, 2, or ≥ 3 lanes, respectively)

Δ_I headway of a bunched vehicle stream in lane group i, ($\Delta = 1.5$ s for a single-lane lane group; $\Delta = 0.5$ s for multilane lane groups)

$\Delta *$ equivalent headway of the bunched vehicle of all lane groups served by the phase $= \frac{\sum_{i=1}^{m} \lambda_i \Delta_i}{\lambda^*}$.

Actuated Phase Times Step 8.

Calculate the average time that the green is extended (g_e) for each phase, which is based on the probability of a vehicle arriving during the MAH.

$$g_e = \frac{p^2(1 - p^n)}{v^*(1 - p)} \tag{9.56}$$

Actuated Phase Times Step 9.

Calculate the call rate to activate a phase. This is dependent on the arrival rate in movements per second of all movements served by the phase. Since both vehicles and pedestrians can call a phase, a separate call rate is computed for each.

The vehicular call rate (q_v^*) is just the sum of the arrival flows for all vehicles served by the phase, in vehicles per second. Note that if dual entry is enabled, then volume for phase 5 and 6 would also call phase 2, and thus the volumes for phases 5 and 6 would be added to that for phase 2.

Similarly, the pedestrian call rate (q_p^*) is computed as the sum of all pedestrian arrival rates served by the phase, in pedestrians per second. And as with vehicles, if dual entry is enabled, then the pedestrian arrival rate (peds/s) on the opposing through movement are added to the subject through movement.

Actuated Phase Times Step 10.

Calculate the probability of a phase being called. If the phase recall is on, then the probability of the phase being called is 1.0. If the phase recall is set to off, then use Eqs. 9.57, 9.58, and 9.59.

$$p_c = p_v\left(1 - p_p\right) + p_p(1 - p_v) + p_v p_p \tag{9.57}$$

$$p_v = 1 - e^{-q_v^* C} \tag{9.58}$$

$$p_p = 1 - e^{-q_p^* 0.51C} \tag{9.59}$$

where:

p_c probability there is a call for the subject phase

p_v probability there is a call from a vehicle

p_p probability there is a call from a pedestrian

0.51 probability of a pedestrian pressing the pedestrian button to activate the phase.

Actuated Phase Times Step 12.

Calculate unbalanced green duration (G_{unbal}). This first calculation of the green duration does not take into account the ring structure and its constraints and will be adjusted in Step 14.

$$G_{unbal} = G_{VehCall} p_v (1 - p_p) + G_{PedCall} p_p (1 - p_v) + \max(G_{VehCall}, G_{PedCall}) p_v p_p \quad (9.60)$$

$$G_{VehCall} = \max(l_1 + g_q + g_e, G_{min}) \quad (9.61)$$

$$G_{PedCall} = Walk + PC \quad (9.62)$$

where:

G_{unbal}	unbalanced green time for a phase, s
$G_{VehCall}$	average green time when a phase is called by a vehicle, s
$G_{PedCall}$	average green time when a phase is called by a pedestrian, s
G_{min}	minimum green time setting input by user, s
Walk	pedestrian walk time setting, s
PC	pedestrian clearance time setting, s.

Note that if maximum recall is set to on, then $G_{unbal} = G_{\max}$. Pedestrians cannot call protected left-turn phases so p_p would be set to zero when calculations are for a left-turn phase. Rest-in-walk is assumed to not be enabled. A shared left-through lane with protected-permitted phasing has its green interval time set to $G_{\min}$.

Actuated Phase Times Step 13.

Calculate unbalanced phase duration (D_{up}) by adding the yellow plus all-red time to the unbalanced green calculated in the previous step.

$$D_{up} = G_{unbal} + y + ar \quad (9.63)$$

Actuated Phase Times Step 14.

Calculate actual average phase duration. For a sequence of phases within barriers, phase *a* is the phase that occurs first, followed by phase *b*. The duration of phase *a* is:

$$D_a = D_{up,a} \quad (9.64)$$

For the major street phases (phase numbers 1, 2, 5, and 6), use Eq. 9.65. For the minor street phases (phase numbers 3, 4, 7, and 8), use Eq. 9.66.

$$D_{p,b} = \max(D_{up,1} + D_{up,2}, D_{up,5} + D_{up,6}) - D_{p,a} \quad (9.65)$$

$$D_{p,b} = \max\left(D_{up,3} + D_{up,4}, D_{up,7} + D_{up,8}\right) - D_{p,a} \tag{9.66}$$

where:

$D_{p,b}$ duration of phase b, the lag phase of the pair
$D_{p,a}$ duration of phase a, the lead phase of the pair
$D_{up,i}$ unbalanced phase duration for phase i.

Actuated Phase Times Step 15.

Calculate green interval duration by subtracting yellow and all-red time.

$$G = D_p - y - ar \tag{9.67}$$

Actuated Phase Times Step 16.

Compare calculated green times of step 15 with green times used in step 1. If difference ≤ 0.1 s, then continue to Step 17, otherwise return to step 1, using green times calculated in Step 15 as new initial green times.

Actuated Step 17.

Do a complete pretimed methodology using final computed average phase times.

Actuated Step 18.

Calculate the probability of a phase maxing out. This step provides an indicator of how well signal is working. If a phase maxes out too often, then it most likely means not enough capacity for that phase.

$$p_x = p^{n_x} \tag{9.68}$$

$$n_s = \frac{G_{\max} - MAH^* - \left(g_q + l_1\right)}{h} \tag{9.69}$$

$$h = \frac{\Delta^* + \left(\frac{\varphi^*}{\lambda^*}\right) - (MAH^* + \left[\frac{1}{\lambda^*}\right]\varphi^* e^{-\lambda^*(MAH^* - \Delta^*)}}{1 - \varphi^* e^{-\lambda^*(MAH^* - \Delta^*)}} \tag{9.70}$$

where:

p_x probability of a phase maxing out
h average call headway for calls with headways less than MAH^*
n_x number of calls needed to extend to $G_{\max}$.

Table 9.8 Default values for number of parking maneuvers

Street type	Number of spaces within 250 feet	Parking time limit, hrs	Turnover rate, *vph*	Maneuvers per hour
Two-way	10	1	1.0	16
		2	0.5	8
One-way	20	1	1.0	32
		2	0.5	16

Source "Highway Capacity Manual: A Guide for Multimodal Mobility Analysis", Transportation Research Board, 2016 [2], Exhibit 18–31, pp. 18–78, used with permission of the National Academy of Science, Courtesy of National Academies Press

9.4 Default Values in the 2010 HCM

There were some changes to recommended default values in the 2010 HCM:

1. The base saturation flow rate default number of 1900 *pcphpl* is still recommended for metropolitan areas with a population greater than 250,000, but for other locations, the default base saturation flow rate is 1750 *pcphpl.*
2. The default value for percent heavy vehicles was changed from 2 to 3%
3. The default for bus stopping rate is still zero if no buses are expected to stop. If buses are expected to stop, the default for a CBD is 12 buses/h, otherwise, the default is 2 buses/hr.
4. Defaults for the number of parking movements per hour are shown in Table 9.8

9.5 Additional Changes in the 2016 HCM

With the publication of the 2016 HCM [2] came a name change for the entire document. It is now called the "Highway Capacity Manual, Sixth Edition: A Guide for Multimodal Mobility Analysis." In the signalized intersection chapter, there were relatively few changes made. The following sections describes the key changes.

9.5.1 Additional Adjustment Factors to Base Saturation Flow Rate

Three new factors were added to adjust the base saturation flow rate. The new equation is shown in Eq. 9.71

$$s = s_o f_w f_{HV} f_g f_p f_{bb} f_a f_{LU} f_{RT} f_{LT} f_{Lpb} f_{Rpb} f_{wz} f_{ms} f_{sp} \tag{9.71}$$

1. **Work Zone Adjustment Factor, f_{wz}**

The work zone factor accounts for the presence of a work zone within 250 feet of the intersection. The factor was based on research done by Bonneson and Nguyen [9]. The factor adjusts saturation flow rate based on the total width of the open lanes during the work zone, and the number of lanes during the work zone and the number of lanes when no work zone is present. The factor is found using Eqs. 9.72, 9.73, and 9.74.

$$f_{wz} = 0.858 f_{wid} f_{reduce} \leq 1.0 \tag{9.72}$$

$$f_{wid} = \frac{1.0}{1.0 - 0.0057(a_w - 12)} \tag{9.73}$$

$$f_{reduce} = \frac{1.0}{1.0 - 0.0402(n_o - n_{wz})} \tag{9.74}$$

where:

f_{wz} adjustment factor for work zone presence,
f_{wid} adjustment factor for approach width,
f_{reduce} adjustment factor for lanes reduced during work zone,
a_w total approach width during work zone,
n_o number of lanes open under normal conditions, excluding right-turn only lanes,
n_{wz} number of lanes open when work zone is present, excluding right-turn only lanes.

2. **Downstream Lane Blockage Adjustment Factor, f_{ms}**

The lane blockage factor accounts for when there is a downstream lane closure. For an isolated intersection, this value would be 1.0. The procedure for calculating this factor is not discussed in the signalized intersection chapters, but is found in the Urban Street Facilities chapter 29 in the 2016 HCM.

3. **Sustained Spillback Adjustment Factor, f_{sp}**

The factor for adjusting for sustained spillback coming from the downstream section is also given in the Urban Street Facilities chapter 29. For an isolated intersection, the value would be 1.0.

4. **Recommendation for Number of Right Turns Made on Red Removed**

In the 2010 HCM, there was a recommendation that when right turns are made from an exclusive lane, where right turn on red (RTOR) is allowed, and there is a complementary left turn phase on the cross street, the number of RTOR turners could be estimated as equal to the number of shadowed left turns. This recommendation was removed in the 2016 manual because it was agreed that the RTOR would necessarily have to be fewer than the lefts due to the nature of a RTOR (stop and then go). In place of this recommendation, the 2016 manual suggested that the RTOR number should be based on field data. If the analysis is a future case, the manual recommends assuming zero RTOR. This would be a conservative assumption and somewhat increase the calculated delay.

5. **New Planning Method**

A new planning methodology was added to the signalized intersection chapter, which replaced the previous quick estimation method (QEM). This will be presented in Chap. 10.

Appendix

Maximum Allowable Headway Equations for All Lane Group Types

The general equation for MAH for presence detectors, shown earlier in Eq. 9.50, needs to be modified for the various possible lane group types to account for differences in vehicle behavior of turning vehicles. The base Eq. 9.50 is rewritten here as Eq. 9.75 for through vehicles.

$$MAH_{th} = PT_{th} + \frac{L_{ds,th} + L_v}{1.47 S_a} \tag{9.75}$$

where:

PT_{th} passage Time for phase serving through vehicles
L_{ds} length of the stop-line detection zone in the through lanes.

The following equations are used to calculate the MAH for all other lane group types.

For protected left turns from an exclusive left-turn lane (or protected-permitted), use Eq. 9.76

$$MAH_{lt,prot,e} = PT_{lt} + \frac{L_{ds,lt} + L_v}{1.47 S_a} + \frac{E_L - 1}{s_o/3600} \tag{9.76}$$

where:

E_L left-turn equivalent for a protected left-turn = 1.05
s_o base saturation flow rate (pc/h/ln).

For left turns made from a shared lane with protected-permitted phasing, use Eq. 9.77

$$MAH_{lt,prot,s} = MAH_{th} + \frac{E_L - 1}{s_o/3600} \tag{9.77}$$

For permitted left turns from a exclusive lane use Eq. 9.78

$$MAH_{lt,perm,e} = PT_{th} + \frac{L_{ds,lt} + L_v}{1.47S_a} + \frac{3600}{s_l} - t_{fh} \tag{9.78}$$

where:

t_{fh} follow-up headway (4.5 if left turn is made from a shared lane and 2.5 if left turn is made from an exclusive lane).

For permitted left turns in a shared lane, use Eq. 9.79:

$$MAH_{lt,perm,s} = MAH_{th} + \frac{3,600}{s_l} - t_{fh} \tag{9.79}$$

For protected right turns from an exclusive right-turn lane, use Eq. 9.80:

$$MAH_{rt,p,e} = PT_{rt} + \frac{L_{ds,rt} + L_v}{1.47S_a} + \frac{E_R - 1}{s_o/3600} \tag{9.80}$$

where:

E_R protected right-turn equivalent = 1.18.

For permitted right turns made from an exclusive right-turn lane, use Eq. 9.81:

$$MAH_{rt,perm,e} = PT_{th} + \frac{L_{ds,rt} + L_v}{1.47S_a} + \frac{\left(E_R/f_{Rpb}\right) - 1}{s_o/3600} \tag{9.81}$$

where:

f_{Rpb} pedestrian-bicycle saturation flow rate adjustment factor.

For permitted right turns made from a shared lane, use Eq. 9.82:

$$MAH_{rt,perm,s} = MAH_{th} + \frac{\left(E_R/f_{Rpb}\right) - 1}{s_o/3600} \tag{9.82}$$

References

1. Highway Capacity Manual (2010) Special report 209, Transportation Research Board, Washington D.C.
2. Highway Capacity Manual (2016) A guide to multimodal mobility analysis, 6th edn. Transp Res Board (National Research Council, Washington DC)
3. Akcelik and Associates (2011) SIDRA intersection user guide for version 5.1, Australia
4. Canadian Capacity Guide for Signalized Intersections, Canadian Institute of Transportation Engineers (CITE), 3rd edn

5. Webster F (1958) Traffic signal settings. Road research paper no. 39, Road Research Laboratory, Her Majesty's Stationery Office, London, U.K.
6. Strong D, Rouphail N, Courage K (2006) New calculation method for existing and extended HCM delay estimation procedures. In: Proceedings 85th annual meeting TRB, Washington DC
7. Strong D, Rouphail N (2006) Incorporating the effects of traffic signal progression into the proposed Incremental Queue Accumulation (IQA) method. In: Proceedings 85th annual TRB meeting, Washington DC
8. Bonneson J (1998) Lane volume and saturation for rate for a multilane intersection approach. J Transp Eng 124(3):240–246 (American Society of Civil Engineers, New York)
9. Bonneson J, Nguyen K (2012) HCM urban streets methodology enhancements—saturation flow rate adjustment factor for work zone presence. Working Paper No. 6, SHRP 2 Project LO8

Chapter 10
Signalized Intersections—Planning Methodology

The first appearance of a formal planning methodology for signalized intersections appears in the 1985 Highway Capacity Manual [1]. *It was introduced because in 1985, doing a quick hand-computed assessment of a signalized intersection became extremely difficult and time-consuming. By 2000 and beyond, it became virtually impossible. The model for the initial planning approach became the signalized intersection methodology introduced in TRB Circular 212 in 1980* [2]. *To this day, there are some agencies that continue to use the Circular 212 methodology because of its simplicity and ease of use.*

10.1 The 1985 HCM

Following the form of the Circular 212 methodology, the planning approach of the 1985HCM does not even consider the details of signal timing or design. Inputs include only the number of lanes, lane usage, and volumes. The method stops at capacity analysis, with a result of either: *Under Capacity*, *Near Capacity*, or *Over Capacity*. No level of service result is provided. The methodology is completed using either one or two worksheets, depending upon whether or not exclusive turn lanes exist.

The steps in the planning application are relatively straightforward, involving little adjustment, and worksheets are provided for easy use. The planning application steps are as follows:

1. Identify the geometrics of the intersection, including lane usage.
2. Identify all movement volumes in vehicles per hour.
3. Assign lane volumes as follows:
   - If a left-turn lane or lanes exist, all left-turn volume is assigned there. If a right-turn lane or lanes exist, all right-turn volume is assigned there. Where two turning lanes exist for a given movement, the volume is split evenly between the lanes.

E. S. Prassas and R. P. Roess, *The Highway Capacity Manual: A Conceptual and Research History Volume 2*, Springer Tracts on Transportation and Traffic 12,
https://doi.org/10.1007/978-3-030-34480-1_10

Table 10.1 Passenger car equivalents for left turners in shared lane

Opposing volume, veh/h	Passenger Car Equivalent (PCE)
0–199	1.1
200–599	2.0
600–799	3.0
800–999	4.0
≥1000	5.0

Source "Highway Capacity Manual," *Special Report 209*, Transportation Research Board, Washington D.C., 1985 [1], pp. 9–21, used with permission of the National Academy of Science, Courtesy of National Academies Press

- If no left-turn lane is provided, Table 10.1 is used to convert the left-turn volume in veh/h to an equivalent in pc/h. Remaining through and right-turning vehicles are then assumed to be in pc/h. The sum of all pc/h volumes is divided equally amongst available lanes. In doing so, *all* left-turning vehicles must be assigned to the left lane, and *all* right-turning vehicles must be assigned to the right lane. When this assignment is completed, left-turn volumes in pc/h are converted back to veh/h.

4. Find sum of critical lane volumes
   Sum of the critical lane volumes is the maximum of the left-turn volume per lane plus the opposing through/right volume per lane. This is found for both the east/west and north/south approaches.

5. Special Procedure for Single-lane Approaches
   For the special case of single-lane approaches, two things are different when assigning lane volumes and the volumes used to find the sum of the critical lanes. First, the volumes in pc/h are not converted back into veh/h. Second, the number of conflicting left turns used to find the total critical volumes is the difference between subject and opposing left-turn volume.

6. Check Intersection Capacity
   The total sum found is step 4 and/or 5 is then compared against the criteria in Table 10.2.

Table 10.2 Capacity criteria: 1985 HCM

Critical volume for intersection, veh/h	Relationship to probable capacity
0–1200	Under capacity
1201–1400	Near capacity
≥1201	Over capacity

Source "Highway Capacity Manual", *Special Report 209*, Transportation Research Board, Washington DC, 1985 [1], Table 9-14, used with permission of the National Academy of Science, Courtesy of National Academies Press

10.2 The 1994 Update [3]

A completely new planning method was created for the 1994 update. The main output is still the relationship to capacity, but there are more inputs, many based on default values, more calculations are required, and there are optional outputs that allow calculating a signal timing and assigning level of service. This methodology is more in line with the operational method and uses many of the same factors. Three worksheets are provided for performing the methodology. The first worksheet is for the inputs. The second worksheet is the lane volume worksheet used to determine critical lane volumes. The third worksheet is the signal operations worksheet for results.

It is possible to complete all the worksheets by hand, but the methodology is long, and more easily and most often done by computer.

10.2.1 Planning Method Inputs (1994 Update)

Inputs required for the planning method include volumes and geometry, as usual, but also include area type (CBD or not), PHF, and cycle length. If the cycle length is not known, a cycle length range (minimum and maximum value) is input, and three questions must be answered for each approach: (1) is parking allowed? (yes/no), (2) will there be coordination? (yes/no), and (3) what is the left-turn treatment? (permitted, protected, or not opposed).

Defaulted into the methodology are the assumptions that there is no grade, 2% heavy vehicles, no bus blockage, no adjustment for unequal lane utilization, no pedestrians, and if there is coordination then arrival type 4 is used. If more detailed information is known, then those values may be used. The worksheet for the methodology inputs is shown in Fig. 10.1.

10.2.2 Lane Volume Worksheet (1994 HCM)

Figure 10.2 shows the lane volume worksheet for one direction. A separate lane volume worksheet is completed for each direction. The steps on this sheet are numbered 1 through 20. Only those steps that are not self-explanatory are discussed here.

- *Step 2*. Opposing Main line volume equals the total opposing volume minus lefts from an exclusive lane or a single-lane approach.
- *Step 4*. Left-turn adjustment factor for exclusive lanes is either 0.95 or 0.92 for one or two lanes, respectively. The 0.95 factor is the same as for the operational methodology. The 0.92 factor subtracts an additional 0.03 reduction (0.95 – 0.03 = 0.92) from the single-lane case to account for not increasing the left-turn volume by 1.03 with the lane utilization factor.

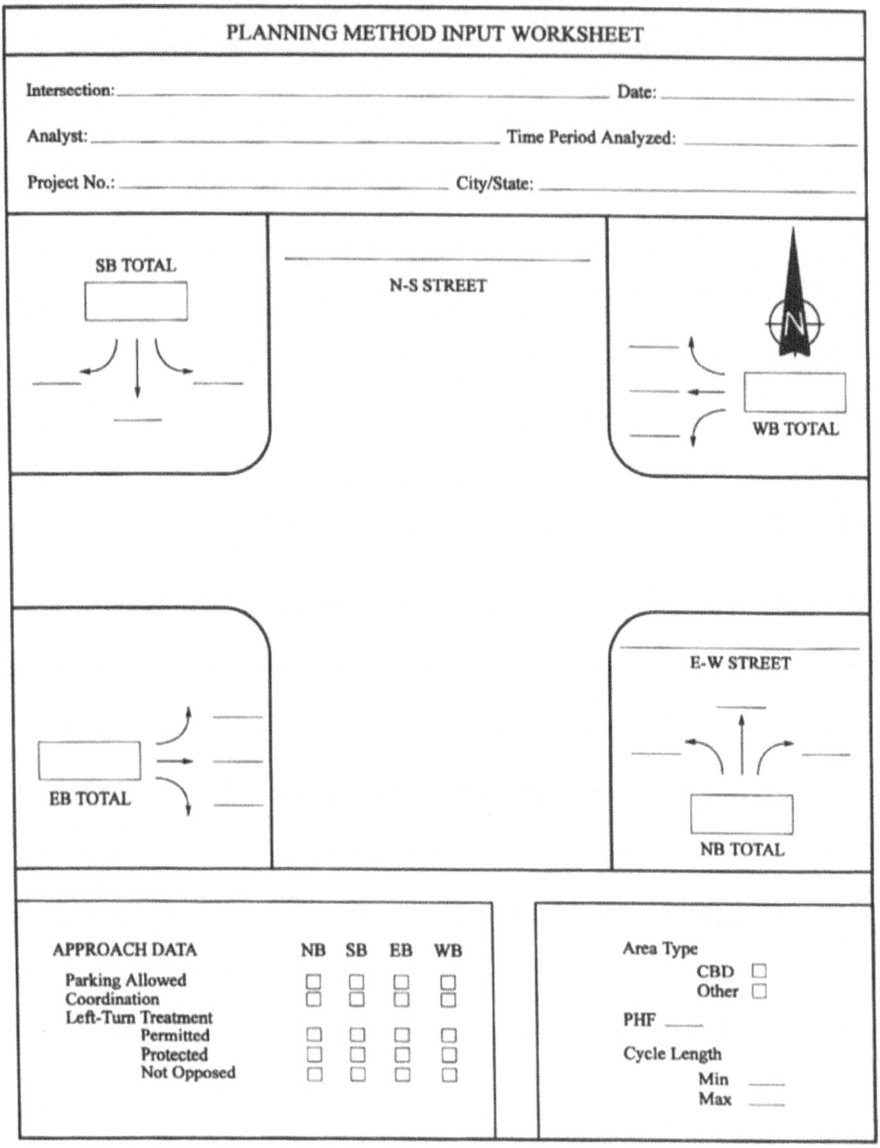

Fig. 10.1 Input worksheet for planning method—1994 update. *Source* "Highway Capacity Manual", *Special Report 209*, Transportation Research Board, Washington D.C., pp. 9–118, 1994 [3], used with permission of the National Academy of Science, Courtesy of National Academies Press

Location: ______________________ Direction ______________________

Left Turn Movement		**Right Turn Movement**	Exclusive RT Lane	Shared RT Lane
1. LT volume	______	6. RT volume	______	______
2. Opposing mainline volume	______	7. RT Lanes	______	1
3. No of exclusive LT lanes	______	8. RT adjustment factor	______	______
4. LT adjustment factor	______	RT lane vol:	[9] ______	[10] ______
(See instructions)				

Cross product: [2] * [1] ______ --->

	Permitted	Protected	Not Opposed
5. LT lane volume: [1] / ([3] * [4])	0	______	______
Through Movement			
11. Through volume	______	______	______
12. Parking adjustment factor	______	______	______
13. No. of through lanes including shared lanes	______	______	______
----------- **Exclusive LT lane computations** -----------			
14. Total approach volume: ([10] + [11]) /[12]	______	______	______
16. Left turn equivalence: (Figure 9-7)	______	XXXXXXXXX	XXXXXXXXX
18. Through lane volume: [14] / [13]	______	______	______
19. Critical lane volume: (See instructions)	______	______	______
----------- **Shared LT lane computations** -----------			
14. Total approach volume: (See instructions)	______	______	______
15. Proportion of left turns in the lane group	______	XXXXXXXXX	XXXXXXXXX
16. Left turn equivalence: (Figure 9-7)	______	XXXXXXXXX	XXXXXXXXX
17. Left turn adjustment factor: (Table 9-15)	______	______	1.0
18. Through lane volume: [14] / ([13] * [17])	______	______	______
19. Critical lane volume: Max([9],[18])	______	______	______
Left Turn Check (if [16] > 8)			
20. Permitted left turn sneaker capacity: 7200 / C_{max}	______	XXXXXXXXX	XXXXXXXXX

Fig. 10.2 Lane volume worksheet for planning method—1994 update. *Source* "Highway Capacity Manual", *Special Report 209*, Transportation Research Board, pp. 119, 1994 [3], used with permission of the National Academy of Science, Courtesy of National Academies Press

Left-turn factors for shared left-turn lanes are found later in step 17.

If it is unknown whether left turns are protected, several checks are performed. If the left-turn volume is greater than 90 veh/h *and* the product of step 1 and step 2 is greater than 50,000 for one opposing lane, 90,000 for two opposing lanes, or 110,000 for three opposing lanes, then a protected left-turn phase is needed. Also, if the left-turn volume is greater than 240 veh/h or if there is more than one exclusive left-turn lane, protection is required.

- *Step 5*. Enter the left-turn volume in the appropriate column (permitted, protected, or not opposed). For permitted lefts, a zero is always entered. Enter the unadjusted left volume from *Step 1* when there is no exclusive lane. Enter the adjusted left-turn if there is an exclusive lane.
- *Step 6*. Enter right-turn volume in the appropriate column, subtracting any right-turn-on-red volume.
- *Step 8*. The right-turn adjustment factor is 0.85 for either a single or a shared lane. It is 0.75 for a double exclusive right-turn lane.
- *Step 9 or 10*. Enter volume from step 6 divided by product of steps 7 and 8. [6]/([7] * [8]).
- *Step 12*. The parking adjustment factor is 1.0 if there is no parking. It is 0.8, 0.9, or 0.933 for one, two, or three lanes, respectively. These factors use the default value of 20 parking movements per hour.

Steps 14–19 are done in either the exclusive left-turn lane section or the shared left-turn section. Note that not all steps are applicable to the exclusive left calculations, as can be seen on the worksheet.

For an exclusive LT Lane:

- *Step 16*. The left-turn equivalence factor is taken from the table for E_L used in the 1994 operational methodology, as shown in Table 10.3.
- *Step 19*. Critical lane volume is equal to the volume of *Step 18*, if no exclusive right-turn lane exists and/or left turners are not opposed. If there is an exclusive right-turn lane and lefts are not opposed, then critical lane volume equals Max ([*Step 18*], [*Step 9*]). If there is no exclusive right lane, but lefts are not opposed, then the critical lane volume is Max ([*Step 18*], [*Step 5*]). If there is both an exclusive right lane and the lefts are not opposed, then the critical lane volume is Max ([*Step 18*], [*Step 9*], [*Step 5*]).

For shared LT lanes:

Table 10.3 Left-turn equivalents, E_{L1}, for permitted left turns from an exclusive lane during unsaturated green, g_u—1994 update

Type of left turn lane	No. of opposing lanes	Opposing flow, v_o						
		0	200	400	600	800	1000	≥1200
Exclusive, with 2 phase	1	1.05	1.7	2.6	4.7	10.4*	10.4*	10.4*
	2	1.05	1.6	2.2	2.9	4.1	6.2	10.4*
	≥3	1.05	1.6	2.1	2.8	3.6	4.8	10.4*
Exclusive, with >2 phase	1	1.05	1.8	3.3	8.2*	8.2*	8.2*	8.2*
	2	1.05	1.7	2.4	3.6	5.9	8.2*	8.2*
	≥3	1.05	1.7	2.4	3.3	4.6	6.8	8.2*

Source Edited from "Highway Capacity Manual", *Special Report 209*, Transportation Research Board, Washington D.C., 1994 [3], Table 9-7, pp. 9–19, used with permission of the National Academy of Science, Courtesy of National Academies Press

*Only sneaker capacity is available

- *Step 14*. If the left-turn is not opposed, then total approach volume is ([*Step 10*] + [*Step 11*] + [*Step 5*])/[*Step 12*] else it is ([*Step 10*] + [*Step 11*])/[*Step 12*].
- *Step 15*. The proportion of lefts in the lane group is only used for calculating the left-turn adjustment for single-lane approaches or when there is protected plus permitted phasing on only one of the approaches. Thus, for a single lane, it would be calculated as [*Step 1*]/([*Step 1*] + [*Step 10*] + [*Step 11*]).
- *Step 16*. The left-turn equivalent is the same as from the operational methodology, found from Table 10.4.
- *Step 17*. The left-turn adjustment factor is found from Table 10.5.

Step 19 is a second check if any permitted only left turns need a protected phase. The check is made if the left-turn equivalency factor, E_L, is greater than 8.0, which

Table 10.4 Left-turn equivalents, E_{L1}, for permitted left turns from a shared lane during unsaturated green, g_u—1994 update

Type of left-turn lane	No. of opposing lanes	Opposing flow, v_o						
		0	200	400	600	800	1000	≥1200
Shared, 2 phase	1	1.05	2.0	3.3	6.5	16*	16*	16*
	2	1.05	1.9	2.6	3.6	6.0	16*	16*
	≥3	1.05	1.8	2.5	3.4	4.5	6.0	16*
Shared >2 phase	1	1.05	2.2	4.5	11*	11*	11*	11*
	2	1.05	2.0	3.1	4.7	11*	11*	11*
	≥3	1.05	2.0	2.9	4.2	6.0	11*	11*

Source Edited from "Highway Capacity Manua"l, *Special Report 209*, Transportation Research Board, Washington D.C., 1994 [3], Table 9-7, pp. 9–19, used with permission of the National Academy of Science, Courtesy of National Academies Press
*Only sneaker capacity is available

Table 10.5 Planning level factors for permitted left turns

Permitted left turn
Lane groups with ≥ 2 lanes {[Step 13] − 1 + $e^{(-[Step 13] * [Step 1] * [Step 16]/600}$}/[13] ≥ {[Step 13] − 1 + $e^{(-[Step 1] * Cmax/3600}$}/[Step 13]
Lane groups with 1 lane $e^{(-0.02 * \{[Step 16]+10 * [Step 15]\} + [Step 1] * Cmax/3600)}$
Protected plus permitted left turn (one direction only)
If [Step 2] < 1220 1/{1 + [(235 + 0.435 * [Step 2]) * [Step 15]]/(1400 − [Step 2])}
If [Step 2] ≥ 1220 1/(1 + 4.525 * [Step 15])

Source "Highway Capacity Manual", *Special Report* 209, Transportation Research Board, Table 9-15, pp. 9–53, used with permission of the National Academy of Science, Courtesy of National Academies Press signifies that the only capacity available for left turners is the sneaker capacity. Sneaker capacity is $7200/C_{max.}$ If the left-turn volume from Step [1] is greater than sneaker capacity, it can be assumed that there will not be enough capacity without a protected left phase.

10.2.3 Signal Operations Worksheet (1994 HCM)

Figure 10.3 shows the signal operations worksheet. The top of the worksheet is the critical movement information transferred from the lane volume worksheets and other previously given data.

There are six possible alternatives for a phase plan, shown in Table 10.6. As is always the case, the phase plan is chosen based on whether there is a need for a protected left-turn phase. A phase plan is chosen separately for each cross street. The choices in Table 10.6 are numbered 1–4. This is because phase plans 2 and 3 have sub-choices based on which (or both) opposing lefts need protection. The choices are:

1. Only permitted left turns
2. Only one opposing left needs protection; dominant left gets lead phase
3. Both lefts need protection; protected left phase is followed by dominant left overlap phase, followed by thru movements
4. Split phase.

There was some feedback from users at the time of this publication concerning confusion over choice 2. Note that the non-dominant left-turn movement is not shown in the phase plan. The methodology does not consider protected-permitted phasing, and at this point in the analysis, only the critical movements are needed. For case two, this will be the sum of the dominant left and opposing through; the non-dominant left is not a possible critical movement.

In Table 10.6, the possible choices for critical movements are given. The largest sum of volumes through the phase is chosen. Also given in Table 10.6 is the lost time for each portion of the phase. A default lost time of 3 s per phase is assumed for any movement that starts and stops in the phase (start-up and clearance lost times). Thus, Plan 1 has 3 s of lost time, and the remaining plans have 6 s of lost time. Note that the 3-phase plans (3a and 3b) also have 6 s of lost time because none of the movements in the middle phase both start and stop. The total lost time for the intersection will be the sum of the lost times for both cross streets.

The sum of the critical movement volumes is then compared to the capacity value at the maximum cycle length considered acceptable. This ratio is the v/c for the intersection, X_{cm}, found from Equation [10.1].

$$X_{cm} = \frac{CS}{\left(1 - \frac{TL}{C_{\max}}\right) * 1900 * CBD * PHF} \tag{10.1}$$

Phase Plan Selection from Lane Volume Worksheets

		EASTBOUND	WESTBOUND	NORTHBOUND	SOUTHBOUND
Critical Through-RT lane volume: [19]		___	___	___	___
LT lane volume: [5]		___	___	___	___
Left turn protection: (Perm, Prot, N/O)		___	___	___	___
Dominant left turn: (Indicate by '*')		___	___	___	___
Selection Criteria based on the specified left turn treatment:	Plan 1:	Perm	Perm	Perm	Perm
		Perm	N/O	Perm	N/O
		N/O	Perm	N/O	Perm
	Plan 2a:	Perm	Prot	Perm	Prot
	Plan 2b:	Prot	Perm	Prot	Perm
* Indicates the dominant left turn for each opposing pair	Plan 3a:	*Prot	Prot	*Prot	Prot
	Plan 3b:	Prot	*Prot	Prot	*Prot
	Plan 4:	N/O	N/O	N/O	N/O

Phase plan selected (1 to 4) ___ ___

Min. cycle [C_{min}] ___ Max cycle [C_{max}] ___ [PHF] (From Input Worksheet) ___

Phasing Plan From Table 9-16

	Note	Value	EAST-WEST Phase 1	EAST-WEST Phase 2	EAST-WEST Phase 3	NORTH-SOUTH Phase 1	NORTH-SOUTH Phase 2	NORTH-SOUTH Phase 3
Movement codes			___	___	___	___	___	___
Critical Phase Volume [CV]			___	___	___	___	___	___
Critical Sum [CS]	1	___						
Lost time/phase [PL]			___	___	___	___	___	___
Lost time/cycle [TL]	2	___						
CBD adjustment [CBD]	3	___						
Critical v/c ratio [X_{cm}]	4	___						
Intersection status	5	___						
Optional Timing Plan Computation								
Reference Sum [RS]	6	___						
Cycle length [CYC]	7	___						
Green time	8		___	___	___	___	___	___

Notes

1. Critical sum = Sum of critical phase volumes [CV's] for all phases.
2. Lost time/cycle = Sum of all lost times/phase, [PL's].
3. CBD adjustment = .9 within CBD, 1.0 elsewhere.
4. Critical v/c ratio = CS /((1-[TL]/C_{max}) * 1900 * [CBD] * [PHF]).
5. Status: (See instructions).
6. Reference Sum = 1710 * [PHF] * [CBD].
7. Cycle length = [TL] / (1-(Min([CS],[RS]) / [RS])), Subject to [C_{min}] and [C_{max}].
8. Green time = ([CYC]-[TL]) * ([CV]/[CS]) + [PL].

Fig. 10.3 Signal operations worksheet for planning method 1994 update. *Source* "Highway Capacity Manual", *Special Report 209*, Transportation Research Board, Washington D.C., 1994 [3], pp. 9–120, used with permission of the National Academy of Science, Courtesy of National Academies Press

Table 10.6 Phase plan summary for planning analysis—1994 update

Phase plan	Phase No.	Lost time	East-West		North-South	
			Movement code	Critical sum	Movement code	Critical sum
1	1	3	EWT	Max (ET,EL,WT,WL)	NST	Max (NT,NL,ST,SL)
2a	1	3	WTL	WL	STL	SL
	2	3	EWT	Max (WT-WL, ET)	NST	Max (ST-SL, NT)
2b	1	3	ETL	EL	NTL	NL
	2	3	EWT	Max (ET-EL, WT)	NST	Max (NT-NL, ST)
3a	1	3	EWL	WL	NSL	SL
	2	0	ETL	EL-WL	NTL	NL-SL
	3	3	EWT	Max (WT,ET-(EL-WL))	NST	Max (ST,NT-(NL-SL))
3b	1	3	EWL	EL	NSL	NL
	2	0	WTL	WL-EL	STL	SL-NL
	3	3	EWT	Max (ET,WT-(WL-EL))	NST	Max (NT,ST-(SL-NL))
4	1	3	ETL	Max (ET,EL)	NTL	Max (NT,NL)
	2	3	WTL	Max (WT,WL)	STL	Max (ST,SL)

Note EWT = eastbound and westbound through; ETL = eastbound through and left; WTL = westbound through and left; NST = northbound and southbound through; STL = southbound through and left; NTL = northbound through and left; ET = eastbound through; EL = eastbound left; WT = westbound through; WL = westbound left; NT = northbound through; NL = northbound left; ST = southbound through; SL = southbound left

Source "Highway Capacity Manual", *Special Report 209*, Transportation Research Board, Washington D.C., 1994 [3], Table 9-16, pp. 9–56, used with permission of the National Academy of Science, Courtesy of National Academies Press

where:

X_{cm} critical v/c ratio at maximum cycle length
CS sum of the critical volumes
TL total lost time per cycle
C_{max} maximum acceptable cycle length.
CBD adjustment for area type (=0.9 for CBD, 1.0 elsewhere).

Table 10.7 gives the relationship of the X_{cm} value to capacity.

Table 10.7 Intersection status criteria for signalized intersection planning analysis 1994 update

Critical v/c ratio, X_{cm}	Relationship to probable capacity
$X_{cm} \leq 0.85$	Under capacity
$0.85 < X_{cm} \leq 0.95$	Near capacity
$0.95 < X_{cm} \leq 1.0$	At capacity
$X_{cm} > 1.0$	Over capacity

Source "Highway Capacity Manual", *Special Report 209*, Transportation Research Board, Washington D.C., 1994 [3], Table 9-14, pp. 9–31, used with permission of the National Academy of Science, Courtesy of National Academies Press

10.2.4 Optional Development of a Timing Plan

The 1994 planning methodology allows a timing plan to estimated, if it is not known or otherwise specified. Developing a timing plan is only needed if a level of service designation (from the delay formulas in the operational methodology) is desired. If the design cycle length is unknown, a cycle length must be calculated before the time can be allocated to each phase. The formula given in the 1994 HCM is shown in Equation [10.2].

$$C = \frac{TL}{1 - \frac{\text{Min}(CS, RS)}{RS}} \tag{10.2}$$

where RS is the Reference Sum (the saturation flow rate). All other variables are as previously defined.

The saturation flow rate recommended is (1710 * PHF * CBD). This formula assumes a target v/c ratio for the intersection of 0.9 (1900 * 0.9 = 1710). Taking the minimum of CS or RS was to remind the user that a volume over capacity cannot be accommodated by the current geometry of the intersection. However, a volume at capacity can also not be accommodated and would give a division by zero error.

After finding the cycle length, the usable (or effective) green time must be allocated to each phase in proportion to the critical volume for each phase (see Eq. [10.3]). The usable green time of the cycle is the cycle length minus the total lost time per cycle, TL.

$$g_i = (C - TL)\frac{(Critical\ Volume)_i}{CS} \tag{10.3}$$

where g_i is the effective green time for phase I (s). and all other variables are as previously defined.

Although it should be done with caution, it is possible to take the output of the signal timing, along with the original inputs, and do an operational analysis to get a level of service for the intersection. In doing so, it must be recognized that many inputs have depended upon default values, and that signal timing was only roughly estimated. Both of these can seriously impact the accuracy of the delay prediction which results.

10.3 The 1997 Update [4]

There were no major changes made to the planning methodology in the 1997 HCM. However, with the change in the left-turn equivalence table from the 1994 to the 1997 Update (see Chap. 9), the sneaker check in Step 20 of the lane volume worksheet uses an E_{L1} value of 3.5.

10.4 The 2000 HCM [5]

In the 2000 HCM, the planning method was removed from the signalized intersection chapter (Chap. 16) and placed in the Urban Street Concepts Chapter (Chap. 10). At the same time, the name was changed from the Signalized Intersection Planning Methodology to the Quick Estimation Method (QEM).

The basic methodology did not change, but along with the name change, the worksheets were reorganized to be made more self-explanatory, a new worksheet for left-turn treatment was added, and there were some changes to the details of the methodology.

10.4.1 QEM Input Worksheet (2000 HCM)

Figure 10.4 shows the input worksheet for the QEM methodology. Lost time per phase is added as an input, giving the option of using a value other than 3 s/phase. Additionally, there are inputs for the proportion of left and right turns in the lane group. There is also direction not to include RTOR volume in the right-turn volume.

10.4.2 Left-Turn Treatment Worksheet (2000 HCM)

Figure 10.5 shows the new left-turn check worksheet added to 2000 HCM. This sheet performs the same tests as in the previous HCM, but with all directions performed on the one sheet, in a more self-explanatory manner.

QUICK ESTIMATION INPUT WORKSHEET

General Information		Site Information	
Analyst	______	Intersection	______
Agency or Company	______	Area Type	☐ CBD ☐ Other
Date Performed	______	Jurisdiction	______
Analysis Time Period	______	Analysis Year	______

Intersection Geometry

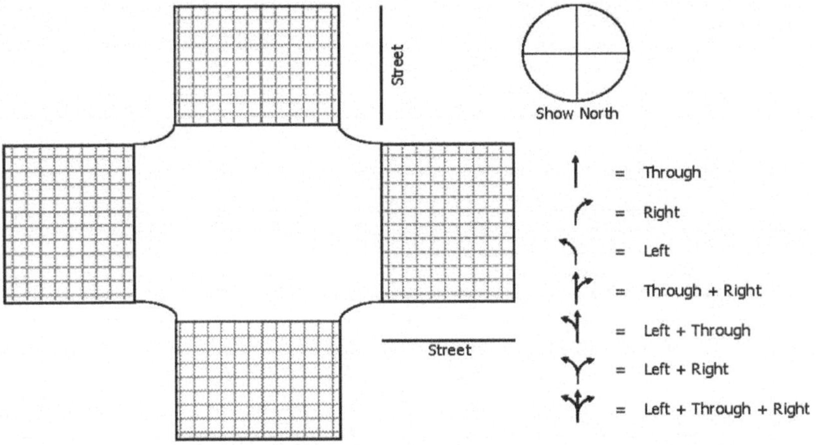

Volume and Signal Input

	EB			WB			NB			SB		
	LT	TH	RT[1]	LT	TH	RT[1]	LT	TH	RT[1]	LT	TH	RT[1]
Volume, V (veh/h)												
Proportion of LT or RT (P_{LT} or P_{RT})[2]		-			-			-			-	
Parking (Yes/No)												
Left-turn treatment (permitted, protected, not opposed) (if known)												

Peak-hour factor, PHF ______

Cycle length Minimum, C_{min} ____ s Maximum, C_{max} ____ s or Given, C ____ s

Lost time/phase ______ s

Notes

1. RT volumes, as shown, exclude RTOR.
2. P_{LT} = 1.000 for exclusive left-turn lanes, and P_{RT} = 1.000 for exclusive right-turn lanes. Otherwise, they are equal to the proportions of turning volumes in the lane group.

Fig. 10.4 Input worksheet for the 2000 HCM QEM. *Source* "Highway Capacity Manual", Transportation Research Board, Washington D.C., 2000 [5], Exhibit A10-2, pp. 10–38, used with permission of the National Academy of Science, Courtesy of National Academies Press

QUICK ESTIMATION LEFT-TURN TREATMENT WORKSHEET

General Information

Description ______________________________

Check # 1. Left-Turn Lane Check

Approach	EB	WB	NB	SB
Number of left-turn lanes				
Protect left turn (Y or N)?				

If the number of left-turn lanes on any approach exceeds 1, then it is recommended that the left turns on that approach be protected. Those approaches with protected left turns need not be evaluated in subsequent checks.

Check # 2. Minimum Volume Check

Approach	EB	WB	NB	SB
Left-turn volume				
Protect left turn (Y or N)?				

If left-turn volume on any approach exceeds 240 veh/h, then it is recommended that the left turns on that approach be protected. Those approaches with protected left turns need not be evaluated in subsequent checks.

Check # 3. Minimum Cross-Product Check

Approach	EB	WB	NB	SB
Left-turn volume, V_L (veh/h)				
Opposing mainline volume, V_o (veh/h)				
Cross-product ($V_L * V_o$)				
Opposing through lanes				
Protected left turn (Y or N)?				

Minimum Cross-Product Values for Recommending Left-Turn Protection

Number of Through Lanes	Minimum Cross-Product
1	50,000
2	90,000
3	110,000

If the cross-product on any approach exceeds the above values, then it is recommended that the left turns on that approach be protected. Those approaches with protected left turns need not be evaluated in subsequent checks.

Check # 4. Sneaker Check

Approach	EB	WB	NB	SB
Left-turn volume, V_L (veh/h)				
Sneaker capacity, c_s (veh/h) $c_s = 7200/C$				
Equivalence factor, E_{L1}				
Protected left turn (Y or N)?				

If the equivalence factor is 3.5 or higher (computed in the Quick Estimation Lane Volume Worksheet) and the unadjusted left turn is greater than the sneaker capacity, then it is recommended that the left turns on that approach be protected.

Notes

1. If any approach is recommended for left-turn protection but the analyst evaluates it as having permitted operation, then this quick estimation method may give overly optimistic results. The analyst should instead use the methodology described in Chapter 18, Signalized Intersections.
2. All volumes used in this worksheet are unadjusted hourly volumes.

Fig. 10.5 Left-Turn Treatment Worksheet—2000 HCM. *Source* "Highway Capacity Manual", Transportation Research Research Board, Washington D.C., 2000 [5], Exhibit A10-3, pp. 10–39, used with permission of the National Academy of Science, Courtesy of National Academies Press

10.4.3 QEM Lane Volume Worksheet (2000 HCM)

Figure 10.6 shows the QEM lane volume worksheet. This worksheet was also reorganized to make it clearer. Instead of referring to numbered steps, the specific variables needed are shown, and many more helpful notes are given.

There was an oversight in the 2000 HCM with the parking adjustment factor. No direction is given as to what number should be used nor where to find the equation. In the 1997 HCM, values were given for one, two, and three lane approaches based on a default value of 20 parking maneuvers per hour. In the 2000 HCM, the equation to find f_p is in HCM Chap. 16, which a user would need to know. But no default number of parking movements per hour is suggested.

10.4.4 QEM Control Delay and LOS Worksheet (2000 HCM)

Figure 10.7 shows the QEM Control Delay and LOS Worksheet. This worksheet replaced the Signal Operations Worksheet of the 1994 and 1997 HCMs. Although, the 2000 HCM includes the same calculations as in the previous versions, there were changes. Explicit calculations for delay were also added to the worksheet. This drastically changed the emphasis of this final result worksheet. In earlier versions, calculating green times and LOS were described as optional, this direction was removed in the 2000 HCM.

A change in the formula for the critical v/c ration, X_{cm}, was made. Equation 10.4 is the equation from the 1994 and 1997 HCMs. Equation 10.5 is the 2000 HCM equation.

$$X_{cm} = \frac{CS}{\left(1 - \frac{TL}{C_{\max}}\right) * 1900 * CBD * PHF} \tag{10.4}$$

$$X_{cm} = \frac{CS}{RS\left(1 - \frac{L}{C}\right)} \tag{10.5}$$

In the 2000 HCM, the actual cycle length replaced the maximum cycle length. Additionally, the saturation flow rate was changed from 1900 to the reference sum, which is 1700 * PHF * f_a. The f_a term is equivalent to the CBD term of the previous editions.

The bottom half of the worksheet gives the complete enumeration for calculating green times, delay, and level of service. In the 1994 and 1997 HCM, these calculations were optional and there was direction to use the operational method if delay and LOS was desired. As stated previously, this drastically changed the emphasis of the planning methodology to a LOS result instead of a relationship to capacity result.

QUICK ESTIMATION LANE VOLUME WORKSHEET

General Information

Description/Approach ______________________________

Right-Turn Movement

	Exclusive RT Lane	Shared RT Lane
RT volume, V_R (veh/h)		
Number of exclusive RT lanes, N_{RT}		use 1
RT adjustment factor,[1] f_{RT}		
RT volume per lane, V_{RT} (veh/h/ln) $V_{RT} = \frac{V_R}{(N_{RT} \times f_{RT})}$		

Left-Turn Movement

LT volume, V_L (veh/h)	
Opposing mainline volume, V_o (veh/h)	
Number of exclusive LT lanes, N_{LT}	
LT adjustment factor,[2] f_{LT}	
LT volume per lane,[3] V_{LT} (veh/h/ln) $V_{LT} = \frac{V_L}{(N_{LT} \times f_{LT})}$	Permitted LT, use 0 ____ Protected LT ____ Not Opposed LT ____

Through Movement

	Permitted LT	Protected LT	Not Opposed LT
Through volume, V_T (veh/h)			
Parking adjustment factor, f_p			
Number of through lanes, N_{TH}			
Total approach volume,[4] V_{tot} (veh/h) $V_{tot} = \frac{V_{RT}\,(\text{shared}) + V_T + V_{LT\,(\text{not opp})}}{f_p}$			

Through Movement with Exclusive LT Lane

Through volume per lane, V_{TH} (veh/h/ln) $V_{TH} = \frac{V_{tot}}{N_{TH}}$			
Critical lane volume,[5] V_{CL} (veh/h) Max[V_{LT}, V_{RT} (exclusive), V_{TH}]			

Through Movement with Shared LT Lane

Proportion of left turns, P_{LT}		Does not apply	Does not apply
Equivalence factor, E_{L1}		Does not apply	Does not apply
Shared lane LT adjustment factor, f_{DL}			use 1.0
Through volume per lane, V_{TH} (veh/h/ln) $V_{TH} = \frac{V_{tot}}{(N_{TH} \times f_{DL})}$			
Critical lane volume,[5] V_{CL} (veh/h) Max[V_{RT} (exclusive), V_{TH}]			

Notes

1. For RT shared or single lanes, use 0.85. For RT double lanes, use 0.75.
2. For LT single lanes, use 0.95. For LT double lanes, use 0.92. For a one-way street or T-intersection, use 0.85 for one lane and 0.75 for two lanes.
3. For unopposed LT shared lanes, $N_{LT} = 1$.
4. For exclusive RT lanes, V_{RT} (shared) = 0. If not opposed, add V_{LT} to V_T and set $V_{LT(\text{not opp})} = 0$.
5. V_{LT} is included only if LT is unopposed. V_{RT} (exclusive) is included only if RT is exclusive.

Fig. 10.6 Lane volume worksheet—2000 HCM. *Source* "Highway Capacity Manual", Transportation Research Board, Washington D.C., 2000 [5], Exhibit A10-4, pp. 10–40, used with permission of the National Academy of Science, Courtesy of National Academies Press

QUICK ESTIMATION CONTROL DELAY AND LOS WORKSHEET

General Information

Description____

East-West Phasing Plan

Selected plan (Exhibit A10-8) ________	Phase No. 1	Phase No. 2	Phase No. 3
Movement codes			
Critical phase volume, CV (veh/h)			
Lost time/phase, t_L (s)			

North-South Phasing Plan

Selected plan (Exhibit A10-8) ________	Phase No. 1	Phase No. 2	Phase No. 3
Movement codes			
Critical phase volume, CV (veh/h)			
Lost time/phase, t_L (s)			

Intersection Status Computation

Critical sum, CS (veh/h) $CS = \Sigma CV$	
Lost time/cycle, L (s) $L = \Sigma t_L$	
Reference sum flow rate RS (veh/h)[1]	
Cycle length, C (s) $C_{min} \le C \le C_{max}$ $C = \dfrac{L}{1 - \left[\dfrac{\min(CS, RS)}{RS}\right]}$	
Critical v/c ratio, X_{cm} $X_{cm} = \dfrac{CS}{RS\left(1 - \dfrac{L}{C}\right)}$	
Intersection status (Exhibit A10-9)	

Green Time Calculation

East-West Phasing	Phase No. 1	Phase No. 2	Phase No. 3
Green time, g (s) $g = \left[(C - L)\left(\dfrac{CV}{CS}\right) + t_L\right]$			
North-South Phasing	**Phase No. 1**	**Phase No. 2**	**Phase No. 3**
Green time, g (s) $g = \left[(C - L)\left(\dfrac{CV}{CS}\right) + t_L\right]$			

Control Delay and LOS

	EB	WB	NB	SB
Lane group				
Lane group adjusted volume from lane volume worksheet, V (veh/h)				
Green ratio, g/C				
Lane group saturation flow rate, s (veh/h) s = RS * number of lanes in lane group				
v/c ratio, X $X = \dfrac{V/s}{g/C}$				
Lane group capacity, c (veh/h) $c = \dfrac{V}{X}$				
Progression adjustment factor, PF (Exhibit 16-12)				
Uniform delay, d_1 (s/veh) (Equation 16-11)				
Incremental delay, d_2 (s/veh) (Equation 16-12)				
Initial queue delay, d_3 (s/veh) (Appendix F, Ch. 16)				
Delay, $d = d_1(PF) + d_2 + d_3$ (s/veh)				
Delay by approach, d_A (s/veh) $\dfrac{\Sigma(d)(V)}{\Sigma V}$				
Approach flow rate, V_A (veh/h)				
Intersection delay, d_I (s/veh) $d_I = \dfrac{\Sigma(d_A)(V_A)}{\Sigma V_A}$		Intersection LOS (Exhibit 16-2)		

Notes

1. $RS = 1710(PHF)(f_a)$, where f_a is area adjustment factor (0.90 for CBD and 1.0 for all others).

Fig. 10.7 Control delay and LOS worksheet—2000 HCM. *Source* "Highway Capacity Manual", Transportation Research Board, Washington D.C., 2000 [5], Exhibit A10-5, pp. 10–41, used with permission of the National Academy of Science, Courtesy of National Academies Press

10.5 The 2010 HCM [6]

In the 2010 HCM, the QEM planning method was moved again, this time from Chapter 10 (of the HCM) to the online-only Chap. 31: Signalized Intersections Supplementary Material.

The basic methodology did not change, but there were some changes made.

10.5.1 QEM Input Worksheet (2010 HCM)

The input worksheet did not change.

10.5.2 Left-Turn Treatment Worksheet (2010 HCM)

The left-turn treatment worksheet did not change.

10.5.3 QEM Lane Volume Worksheet (2010 HCM)

The lane Volume Worksheet itself did not change, but an equation is given in the text for calculating the parking factor, f_p. The equation is the same as in the operational methodology, shown in Eq. 10.6.

$$f_p = \frac{N_{TH} - 0.1 - \frac{18N_m}{3600}}{N_{TH}} \geq 0.05 \tag{10.6}$$

where:

N_m the number of parking movements per hour, and
N_{TH} number of lanes that include through vehicles (including shared lanes).

No default value for N_m is given.

10.5.4 QEM Control Delay and LOS Worksheet (2010 HCM)

Figure 10.8 shows the 2010 QEM Control Delay and LOS Worksheet. There were a few small but significant changes to this worksheet.

The top two sections of the worksheet stayed the same. In the Intersection Status Computation section, the reference sum (RS) used to calculate cycle length was

QUICK ESTIMATION CONTROL DELAY AND LOS WORKSHEET

General Information

Description ____________________

East-West Phasing Plan

Selected plan ______	**Phase No. 1**	**Phase No. 2**	**Phase No. 3**
Movement codes			
Critical phase volume, CV (veh/h)			
Lost time/phase, t_L (s)			

North-South Phasing Plan

Selected plan ______	**Phase No. 1**	**Phase No. 2**	**Phase No. 3**
Movement codes			
Critical phase volume, CV (veh/h)			
Lost time/phase, t_L (s)			

Intersection Status Computation

Critical sum, CS (veh/h) $CS = \Sigma CV$	
Lost time/cycle, L (s) $L = \Sigma t_L$	
Reference sum flow rate RS (veh/h)[1]	
Cycle length, C (s) $C_{min} \leq C \leq C_{max}$ $C = C_{max}$ when $CS \geq RS$ $C = \dfrac{L}{1-\dfrac{CS}{RS}}$	
Critical v/c ratio, X_c $X_c = \dfrac{CS}{1700\ PHF\ f_a\left(1-\dfrac{L}{C}\right)}$	
Intersection status (relationship to capacity)	Under ___ Near ___ At ___ Over ___

Green Time Calculation

East-West Phasing	**Phase No. 1**	**Phase No. 2**	**Phase No. 3**
Green time, g (s) $g = (C-L)\left(\dfrac{CV}{CS}\right) + t_L$			
North-South Phasing	**Phase No. 1**	**Phase No. 2**	**Phase No. 3**
Green time, g (s) $g = (C-L)\left(\dfrac{CV}{CS}\right) + t_L$			

Control Delay and LOS

	EB	WB	NB	SB
Lane group				
Lane group adjusted volume from Lane Volume worksheet, V (veh/h)				
Green ratio, g/C				
Lane group saturation flow rate, s (veh/h) s = RS * number of lanes in lane group				
v/c ratio, X $X = \dfrac{V/s}{g/C}$				
Lane group capacity, c (veh/h) $c = \dfrac{V}{X}$				
Progression adjustment factor, PF				
Uniform delay, d_1 (s/veh)				
Incremental delay, d_2 (s/veh)				
Control delay, $d = d_1(PF) + d_2$ (s/veh)				
Delay by approach, d_A (s/veh) $\dfrac{\Sigma(d)(V)}{\Sigma V}$				
Approach flow rate, V_A (veh/h)				
Intersection delay, d_I (s/veh) $d_I = \dfrac{\Sigma(d_A)(V_A)}{\Sigma V_A}$		Intersection LOS (Exhibit 18-4)		

Notes

1. RS = 1530 x PHFx f_a, where f_a is area adjustment factor (= 0.90 for CBD and 1.0 for all other area types).

Fig. 10.8 QEM control and delay worksheet—2010 HCM. *Source* "Highway Capacity Manual", Transportation Research Board, Washington D.C., 2010 [6], Exhibit 31–42, pp. 31–90, 2010, used with permission of the National Academy of Science, Courtesy of National Academies Press

Table 10.8 Progression adjustment factors—2010 HCM

Arrival type	Progression adjust factor *PF* as a function of green ratio					
	0.2	0.3	0.4	0.5	0.6	0.7
Uncoordinated	1.00	1.00	1.00	1.00	1.00	1.00
Coordinated	0.92	0.86	0.78	0.67	0.50	0.22

Note $PF = (1 - [1.33\ g/C])/(1 - g/C)$.
Source "Highway Capacity Manual", Transportation Research Board, Washington D.C., 2010 [6], Exhibit 31–46, used with permission of the National Academy of Science, Courtesy of National Academies Press

changed from Eqs. 10.7 to 10.8.

$$RS = 1700 * PHF * f_a \quad (2000\,HCM) \tag{10.7}$$

$$RS = 1530 * PHF * f_a. \quad (2010\,HCM) \tag{10.8}$$

The 1530 number represents a value that is about 90% of adjusted saturation flow rate (1700 * 0.9). The objective is to calculate a cycle length based on having a v/c ratio of 0.90 for all the critical movements.

The error from the 1994 edition of being able to get a division by zero when RS < CS was finally corrected by setting $C = C_{max}$ when $CS \geq RS$.

When calculating the critical volume, the old value for reference sum is used, as shown in Equation [10.9]. Using the full adjusted saturation flow rate allows the critical v/c ratio to be at 90% rather than 100% when the minimum cycle length is reached.

$$X_c = \frac{CS}{1700 * PHF * f_a\left(1 - \frac{L}{C}\right)} \tag{10.9}$$

In the calculation of delay, the d_3 term was removed in the 2010 HCM. Also, the progression factor was simplified, as shown in Table 10.8. (In the 2000 HCM, progression factor was taken from the operational methodology, Chapter 16 of the HCM). The factors for Uncoordinated are those for Arrival Type 3 in the operational methodology and the factors for Coordinated are those for Arrival Type 4. Using the formula given, the PF can be calculated, which would give the values for the other arrival types, however. The remainder of the delay calculations remained the same.

10.6 The 2016 HCM [6]

Between the publication of the 2010 and the 2016 HCMs, the number one concern of the subcommittee was to change the planning method that had been in place since

1994 (with minor updates and changes) to a more simple procedure that returned to using critical movement analysis, as in Circular 212. The new method is divided into two distinct parts. The first part is a critical movement analysis (CMA) that goes back to the 1985 planning outputs of under, near, or over capacity. The method is more detailed than the 212 circular, and incorporates additional adjustment factors based on the intersection characteristics, but detailed signal timing is not needed. The second part of the methodology has level of service as its output, based on delay calculations and thus detailed signal timing is needed. The analyst can use only Part I or continue and perform Part II calculations.

10.6.1 Part I of the Planning Method, Critical Movement Analysis

Critical movement analysis (CMA), unlike the operational analysis of Chap. 9, does not make adjustments to the base saturation flow rate. Rather, adjustments are made to the input demand volume. So rather and decreasing the saturation flow rate for conditions that differ from the base, the demand volume is increased by converting the volume to equivalent through-only passenger car units.

Step 1. Input Data

(a) Required Input Data. The following data is required to complete Part I of the methodology. Table 10.9 shows the data needed for one approach. The same data is needed for each approach on the intersection.

In addition, the analyst may enter a peak-hour factor, cycle length, and base saturation flow rate (pc/h/ln). PHF is defaulted to 1.0; base saturation flow rate is defaulted to 1900 pc/h/ln for metropolitan areas with population $\geq$250,000 and 1750 for areas with population <250,000. If Cycle length is not known, it is estimated later in the methodology based on the number of critical movements.

Table 10.9 Required data needed for Part I—2016 HCM

Movement	Left	Thru	Right
Volume, veh/h			
Number of lanes	#Exclusive left lanes	#thru only and shared lanes	#exclusive right lanes
Lane use (exclusive or shared)			

Source "Highway Capacity Manual": A Guide to Multimodal Mobility Analysis, Transportation Research Board, Washington D.C., 2016 [7], Exhibit 31–32, pp. 31–80, used with permission of the National Academy of Science, Courtesy of National Academies Press

(b) Optional Input Data. The analyst may also enter a limited number of characteristics about the intersection and its traffic composition. For each approach, the following may be entered for:

- Percent heavy vehicles, P_{HV}
- Parking activity (none, low, medium, high)
- Pedestrian activity (none, low, medium, high, extreme)
- Left-turn operation (protected, permitted, protected/permitted)
- Phase sequence, if protected left-turn operation (lead or lag).

Step 2. Determine Left-Turn Operation if Not Entered in Step 1

If the left-turn operation is unknown, then the following three rules are applied to determine if there is protected left-turn phasing. If any of the three rules are met, then protected left-turn phasing is assigned to that approach. Note that if one approach requires protected phasing and the opposing approach does not, the method assumes that both opposing approaches have protected left-turn phasing.

1. Left-turn volume is greater than or equal to 240 veh/h.
2. The cross product of the subject approach left-turn volume and the opposing approach through volume is greater than or equal to 50,000, 90,000, or 110,000 for an opposing approach with one, two, or three or more through lanes, respectively.
3. There is more than one exclusive left-turn lane on the subject approach.

Step 3. Convert Input Demand Volumes to Through-Passenger-Car Equivalents

Equation 10.10 is used to convert the input demand volumes to through-passenger car equivalents, which are given the symbol *tcu* (through car units). Table 10.10 gives the equivalency values to be used in the equation.

$$v_{adj} = V E_{HV} E_{PHF} E_{LT} E_{RT} E_p E_{LU} E_{other} \tag{10.10}$$

where:

v_{adj} adjusted volume; input volumes converted to equivalent through passenger cars per hour (tcu/h)

Table 10.10 Equivalency factors for right turns—2016 HCM

Pedestrian activity level	Pedestrian volume (peds/h in conflicting xwalk)	E_{RT}
None or low	0–199	1.2
Moderate	200–399	1.3
High	400–799	1.5
Extreme	≥800	2.1

Source "Highway Capacity Manual": *A Multimodal Guide to Mobility Analysis*, Transportation Research Board, Washington D.C., 2016 [7], Exhibit 31–34, pp. 31–83, used with permission of the National Academy of Science, Courtesy of National Academies Press

Table 10.11 Equivalency factors for left turns—2016 HCM

	E_{LT}
Protected-only phasing	1.05
Permitted-only phasing	
Opposing through + right-turn volume	
<200	1.1
200–599	2.0
600–799	3.0
800–999	4.0
≥1000	5.0
Protected-permitted phasing (see text in Step 3 below)	$E_{LTC} = \dfrac{E^{*}_{LTprot} g_{prot} + E_{LTperm^{*}} g_{perm}}{g_{prot} + g_{perm}}$

Source "Highway Capacity Analysis": *A Guide for Multimodal Mobility Analysis*, Transportation Research Board, Washington D.C., 2016, Exhibit 31–33, pp. 31–83, used with permission of the National Academy of Science, Courtesy of National Academies Press

V volume input by analyst per movement, veh/h
E_{HV} equivalency factor for heavy vehicles $= 1 + 0.01P_{HV}(E_T - 1)$; $E_T = 2.0$
E_{PHF} equivalency factor for peak-hour characteristics $= 1/PHF$
E_{RT} equivalency factor for right turns, see Table 10.10
E_{LT} equivalency factor for left turns, see Table 10.11
E_p equivalency factor for parking, see Table 10.12
E_{LU} equivalency factor for lane utilization, see Table 10.13
E_{other} equivalency factor for other conditions not detailed in previous factors, such as buses, grade, or area type.

Step 3 Special Case: Protected-Permitted Left-Turn Phasing Equivalency

In the operational methodology, when there is protected-permitted phasing, the two portions of the phase are analyzed separately, and each portion of the phase gets its own left-turn equivalency value. The planning methodology uses a simpler approach and combines the phases, using one left-turn equivalency value. The equivalency for

Table 10.12 Equivalency factors for parking activity—2016 HCM

Parking activity level	Number of lanes in Lane group	E_p
No parking lane	All	1.00
Adjacent parking allowed	1	1.20
	2	1.10
	3	1.05

Source "Highway Capacity Analysis": *A Guide for Multimodal Mobility Analysis*, Transportation Research Board, Washington D.C., 2016 [7], Exhibit 31–35, pp. 31–83, used with permission of the National Academy of Science, Courtesy of National Academies Press

Table 10.13 Equivalency factors for lane utilization—2016 HCM

Lane group	Number of lanes in lane group	E_{LU}
Through and/or shared lane	1	1.00
	2	1.05
	≥3	1.10
Exclusive left	1	1.00
	≥2	1.03
Exclusive right	1	1.00
	≥2	1.13

Source "Highway Capacity Analysis": *A Guide for Multimodal Mobility Analysis*, Transportation Research Board, Washington D.C., 2016 [7], Exhibit 31–36, pp. 31–84, used with permission of the National Academy of Science, Courtesy of National Academies Press

the compound left-turn phase is found by weighting the equivalency for each portion of the phase in proportion to the amount of effective green time available for each portion. Because of this, when there is compound phasing, in addition to the all the previous inputs, signal timing must also be entered for the Part I methodology. The weighted average is found using Eq. 10.11.

$$E_{LTC} = \frac{E^*_{LTprot}g_{prot} + E^*_{LTperm}g_{perm}}{g_{prot} + g_{perm}} \tag{10.11}$$

Step 4. Find Flow Rate per Lane for each Lane Group

For the planning method, lane groups are defined as they were in the 2000 HCM operational methodology. Exclusive turning lanes are separate lane groups and all other lane types are combined into one lane group. To find the flow rate per lane for each lane group, divide by the number of lanes in that lane group, using Eq. 10.12.

$$v_i = \frac{v_{adj,i}}{N_i} \tag{10.12}$$

where:

v_i through passenger cars per hour per lane for lane group i
$v_{adj,i}$ adjusted equivalent through flow rate for lane group i
N_i number of lanes in lane group i.

Note: If a lane group with multiple lanes includes a shared lane and the flow rate of the turning traffic (left or right) is greater than v_i, then a defacto turn lane should be assumed and the flow rates reassigned.

Note: If there is an exclusive right-turn lane and a protected left-turn on the cross street, then the right-turn flow rate should be reduced by the flow rate of the protected lefts on the cross street. For example, if there is an exclusive lane on the eastbound approach with 200 tcu/h/ln and a protected left on the northbound approach of 50 tcu/h/ln, then 50 tcu/h/ln should be deducted from the eastbound 200 tcu/h/ln, for a result of 150 tcu/h/ln in the eastbound exclusive right-turn lane.

Step 5. Calculate Sum of Critical Lane Flow Rates

Find the critical lane groups for the east/west and then north/south approaches. Sum the per lane flow rates of each critical lane.

Step 6. Calculate Capacity of the Intersection

In order to find capacity, you will need a cycle length. If the analyst did not enter a cycle length, then it can be estimated as the number of critical lane groups times 30 s.

To calculate intersection capacity, use Eq. 10.13.

$$c_I = s_o \frac{C - (n_{cp} l_t)}{C} \tag{10.13}$$

where:

c_I capacity of the intersection, tcu/h/ln
s_o base saturation flow rate, pc/h/ln
C cycle length, s
n_{cp} number of critical phases
l_t total lost time per phase, s, defaulted to 4 s/phase.

Step 7. Calculate Intersection Sufficiency

Intersection sufficiency is based on the intersection volume-to-capacity ratio, which is found using Eq. 10.14.

$$X_c = \frac{V_c}{c_I} \tag{10.14}$$

where:

X_c critical intersection volume-to-capacity ratio
V_c sum of the critical lane flow rates, tcu/h/ln
c_I intersection capacity, tcu/h/ln.

The intersection sufficiency is defined as either under, near, or over capacity and describes the ability of the intersection to process the given demand. Table 10.14 provides the sufficiency levels and descriptions of what each represents. The analyst may stop here or continue on to Step 8 in Part II of the planning method.

Table 10.14 Intersection sufficiency—2016 HCM

Critical v/c ratio, X_c	Capacity assessment	Description
<0.85	Under	All demand may be accommodated with low to moderate delays
0.85–0.98	Near	Demand for critical lane groups is near capacity and some lane groups require more than one cycle to clear the intersection; all demand is able to be processed within the analysis period; delays are moderate to high
>0.98	Over	Demand for critical lane groups is just able to be accommodated with a cycle but oftentimes requires multiple cycles to clear the intersection, with long queues and high delays

Source "Highway Capacity Manual", *6th Edition*: *A Multimodal Guide for Mobility Analysis*, Transportation Research Board, Washington DC, 2016 [7], Exhibit 31–37, pp. 31–92, used with permission of the National Academy of Science, Courtesy of National Academies Press

10.6.2 Part II of the Planning Method, Delay and LOS

Part II of the planning method estimates delay and assigns a level of service grade based on that delay. But in order to calculate delay, detailed signal timing of the phases must be known or calculated.

Step 8. Calculate Effective Green Times, if Unknown

If the signal timing is unknown, then it is found by dividing the usable time available of the cycle length in proportion to the critical lane volumes. The time available of the cycle length is found by subtracting the total lost time, as shown in Eq. 10.15

$$g_{tot} = C - L \tag{10.15}$$

where:

g_{tot} total effective green time in the cycle, s
C cycle length, s
L total lost time in cycle length, s $= n_{cp} l_t$.

Step 9. Find Capacity and v/c Ratios of the Lane Groups

Lane group capacity and v/c ratio are found using Eqs. 10.16 and 10.17.

$$c_{i=} s_o N_i \frac{g_i}{C} \tag{10.16}$$

$$X_i = \frac{N_i v_i}{c_i} \tag{10.17}$$

where:

c_i capacity of lane group i, tcu/h
s_o base saturation flow rate, tcu/hg/ln
N_i number of lanes in lane group i
g_i effective green time for lane group i, s
C signal cycle length, s
v_i lane group flow rate per lane, tcu/h/ln
X_i v/c ratio for lane group i.

The intersection capacity and intersection v/c ratio are then computed from Eqs. 10.18 and 10.19.

$$c_{sum} = s_o \frac{\sum_{i=1}^{n_{cp}} g_{c,i}}{C} \quad (10.18)$$

$$X_c = \frac{V_c}{c_{sum}} \quad (10.19)$$

where:

c_{sum} Intersection capacity per lane, tcu/h/ln
X_c Intersection v/c ratio.

Step 10: Calculate Delay and Assign LOS

The calculations for delay are simplified equations, similar to those used in the 2000 HCM, before IQA was introduced in 2010, and with no delay calculation for pre-existing queues and with simplified values for progression factor. Delay is calculated using Eqs. 10.20 and the progression factors are found in Table 10.15. In Table 10.15, it can be seen that only very general description of the progression must be entered by the analyst. The level of service definitions are the same as those for the operational method and are found in Table 10.16.

$$d_i = d_{1i}\, PF + d_{2i}$$

$$d_{1i} = \frac{0.5\, C\left[1 - \left(\frac{g_i}{C}\right)\right]^2}{1 - \left[\min(1, X_i)^* \left(\frac{g_i}{C}\right)\right]}$$

Table 10.15 Progression factors—2016 HCM

Quality of progression	Progression factor, *PF*
Good	0.70
Random	1.00
Poor	1.25

Source "Highway Capacity Analysis": *A Guide for Multimodal Mobility Analysis,* Transportation Research Board, Washington D.C., 2016 [7], Exhibit 31–38, pp. 31–92, used with permission of the National Academy of Science, Courtesy of National Academies Press

Table 10.16 Level of service—2016 HCM

Level of service grade	Delay, s/veh
A	≤10
B	>10–20
C	>20–35
D	>35–55
E	>55–80
F	>80

Source "Highway Capacity Analysis": *A Guide for Multimodal Mobility Analysis,* Transportation Research Board, Washington D.C., 2016 [7], Exhibit 19–8, pp. 19–16, used with permission of the National Academy of Science, Courtesy of National Academies Press

$$d_{2i} = 225\left[(X_i - 1) + \sqrt{(X_i - 1)^2 + \frac{16\,X_i}{c_i}}\right] \tag{10.20}$$

where:

d_i average delay per vehicle in lane group i, s/veh.
d_{1i} average uniform delay per vehicle in lane group i, s/veh.
d_{2i} average overflow delay per vehicle in lane group i, s/veh.

Delay and LOS may also be calculated for the approach and for intersection, if desired. Note that if any v/c ratio is greater than 1.0, then LOS is defined as 'F', even if the delay value is less than 80 s per vehicle.

10.7 Conclusion

The planning method went from a simple methodology in the 1985 HCM to a much more detailed methodology in the 1994 HCM (the QEM method). The QEM methodology remained fundamentally the same through the 2010 HCM, with some small changes in each version. In the 2016 HCM, the signalized intersection subcommittee returned to a much simpler planning analysis, which uses critical movement analysis similar to circular 212, but added an optional delay calculation part. There have been discussions in the signalized subcommittee of continuing to simplify the methodologies, including the current operational methodology to make it look more like this planning method. This issue is further discussed in Chap. 11.

References

1. Highway Capacity Manual (1985) Special report 209. Transportation Research Board, Washington D.C.
2. Interim Materials on Highway Capacity (1980) Circular 212. Transportation Research Board, Washington D.C.
3. Highway Capacity Manual (1994) Special report 209. Transportation Research Board, Washington D.C.
4. Highway Capacity Manual (1997) Special report 209. Transportation Research Board, Washington D.C.
5. Highway Capacity Manual (2000) Special report 209, 4th edn. Transportation Research Board, Washington D.C.
6. Highway Capacity Manual (2010) Special report 209, 5th edn. Transportation Research Board, Washington D.C.
7. Highway Capacity Manual (2016) A guide for multimodal mobility assessment. Transportation Research Board, Washington D.C.

Chapter 11
Comments and Recommendations

This chapter contains thoughts and recommendations of the authors on the future development and use of the Highway Capacity Manual.

In Volume I of this book [1], the authors laid out a series of issues related to the future of the HCM. At the time, the HCQSC was working on an "update" to the 2010 HCM, expected in 2015. The chapter presented a series of fundamental issues that the Committee needed to address before publication of a full 6th edition of the manual.

In fact, the "update" became the full 6th edition, published in 2016. Because the full update was an afterthought, determined when the "update" was nearing completion, the Committee addressed none of the issues laid out in Volume I. Because the "update" had become such a massive effort, it was thought that designation as a full 6th edition was appropriate. The issues raised remain largely unresolved, and in some cases, unexplored. While this chapter will touch on some of the issues raised in Volume I, it will not repeat the entire discussion.

11.1 Complexity, Precision, and Accuracy

It is obvious that as the HCM has developed, its procedures have become more and more complex. Because of the accompanying software (HCS), the need for efficient manual computational procedures has been removed.

The result is that methodologies now consider a wider range of input variables, and produce sensitivities to a broad range of field conditions. The major downside is that most methodologies have become virtual black boxes, and the vast majority of users have little knowledge and understanding of what is in them.

This problem is particularly acute for interrupted flow facilities. The signalized intersection model, the basic building block of virtually all interrupted flow analysis, is only fully described in computer code. We have attempted to present a text description of the basic methodology and its analytic basis, but it has been a difficult task

E. S. Prassas and R. P. Roess, *The Highway Capacity Manual: A Conceptual and Research History Volume 2*, Springer Tracts on Transportation and Traffic 12,
https://doi.org/10.1007/978-3-030-34480-1_11

indeed. In writing this volume, it has become clear that the code adopted for the 2016 HCM contains myriad small (and not so small) revisions (from 2010) that were never explicitly discussed by the Committee or its Signalized Intersection Subcommittee. Methodologies for unsignalized intersections are similarly complex, though more documentation is available than for signalized intersections. The basic fact of the situation, is that for some methodologies, only the developer actually knows what is in them.

As has been noted in Volume 1, complexity has led to a situation where most members of the HCQSC do not have significant knowledge of methodologies they must vote to approve. They depend heavily on input from subcommittees, many of which contain only one or two people with comprehensive knowledge of recommended methodologies.

More important is the fact that complexity has made it more difficult to verify whether methodologies are producing accurate estimates of operating conditions. The principal outputs of the signalized intersection methodology are predictions of saturation flow rates and delays for component movements. Yet, it has been decades since any significant body of field data was used to verify whether these predictions are indeed accurate. More specifically, as each manual produced yet greater complexity, there is little or no evidence that increasing complexity has led to increased accuracy. Individual components have occasionally been calibrated with small data bases (but not always). Sometimes, procedural outputs have been compared to simulation outputs (who verified those models?) and even previous HCM methodologies. If a previous HCM is the standard, why change it?

Just because computational tools allow for more complex models to be implemented, it does not logically follow that more accurate results are guaranteed. Detailed data on interrupted flow facilities, however, is expensive to collect, so the Committee has been forced, in many cases, to proceed without it.

With complexity, the size of the HCM has expanded exponentially. The 2016 HCM now contains well over 3000 pages of material, much of it available only on line. Only a handful of people have read all of it, and most of these are from the contractors who produced various editions of the HCM. The task of the HCQSC to control its principal product has become virtually impossible.

11.2 Who's the Audience?

In 1950, the user community for the HCM was largely limited to facility designers who were searching for some national continuity of standards and approaches. The list of users has greatly expanded since then to include:

- Facility designers
- Traffic control engineers
- Operational analysts
- Planners

- Facility operators
- Environmental analysts
- Other occasional users.

The problem is that each set of users has very different needs in terms of output they are seeking. Designers continue to be focused on basic physical parameters of highway design—how many lanes, use of lanes, shoulders, etc. Traffic control engineers are focused on signal design and timing, and the operation of traffic coordination systems. Planners need many of the same things, but do not have detailed input details, and thus rely on defaults, which need to be carefully selected. Operational analysts continue to want predictions of various operational parameters, such as speed, density, v/c ratios, delays, etc. Facility operators need current information on system operations, and have special interest in the effects of high-occupancy vehicle lanes, user fees and collection methodologies, identification and removal of accidents and incidents, impacts of transportation management strategies, etc. Environmental analysts want traffic parameters that can be used to predict emissions.

The point is that it has become very difficult to provide each of these groups with what they need or want with a single set of methodologies. The risk is that in trying to address all, virtually no one is getting what they really need.

11.3 The Level of Service Predicament

The authors have been calling for a complete review of the level of service concept for many years.

Originally, LOS was introduced as a simple way of communicating quality of traffic operations to decision-makers, and to address situations in which early methodologies produced no numeric operational predictions upon which to base a judgment on operating quality. By 1994, the latter issue had been eliminated, with all methodologies producing numeric operating parameters. Over the years, LOS has retained it function to describe increasingly complex operating scenarios to the public and decision-makers.

LOS criteria for various facility types have dramatically changed over the years. Unfortunately, LOS has been written into numerous laws and regulations by various agencies, often as part of development control programs. Every time the Committee changes the definition of a particular LOS criterion, it changes laws. This is far beyond its job description.

In the 2016 HCM, LOS criteria are defined in several different ways:

- Single service measures (operation parameters such as speed, density, delay, etc.)
- Multiple service measures
- User perception indices.

While not yet strictly defined as LOS criteria, the 2016 HCM contains two additional concepts that various agencies have sought to be included in LOS:

- Travel time reliability
- Parameters describing only truck operations.

It is virtually impossible to develop a consistent set of LOS criteria for all facility types that include all of these as fundamental inputs.

The time has long since passed for the HCQSC to formally discuss and decide how to address this concept in future manuals. LOS has become far more than a convenient communications tool for non-professionals. It is actually a tool for *setting policy* on a local, regional, and national level. Consider the following:

- LOS is a critical tool for determining investment for various state and local highway agencies. Relative LOS on rural and urban facilities can be used, for example, to drive investments towards one or the other. In recent times, a number of agencies have advocated changing rural LOS designations in ways that would degrade them—thus forcing investments more towards rural facilities.
- LOS is a critical driver of development decisions, and critical to regulations which force developers to pay for various mitigations for developments that would lead to decreases in LOS.
- Possible new measures of travel time reliability and truck operations would have dramatic influences on national transportation polices and investment.

While clearly the Committee's role is to provide transportation decision-makers with information that assists them in making policy, it *is not* to provide methodologies which bias policy decisions in one direction or another.

11.3.1 An Argument for Eliminating LOS

The simplest way out of this conundrum is to eliminate the use of the term *level of service* in all future manuals and methodologies.

Current methodologies enable the production of multiple traffic operational parameters, user perception measures (for some facility types), and travel time reliability parameters (for facilities and systems). Since the HCM is not intended for non-professionals, perhaps it could be assumed that users have the required ability to interpret for themselves what the numerical outputs of HCM analyses actually mean.

Doing so would restore some sense of balance between the Committee's responsibility, and the responsibility of various user groups. It's not up to the Committee to decide what type of operations are acceptable or not in various contexts. That is the job of various federal, state, and local agencies. It would free the Committee to work solely on predictive models for a variety of desired outputs without concern for how these might affect policy.

Can the user community be cured of its dependency on LOS? The answer is probably not easily, and probably not quickly. That is why it is so critical for the HCQSC to address this issue significantly in advance of producing a 7th edition of the HCM. Seminars and other forms of communications with user groups would have to be systematically employed to insure that everyone understands what is coming.

11.3.2 Options for Retaining LOS

If the judgment of the HCQSC is that LOS must be retained, there are various avenues it can pursue.

a. The HCM could conceivably define a LOS structure, while leaving it to federal, state, and local agencies to determine where the breakpoints should be—in other words, let agencies define specific criteria for LOS A, B, C, D, and E, with the Committee perhaps providing clear guidance on LOS F and its meaning.
b. From 1965 through 1997, the HCM focused almost entirely on points and uniform segments within the traffic system. From 2000 onward, more attention has been paid to entire facilities and network systems. It is apparent that operational quality would have to be defined in very different ways for each.

   Traditional LOS could be defined for points and segments, using multiple operational parameters as service measures. User perception indices could be provided as supplemental material.

   Long sections of facilities and network systems could focus more on average or representative operational parameters, travel time reliability and perhaps truck-only measures. To avoid confusion, LOS would not be used for these. Perhaps the old term *Quality of Flow* could be revised and applied.
c. The Committee could continue as at present, with incremental changes to LOS. It would have to make special efforts to avoid inconsistencies across methodologies, and should remove, as far as is possible, policy implications of its specified criteria.

11.4 Form of Future Manuals

Given modern technology, it is hard to argue that the HCM needs to be a paper document at all. The one benefit of at least part of the document be available in hard copy is that it requires at least a minimal effort at efficiency and page limitations. It would be nice for the HCM to remain a document that at least some professionals actually read and understand.

It seems to the authors that the traditional split between interrupted and uninterrupted flow in the HCM has outlived its usefulness. The differences in approach, in

LOS definitions and criteria, and in applications, between interrupted and uninterrupted flow are not as stark as those between point and uniform segment analysis and facility and system analysis. Our view is that the HCM should be produced as two volumes, one on points and uniform segments, the other on facilities and systems. If the volumes are produced on paper, then the practice of having supplementary chapters (available only electronically) would continue. It is further our view that introductory material for each would be contained in each volume.

Introductory material should be greatly curtailed. The explosion of such material into a fully separate volume is not needed. It should be assumed that users are generally familiar with basic concepts of traffic flow. Extensive coverage of these basic concepts, and of traffic characteristics could be included in supplementary chapters if deemed necessary in the future.

11.5 A Recommendation on Complexity

In the absence of robust validation of the principal results of HCM models, it is easy to support a recommendation for simplifying many methodologies. While uninterrupted flow models (freeways, weaving, merging, diverging, two-lane and multilane surface highways) have become more complex, all of them can be reasonably applied by hand, and all of the key analytic principles can be adequately explained to users.

The same cannot be said about interrupted flow models. Unsignalized and signalized intersections, and urban street segments and facilities have become hopelessly complex, and defy complete description for users.

The methodology for two-way STOP-controlled intersections is still fundamentally tractable by hand, if it were not for the adjustment for platoon flow on the uncontrolled arterial. The precise methodology for this adjustment is not fully explained anywhere in the HCM, although there are references to it in multiple chapters. For all-way STOP-controlled intersections, the methodology is extremely complex involving multiple nested iterations. At least for roundabouts, the methodology is still relatively simple.

The complications of the signalized intersection model have been previously discussed.

Two questions arise:

- Even application of the HCS software to some cases (particularly signalized intersections) is not simple or straightforward. One might argue that direct use of simulation packages might be a more efficient way to analyze them. The advantage of a deterministic model is that it produces a uniform answer with uniform inputs. How much complexity is this worth?
- For unsignalized intersections, what do users truly need? How much complexity is justifiable when the majority of users just want to know whether a given location will work or not. One might argue that even a capacity determination is of little

value, since operating characteristics at such intersections will have long since become unacceptable.

In 2016, the Signalized Intersection Subcommittee opted to greatly simplify the Planning Method, going back to the principles of *Circular 212* again. The fact that several agencies still use the Circular 212 model is a strong argument that there need to be practical limits on complexity.

The Subcommittee has discussed making the planning method (or something like it) the operational analysis methodology, but has not made a determination to date.

The authors recommend an approach that lies somewhere between the simplicity of *Circular 212* and 2016 Planning Methodology, and the operational analysis methodology of the 2016 HCM. In simplifying the operational methodology, it is recommended that:

- The complete methodology should be explainable in text and equations. Coding *should not* be used as documentation. Such explanations should be possible in a 50–60 page chapter. A supplemental chapter could be added to contain additional details for software developers.
- The methodology should be implementable by hand with an analysis taking no more than 30 min for complex cases.

All methods, including uninterrupted and interrupted flow methodologies, should be simplified to the extent possible, with uninterrupted intersections focusing on whether or not the operation will be stable with given demands.

11.6 A Suggested Outline for a 7th Edition HCM

Table 11.1 shows a specific recommendation for the organization of a 7th edition HCM, based upon the authors' views. These are only recommendations made for discussion.

It would be envisioned that Volume I could be delivered within 500 pages, as would Volume II. If not, the very concept of a paper document would be impossible. The size of the electronic volumes is not subject to any arbitrary limitations, but (except for Volume V) reasonable targets should be set to insure that people actually read the material.

Note that we have recommended (in Volume I) that pedestrian and bicycle methodologies be re-separated into focused chapters. The current presentation of pedestrian and bicycle impacts within each chapter is inefficient, weakens the focus of other chapters, and prevents any cohesive presentation and discussion of pedestrians and bicycles as a more global subject. Many users have expressed this view to the authors in recent years.

Table 11.1 A proposed outline for a 7th edition HCM

Chapter	Form
Volume I: Points and uniform segments	
1. Basic Characteristics of Uninterrupted and Interrupted Flow 2. Concepts of Capacity, Level of Service(?) and v/c Ratio 3. Basic Freeway and Multilane Highway Segments 4. Weaving Areas on Freeways and Multilane Highway Segments 5. Merging and Diverging Areas on Freeways and Multilane Highway Segments 6. Two-Lane Highway Segments 7. Two-Way STOP (and YIELD?) Controlled Intersections 8. All-Way STOP Controlled Intersections 9. Roundabouts 10. Signalized Intersections 11. Pedestrian and Bicycle Segments	Paper, Electronic
Volume II: Facilities and network systems	
12. Facility and Network Concepts 13. Freeway Facilities 14. Multilane and Two-Lane Highway Facilities 15. Freeway, Multilane, and Two-Way Highway Systems 16. Urban Street Segments and Facilities 17. Urban Street Networks 18. Pedestrian and Bicycle Facilities 19. Transit Services	Paper, Electronic
Volume III: Supplemental chapters	
20–38. Supplemental Material for Chaps. 1–19 39. Traffic Characteristics: Uninterrupted and Interrupted Flow 40. Traffic Systems and Demand Management Policies and Impacts	Electronic only
Volume IV: Applications guide	
Chapter division as needed.	Electronic only
Volume V: Temporary chapters	
Chapter updates and interim replacements as needed	Electronic only
Volume VI: Source Documents and Other Reference Material	
As needed	Electronic only

11.7 Concluding Remarks

The *Highway Capacity Manual* has been one of the fundamental documents that have guided the design, planning, and analysis of traffic facilities and the users of such facilities. Nevertheless, a recent trend has been clear: HCM sales are down, and its use is challenged by an ever-increasing set of competing tools, including simulators.

If there is to be a critical role for the HCM, we believe it lies in returning to its roots: set of rational calibrated and validated methodologies that can predict operational outcomes of a wide variety of design, control, and management strategies in a reasonable period of time at reasonable cost.

We further believe that the validation role of the Committee must be greatly strengthened, and that the temptation to compete with the level of detail addressed by simulation and other available modeling tools must be resisted.

We must return to a time when HCQSC members all have a fundamental understanding of the methodologies they are promulgating, and the material in the ultimate product of their efforts: the HCM.

Reference

1. Roess R, Prassas E (2014) The highway capacity manual: a conceptual and research history. In: Uninterrupted flow, vol 1, Springer, 2014

Appendix A
Sample Calculations: TWSC Intersections

Given the general complexity of analysis methodologies for unsignalized and signalized intersections, it is difficult to illustrate complex problems. Computations, particularly when using the 2000, 2010, and 2016 HCMs, generally require software, and display of all computational steps would be extremely difficult. For this reason, this appendix focuses on a relatively simple example that can, for the most part, be implemented and documented by hand. These should assist the reader in understanding the conceptual approaches to analyses as they have developed over the years.

The Problem: A Simple TWSC Intersection

The intersection for this sample calculation is shown in Fig. A.1. It is a TWSC location with four lanes on the major street, and two lanes on the minor street. Left turns from the major street are handled from a short exclusive turn lane that can accommodate 2 stored vehicles.

Because the 1965 HCM really offers no methodology for TWSC Intersections other than to model them as if they were signalized intersections, sample solutions will begin with Circular 212.

Circular 212

Conversion of Demand Volumes to pc/h

The first step in the Circular 212 methodology is to convert demand volumes to flow rates in pc/h. Actually, this would be done in two separate steps, as *all* volumes

E. S. Prassas and R. P. Roess, *The Highway Capacity Manual: A Conceptual and Research History Volume 2*, Springer Tracts on Transportation and Traffic 12,
https://doi.org/10.1007/978-3-030-34480-1

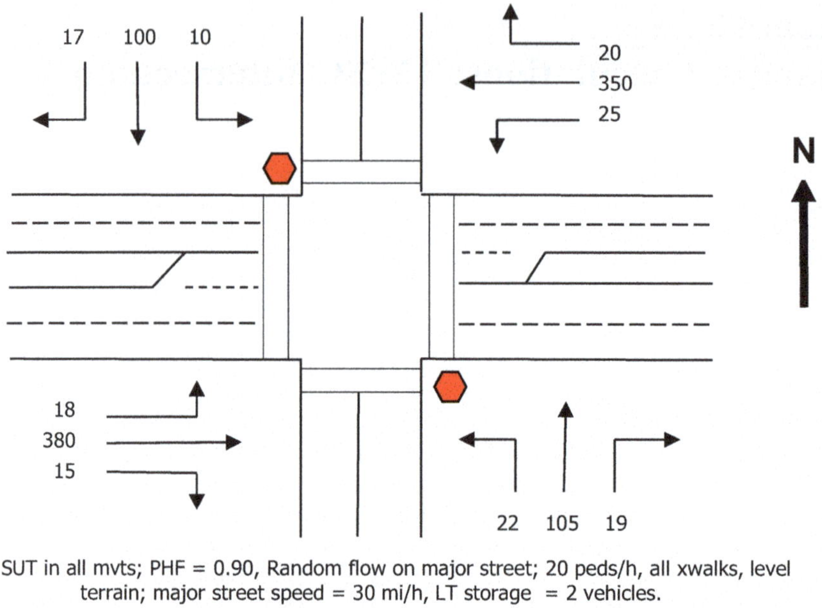

Fig. A.1 TWSC intersection example

should be converted to flow rates, but through and right-turn demands on the major street *do not* have to be converted to pc/h.

Conversion to flow rates in veh/h is accomplished by dividing by the PHF using Eq. 4.4. Conversion from flow rates in veh/h to flow rates in pc/h is accomplished by dividing by the heavy vehicle adjustment factor (f_{HV}). From Table 4.3, the passenger car equivalent (E_{HV}) for single-unit trucks on a level grade is 1.5. Thus, the heavy vehicle adjustment factor is:

$$f_{HV} = \frac{1}{1 + P_{HV}(E_{HV} - 1)} = \frac{1}{1 + 0.10(1.5 - 1)} = 0.952$$

Table A.1 shows these conversions. The tables also assigned movement numbers in accordance with Fig. 4.1. These will be used throughout the various solutions that are illustrated.

Computing Conflicting Volumes

The first computational step involves determining the conflicting volumes for each movement using gaps. This is done in using the equations presented in Fig. 4.2, and uses flow rates in veh/h (*not* pc/h). These computations are shown in Table A.2.

Finding the Critical Gap

Critical gaps for Circular 212 are taken from Table 4.5, for STOP control, a 4-lane major street, and 30 mi/h major street prevailing speed. Critical gaps are as follows:

Table A.1 Conversion to flow rates in pc/h—Circular 212

Mvt	Mvt No	Volume (veh/h)	Flow rate (veh/h)	Flow rate (pc/h)
EBL	1	18	$\div 0.90 = 20$	$\div 0.952 = 21$
EBT	2	380	$\div 0.90 = 422$	$\div 0.952 = 469$
EBR	3	15	$\div 0.90 = 17$	$\div 0.952 = 18$
WBL	4	25	$\div 0.90 = 28$	$\div 0.952 = 29$
WBT	5	350	$\div 0.90 = 389$	$\div 0.952 = 409$
WBR	6	20	$\div 0.90 = 22$	$\div 0.952 = 23$
NBL	7	22	$\div 0.90 = 24$	$\div 0.952 = 25$
NBT	8	105	$\div 0.90 = 117$	$\div 0.952 = 123$
NBR	9	19	$\div 0.90 = 21$	$\div 0.952 = 22$
SBL	10	10	$\div 0.90 = 11$	$\div 0.952 = 12$
SBT	11	100	$\div 0.90 = 111$	$\div 0.952 = 117$
SBR	12	17	$\div 0.90 = 19$	$\div 0.952 = 20$

Notes All results rounded to the nearest integer
All crosswalk volumes (13–16) = 20 peds/h or 20/0.9 = 22 peds/h as a flow rate

Table A.2 Conflicting volumes—Circular 212

Movements
Right Turns from the Minor Street Conflicting volume includes ½ of the major street through movement (from the left) - assuming that there is a 50-50 split in the two major street lanes, and ½ right turn from the major street (from the left)
$v_{c9} = 422/2 + 17/2 = 220\,\text{veh/h}$ $v_{c12} = 389/2 + 22/2 = 206\,\text{veh/h}$
Left Turns from the Major Street Conflicting volumes consist of the opposing through and right turn movements
$v_{c4} = 422 + 17 = 439\,\text{veh/h}$ $v_{c1} = 389 + 22 = 411\,\text{veh/h}$
Minor Street Through Movements Conflicting volumes include all major street flows, except for the right turn (approaching from the left), of which only ½ is counted
$v_{c8} = 17/2 + 20 + 422 + 22 + 389 + 28 = 890\,\text{veh/h}$ $v_{c11} = 22/2 + 389 + 28 + 17 + 422 + 20 = 887\,\text{veh/h}$
Left Turns from the Minor Street Conflicting volumes include all conflicting volume for minor street TH movements, *plus* the opposing minor street through and right-turn movements
$v_{c7} = 890 + 111 + 19 = 1020\,\text{veh/h}$ $v_{c10} = 887 + 117 + 21 = 1019\,\text{veh/h}$

Note All results rounded to the nearest integer

RTs, Minor Street (Mvts 9, 12) = 5.0 s/veh
LTs, Major Street (Mvts 1, 4) = 5.5 s/veh
THs, Minor Street (Mvts 8, 11) = 6.5 s/veh
LTs, Minor Street (Mvts 7, 10) = 7.0 s/veh.

Computing Potential Capacity
Figure 4.4 is used in Circular 212 to determine the potential capacity of each movement using gaps. It is entered with the conflicting volume in veh/h and the critical gap (for each movement) to find the potential capacity in pc/h. Figure A.2 illustrates two of the eight potential capacity determinations. Shown are Mvt 4 (LT from the major street) with a conflicting volume of 439 veh/h and a critical gap of 5.5 s/veh and Mvt 7 (LT from the minor street), with a conflicting volume of 1020 veh/h and a critical gap of 7.0 s. All results are summarized in Table A.3.

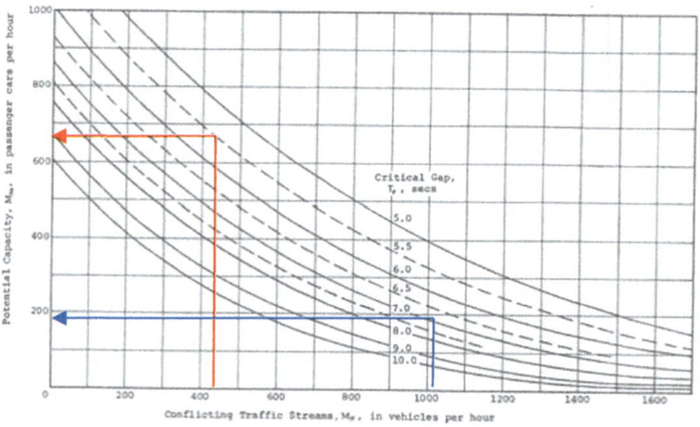

Fig. A.2 Potential capacity determination—Circular 212

Table A.3 Potential capacities for sample calculation A.1—Circular 212

Movement	Conflicting Vol. (veh/h)	Critical gap (s/veh)	Potential capacity (pc/h)*
1 (LT Major)	411	5.5	690
4 (LT Major)	439	5.5	670
9 (RT Minor)	220	5.0	960
12 (RT Minor)	206	5.0	980
8 (TH Minor)	890	6.5	270
11 (TH Minor)	887	6.5	270
7 (LT Minor)	1020	7.0	180
10 (LT Minor)	1019	7.0	180

*Due to the scale of Fig. A.2, potential capacities are read to the nearest 10 pc/h

Finding Impedance and Movement Capacities

Figure 4.6 illustrates the impedance effects that are accounted for in Circular 212. To summarize:

- Right turns from the minor street (Mvts 9, 12) are unimpeded.
- Left turns from the major street (Mvts 1, 4) are unimpeded.
- Through movements from the minor street (Mvts 8, 11) are impeded by *both* left turns from the major street (Mvts 1, 4).
- Left turns from the minor street (Mvts 7, 10) are impeded by *both* left turns from the major street (Mvts 1, 4) and the opposing minor street through and right turn movements (Mvts 11, 12 for Mvt 7, and Mvts 8, 9 for Mvt 10).

Because Movements 1, 4, 9, and 12 are unimpeded, their movement capacities equal their potential capacities:

$$c_{m1} = c_{p1} = 690 \text{ pc/h}$$
$$c_{m4} = c_{p4} = 670 \text{ pc/h}$$
$$c_{m9} = c_{p9} = 960 \text{ pc/h}$$
$$c_{m12} = c_{p12} = 980 \text{ pc/h}$$

Figure 4.6 is used to find individual impedance factors for each impeding movement, based upon the percent of movement capacity used by the impeding movement, i.e., 100 (v_i/c_{mi}). Note that the percentage is based upon *movement* capacities. This, impedance factors can be determined only *after* the movement capacity of each impeding factor is found.

Through movements from the minor street are impeded only by both left turns from the major street. These, however, are themselves not impeded, so we already have determined their movement capacities. Note also that the percent of capacity used must use the demand flow rate (v_i) in units of pc/h for consistency. These values were computed in Table A.1. Percent of capacity used for Mvt 1 is 100 * (21/690) = 3.04%; percent of capacity used for Mvt 4 is 100 * (29/670) = 4.3%. Figure A.3 shows the determination of impedance factors (P) for these values.

From Fig. A.3, $P_1 = 0.99$ and $P_4 = 0.97$. The equations in Fig. 4.6 may then be used to find the movement capacities of Mvts 8 and 11, the through movements from the minor street:

$$c_{m8} = c_{p8}\, P_1\, P_4 = 270^*0.99^*0.97 = 259 \text{ pc/h}$$
$$c_{m11} = c_{p11}\, P_1\, P_4 = 270^*0.99^*0.97 = 259 \text{ pc/h}$$

In terms of the movement capacity of minor street left turns, Mvt 7 is impeded by Mvts 1, 4, 11, and 12; Mvt 10 is impeded by Mvts 1, 4, 8 and 9. Thus, impedance factors must be found for Mvts 11, 12, 8, and 9. The percent of capacity used for these (using movement capacities) is:

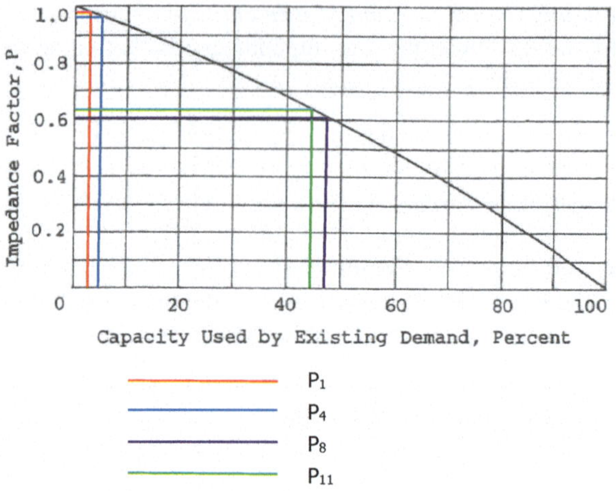

Fig. A.3 Impedance factors—Circular 212

$$\text{Mvt. } 11 = 100^*(117/259) = 45.1\%$$
$$\text{Mvt. } 12 = 100^*(20/980) = 2.0\%$$
$$\text{Mvt. } 8 = 100^*(123/259) = 47.5\%$$
$$\text{Mvt. } 9 = 100^*(22/960) = 2.3\%$$

Impedance factors for these are determined as shown in Fig. A.3. Only Mvts 11 and 8 are shown due to the scale of the figure. The results are:

$$P_8 = 0.61$$
$$P_{11} = 0.64$$
$$P_9 = 0.99$$
$$P_{12} = 0.99$$

Then:

$$c_{m7} = c_{p7} P_1 P_4 P_{11} P_{12} = 180^*0.99^*0.97^*0.64^*0.99 = 110\,\text{pc/h}$$
$$c_{m10} = c_{p10} P_1 P_4 P_8 P_9 = 180^*0.99^*0.97^*0.61^*0.99 = 104\,\text{pc/h}$$

Finding Shared-Lane Capacities
Movement capacities assume that each movement has access to an exclusive lane. In this case, however, Movements 7, 9, and 9 share a one-lane approach, as do Movements 10, 11, and 12. The shared-lane capacity for each of the minor street approached is computed using Eq. 4.18.

$$c_{SHNB} = \frac{25 + 123 + 22}{\left(\frac{25}{110}\right) + \left(\frac{123}{259}\right) + \left(\frac{22}{960}\right)} = \frac{170}{0.227 + 0.475 + 0.023} = \frac{170}{0.725} = 234\,\text{pc/h}$$

Table A.4 Reserve capacity and LOS—Circular 212

Lane or Mvt	Capacity (pc/h)	Demand (pc/h)	Reserve capacity (pc/h)	LOS (Table 4.15)
Mvt 1	690	21	690 – 21 = 669	A
Mvt 4	670	29	670 – 29 = 641	A
NB	234	170	234 – 170 = 64	E
SB	254	149	254 – 149 = 105	D

$$c_{SHSB} = \frac{20 + 117 + 12}{\left(\frac{20}{980}\right) + \left(\frac{117}{259}\right) + \left(\frac{12}{104}\right)} = \frac{149}{0.020 + 0.452 + 0.115} = \frac{149}{0.587} = 254\,\text{pc/h}$$

Movements 1 and 4 have an exclusive LT lane, so there is no shared-lane capacity involved.

Finding Reserve Capacity
In Circular 212, level of service is defined based upon "reserve" or "unused capacity." Reserve capacity is computed using Eq. 4.19, and LOS criteria are found in Table 4.13. Table A.4 summarizes the results.

The Circular 212 analysis shows that the left-turn movements from the major street will have little difficulty and very good LOS. The minor street approaches, however, will experience significant delay, and operate at poor levels of service. The capacity of the minor street approaches is near capacity, and could be severely affected by an increase in either/or minor street demand and/or major street demand.

1985 Highway Capacity Manual

Finding Conflicting Volumes
There were no changes in conflicting volumes in the 1985 HCM. The values computed in Table A.2 remain valid:

$$\begin{array}{ll} v_{c1} = 411\,\text{veh/h} & v_{c4} = 439\,\text{veh/h} \\ v_{c9} = 220\,\text{veh/h} & v_{c12} = 206\,\text{veh/h} \\ v_{c8} = 890\,\text{veh/h} & v_{c11} = 887\,\text{veh/h} \\ v_{c7} = 1020\,\text{veh/h} & v_{c10} = 1019\,\text{veh/h} \end{array}$$

Finding Critical Gaps
Critical gaps for the 1985 HCM are given in Table 4.6. None of the adjustments shown in Table 4.6 apply, so the base values are used directly. The critical gaps are:

RT Minor Street (Mvts 9, 12) = 5.5 s/veh

LT Major Street (Mvts 1, 4) = 5.5 s/veh
TH Minor Street (Mvts 8, 11) = 6.5 s/veh
LT Minor Street (Mvts 7, 10) = 7.0 s/veh.

Finding Potential Capacity
For the 1985 HCM, potential capacity is determined graphically from Fig. 4.5, which is entered with the conflicting volume in veh/h and critical gap to determine the potential capacity in pc/h. Some of these determinations are shown in Fig. A.4.

The resulting values are:

$$c_{p1} = 680 \text{ pc/h } c_{p4} = 670 \text{ pc/h (not shown)}$$
$$c_{p9} = 860 \text{ pc/h } c_{p12} = 870 \text{ pc/h (not shown)}$$
$$c_{p8} = 270 \text{ pc/h } c_{p11} = 270 \text{ pc/h (not shown)}$$
$$c_{p7} = 180 \text{ pc/h } c_{p10} = 180 \text{ pc/h (not shown)}$$

Note that estimates of potential capacity are rounded to the nearest 10 pc/h due to the scale of Fig. 4.5.

Finding Impedance and Movement Capacity
The process for determining impedance factors does not change in the 1985 HCM. However, some of the values of potential capacity have changed from Circular 212, and different impedance factors will, therefore, result, even using the same model and process.

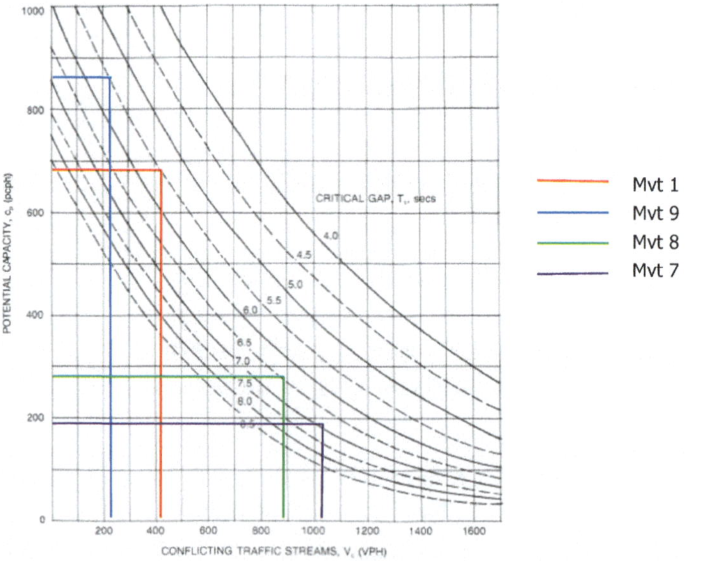

Fig. A.4 Determination of potential capacity—1985 HCM

As previously, left turns from the major street and right turns from the minor street are not impeded. Therefore, their movement capacities are the same as their potential capacities:

$$c_{m1} = 680\,\text{pc/h} \; c_{m4} = 670\,\text{pc/h}$$
$$c_{m9} = 860\,\text{pc/h} \; c_{m12} = 870\,\text{pc/h}$$

Once again, through movements from the minor street are impeded only by both left turns from the major street. As previously, the percent of capacity used uses the demand flow rate (v_i) in units of pc/h for consistency. These values were computed in Table A.1. Percent of capacity used for Mvt 1 is 100 * (21/680) = 3.1%; percent of capacity used for Mvt 4 is 100 * (29/670) = 4.3%. This is a negligible change from Circular 212, so the impedance factors remain $P_1 = 0.99$ and $P_4 = 0.97$. Then:

$$c_{m8} = c_{p8} P_1 P_4 = 270{*}0.99{*}0.97 = 259\,\text{pc/h}$$
$$c_{m11} = c_{p11} P_1 P_4 = 270{*}0.99{*}0.97 = 259\,\text{pc/h}$$

In terms of the movement capacity of minor street left turns, Mvt 7 is again impeded by Mvts 1,4, 11, and 12; Mvt 10 is impeded by Mvts 1,4, 8 and 9. Thus, impedance factors must be found for Mvts 11, 12, 8, and 9. The percent of capacity used for these (using movement capacities) is:

$$\text{Mvt.}11 = 100{*}(117/259) = 45.1\%$$
$$\text{Mvt.}12 = 100{*}(20/870) = 2.3\%$$
$$\text{Mvt.}8 = 100{*}(123/259) = 47.5\%$$
$$\text{Mvt.}9 = 100{*}(22/860) = 2.6\%$$

The values differ from Circular 212 for Mvts 9 and 12 (the minor street right turns). The values for Mvts 8 and 11 are the same as in Circular 212. Impedance factors for Mvts 9 and 12 do not change much either. All are obtained from Fig. A.4. The results are:

$$P_8 = 0.61$$
$$P_{11} = 0.64$$
$$P_9 = 0.98$$
$$P_{12} = 0.99$$

Then:

$$c_{m7} = c_{p7} P_1 P_4 P_{11} P_{12} = 180{*}0.99{*}0.97{*}0.64{*}0.98 = 108\,\text{pc/h}$$
$$c_{m10} = c_{p10} P_1 P_4 P_8 P_9 = 180{*}0.99{*}0.97{*}0.61{*}0.99 = 104\,\text{pc/h}$$

Table A.5 Reserve capacity and LOS—85HCM

Lane or Mvt	Capacity (pc/h)	Demand (pc/h)	Reserve capacity (pc/h)	LOS (Table 4.15)
Mvt 1	680	21	680 – 21 = 659	A
Mvt 4	660	29	660 – 29 = 631	A
NB	232	170	232 – 170 = 62	E
SB	253	149	253 – 149 = 104	D

Finding Shared-Lane Capacities

As previously, Mvts 7, 8, and 9 share a lane (the NB approach), and Mvts 10, 11, and 12 share a lane (the SB approach). The shared-lane capacities for these lanes are:

$$c_{SHNB} = \frac{25+123+22}{\left(\frac{25}{108}\right)+\left(\frac{123}{259}\right)+\left(\frac{22}{860}\right)} = \frac{170}{0.231+0.475+0.026} = \frac{170}{0.732} = 232\ \text{pc/h}$$

$$c_{SHSB} = \frac{20+117+12}{\left(\frac{20}{870}\right)+\left(\frac{117}{259}\right)+\left(\frac{12}{104}\right)} = \frac{149}{0.023+0.452+0.115} = \frac{149}{0.590} = 253\ \text{pc/h}$$

These values represent only a minor change from the Circular 212 results.

Finding Reserve Capacity and Level of Service

The reserve capacity for each lane and the resulting level of service is shown in Table A.5. Reserve capacity is computed using Eq. 4.19, and levels of service are drawn from Table 4.13.

The results do not change radically from Circular 212: The left turn movements from the major street will experience little difficulty. The STOP-controlled NB and SB approaches will not work well, and long delays would be expected.

The only differences in the 1985 HCM (compared to Circular 212) come from the increased critical gap for minor street right turns (from 5.0 to 5.5 s/veh), and slight changes in the figure for determining potential capacities.

The 1994 Update to the 1985 HCM

Finding Conflicting Volume

Conflicting volume computations remain as in Circular 212:

$$\begin{aligned} v_{c1} &= 411\ \text{veh/h} & v_{c4} &= 439\ \text{veh/h} \\ v_{c9} &= 220\ \text{veh/h} & v_{c12} &= 206\ \text{veh/h} \\ v_{c8} &= 890\ \text{veh/h} & v_{c11} &= 887\ \text{veh/h} \\ v_{c7} &= 1020\ \text{veh/h} & v_{c10} &= 1019\ \text{veh/h} \end{aligned}$$

Finding Critical Gaps and Follow-Up Times

The 1994 update is the first edition to use both critical gaps and follow-up times explicitly in gap acceptance model. Critical gaps and follow-up times for the 1994 update are taken from Table 4.7. The results are:

$$\begin{aligned}
&\text{LT Major Street (Mvts 1, 4)}: && t_c = 5.5\,\text{s/veh} \quad t_f = 2.1\,\text{s/veh}\\
&\text{RTMinor Street (Mvts 9, 12)}: && t_c = 5.5\,\text{s/veh} \quad t_f = 2.6\,\text{s/veh}\\
&\text{TH Minor Street (Mvts 8, 11)}: && t_c = 6.5\,\text{s/veh} \quad t_f = 3.3\,\text{s/veh}\\
&\text{LT Minor Street: (Mvts 7, 10)}: && t_c = 7.0\,\text{s/veh} \quad t_f = 3.4\,\text{s/veh}
\end{aligned}$$

All values are selected from Table 4.7 for a four-lane major street.

Finding Potential Capacity

The 1994 update abandons the graphic approach to determining potential capacity in favor of Eq. 4.7:

$$c_{px} = \frac{3600}{t_f} e^{-\left(\frac{v_{cx}\, t_o}{3600}\right)}$$

The term t_o is computed as $t_c - (t_f/2)$. For the four types of movements, the value of t_o is:

$$\begin{aligned}
&\text{LTMajor Street (Mvts 1, 4)}: && t_o = 5.5 - (2.1/2) = 4.45\ \text{s/veh}\\
&\text{RT Minor Street (Mvts 9, 12)}: && t_o = 5.5 - (2.6/2) = 4.20\,\text{s/veh}\\
&\text{TH Minor Street (Mvts 8, 11)}: && t_o = 6.5 - (3.3/2) = 4.85\ \text{s/veh}\\
&\text{LT Minor Street (Mvts 7, 10)}: && t_o = 7.0 - (3.4/2) = 5.30\,\text{s/veh}
\end{aligned}$$

Then:

$$\begin{aligned}
c_{p1} &= \frac{3600}{2.1} e^{-\left(\frac{411*4.45}{3600}\right)} = 1031\ \text{pc/h}\\
c_{p4} &= \frac{3600}{2.1} e^{-\left(\frac{4.39*4.45}{3600}\right)} = 996\ \text{pc/h}\\
c_{p9} &= \frac{3600}{2.6} e^{-\left(\frac{220*4.20}{3600}\right)} = 1071\ \text{pc/h}\\
c_{p12} &= \frac{3600}{2.6} e^{-\left(\frac{206*4.20}{3600}\right)} = 1089\ \text{pc/h}\\
c_{p8} &= \frac{3600}{3.3} e^{-\left(\frac{890*4.85}{3600}\right)} = 329\ \text{pc/h}\\
c_{p11} &= \frac{3600}{3.3} e^{-\left(\frac{887*4.85}{3600}\right)} = 330\ \text{pc/h}\\
c_{p7} &= \frac{3600}{3.4} e^{-\left(\frac{1020*5.3}{3600}\right)} = 236\ \text{pc/h}\\
c_{p10} &= \frac{3600}{3.4} e^{-\left(\frac{1019*5.3}{3600}\right)} = 236\ \text{pc/h}
\end{aligned}$$

All of these values are significantly higher than the results of previous editions (Circular 212, 85 HCM). This is primarily due to the fact that follow-up times are specifically recognized. Follow-up time allows for multiple minor street vehicles to use the *same* gap when it is sufficiently long.

Finding Impedance and Movement Capacity
The basic impedances do not change from Circular 212 or the 1985 HCM. Left turns from the major street and right turns from the minor street continue to be unimpeded, and their movement capacities are the same as their potential capacities:

$$c_{m1} = 1031\,\text{pc/h}\ c_{m4} = 996\,\text{pc/h}$$
$$c_{m9} = 1071\,\text{pc/h}\ c_{m12} = 1089\,\text{pc/h}$$

Through movements from the minor street (Mvts 8 and 11) are impeded by both left turns from the major street. In the 1994 update, impedance factors are no longer graphically determined, but are computed using Eq. 4.10. Then:

$$P_1 = 1 - (v_1/c_{m1}) = 1 - (21/1031) = 0.98$$
$$P_4 = 1 - (v_4/c_{m4}) = 1 - (29/996) = 0.97$$

Then:

$$c_{m8} = c_{p8}\, P_1\, P_4 = 329{*}0.98{*}0.97 = 313\,\text{pc/h}$$
$$c_{m11} = c_{p11}\, P_1\, P_4 = 330{*}0.98{*}0.97 = 314\,\text{pc/h}$$

At this point, the procedure for impedance factor determination gets a bit more complicated. As previously, left turns from the minor street are impeded by both major street left turns, and the opposing minor street through movement and right-turn movement. This would then result in the following:

$$c_{m7} = c_{p7} P_1\, P_4\, P_{11}\, P_{12}$$
$$c_{m10} = c_{p10} P_1\, P_4\, P_8\, P_9$$

The 1994 update, however, recognizes that the impedances for the major street left turns and the opposing through movement somewhat overlap. This is because it is possible for some opposing through movements and major street left turns to operate through the *same* gap. Thus, the net effect of those impedances should be somewhat reduced to reflect this.

Before this can be done, the impedance factors for the minor street through movements (Mvts 8 and 11) must be computed using Eq. 4.10:

$$P_8 = 1 - (v_8/c_{m8}) = 1 - (123/313) = 0.61$$
$$P_{11} = 1 - (v_{11}/c_{m11}) = 1 - (117/314) = 0.63$$

In determining the movement capacity for Mvt 7, impedances for movements 1, 4, and 11 overlap. Normally, when multiplied, these three impedances would yield 0.98 * 0.97 * 0.63 = 0.60. In determining the movement capacity for Mvt 10, impedances for movements 1, 4, and 7 overlap. These three would normally yield 0.98 * 0.97 * 0.61 = 0.58.

The 1994 update indicates that these products should be modified as follows:

$$P' = 0.65\,P'' - \left(\frac{P''}{P''+3}\right) + 0.6\sqrt{P''}$$

where P″ is the product of the three overlapping impedances (see previous paragraph). The modified values are:

$$P'_7 = 0.65(0.60) - \left(\frac{0.60}{0.60+3}\right) + 0.6\sqrt{0.60} = 0.390 - 0.167 + 0.465 = 0.69$$

$$P'_{10} = 0.65(0.58) - \left(\frac{0.58}{0.58+3}\right) + 0.6\sqrt{0.58} = 0.377 - 0.162 + 0.457 = 0.67$$

Impedance factors for the minor street right-turn movements must also be determined using Eq. 4.10:

$$P_9 = 1 - (v_9/c_{m9}) = 1 - (22/1071) = 0.98$$
$$P_{12} = 1 - (v_{12}/c_{m12}) = 1 - (10/1089) = 0.99$$

Then:

$$c_{m7} = c_{p7}P'_7P_{12} = 236{*}0.69{*}0.99 = 161\text{ pc/h}$$
$$c_{m10} = c_{p10}P'_{10}P_9 = 236{*}0.67{*}0.98 = 155\text{ pc/h}$$

Finding Shared-Lane Capacity

As with previous solutions, Mvts 7, 8, and 9 share one minor street approach lane (NB), and Mvts 10, 11, and 12 share one minor street approach lane (SB). Shared-lane capacities for these are computed as:

$$c_{SHNB} = \frac{25+123+22}{\left(\frac{25}{161}\right)+\left(\frac{123}{313}\right)+\left(\frac{22}{1071}\right)} = \frac{170}{0.255+0.393+0.021} = \frac{170}{0.669} = 254\text{ pc/h}$$

$$c_{SHSB} = \frac{20+117+12}{\left(\frac{20}{1089}\right)+\left(\frac{117}{314}\right)+\left(\frac{12}{155}\right)} = \frac{149}{0.018+0.373+0.077} = \frac{149}{0.468} = 318\text{ pc/h}$$

Finding Delay and Level of Service

The 1994 update is the first to use control delay as the service measure to determine level of service. Delay is estimated using Eq. 4.20:

$$D_x = \frac{3600}{c_{mx}} + 900T\left[\frac{v_x}{c_{mx}} - 1 + \sqrt{\left(\frac{v_x}{c_{mx}} - 1\right)^2 + \frac{\left(\frac{3600}{c_{mx}}\right)\left(\frac{v_x}{c_{mx}}\right)}{450T}}\right]$$

There are *four* delay values to estimate in this problem: The two major street left turns (operating out of an exclusive lane), and the two shared-lanes on the STOP-controlled approaches. Note that the length of the analysis period, T, is taken to be the standard 15 min, or T = 0.25 h. Then:

$$D_1 = \frac{3600}{1031} + 900(0.25)\left[\frac{21}{1031} - 1 + \sqrt{\left(\frac{21}{1031} - 1\right)^2 + \frac{\left(\frac{3600}{1031}\right)\left(\frac{21}{1031}\right)}{450{*}0.25}}\right]$$

$$D_1 = 3.49 + 225\left[-0.980 + \sqrt{0.960 + 0.001}\right] = 3.49 + 225[-0.980 + 0.980]$$

$$= 3.5 \text{ s/veh}$$

$$D_4 = \frac{3600}{996} + 900(0.25)\left[\frac{29}{996} - 1 + \sqrt{\left(\frac{29}{996} - 1\right)^2 + \frac{\left(\frac{3600}{996}\right)\left(\frac{29}{996}\right)}{450{*}0.25}}\right]$$

$$D_4 = 3.61 + 225\left[-0.971 + \sqrt{0.943 + 0.001}\right] = 3.61 + 225[-0.971 + 0.972]$$

$$= 3.8 \text{ s/veh}$$

$$D_{NB} = \frac{3600}{254} + 900(0.25)\left[\frac{170}{254} - 1 + \sqrt{\left(\frac{170}{254} - 1\right)^2 + \frac{\left(\frac{3600}{254}\right)\left(\frac{170}{254}\right)}{450{*}0.25}}\right]$$

$$D_{NB} = 14.17 + 225\left[-0.331 + \sqrt{0.110 + 0.0441}\right]$$

$$= 14.17 + 225\,[-0.331 + 0.441]$$

$$= 38.9 \text{ s/veh}$$

$$D_{SB} = \frac{3600}{318} + 900(0.25)\left[\frac{149}{318} - 1 + \sqrt{\left(\frac{149}{318} - 1\right)^2 + \frac{\left(\frac{3600}{318}\right)\left(\frac{149}{318}\right)}{450{*}0.25}}\right]$$

$$D_{SB} = 11.32 + 225\left[-0.531 + \sqrt{0.282 + 0.047}\right]$$

$$= 11.32 + 225[-0.531 + 0.574]$$

$$= 21.0 \text{ s/veh}$$

Levels of service for the 1994 update are given in Table 4.14. The delay and LOS results are summarized in Table A.6.

While the 1994 update introduced a number of new features, the most important of which was the use of control delay to determine LOS, the final results (in terms of LOS) do not change from either Circular 212 or the 1985 HCM. The major

Table A.6 Levels of service—1994 update

Movement (Lane)	Delay (s/veh)	LOS
1	3.5	A
4	3.8	A
NB	38.9	E
SB	21.0	D

street left turns continue to operate very well, while the STOP-controlled approaches experience poor levels of service.

The 1997 Update to the 95 HCM and the 2000 HCM

The 1997 update introduces the specific consideration of pedestrians into the methodology for analysis of TWSC intersections. This affects the conflicting volumes used in the methodology, and also impedance. The 2000 HCM method is the same as in the 1997 update.

Finding Conflicting Volumes
Conflicting volumes for the 1997 update/2000 HCM are illustrated in Fig. 4.3. Pedestrians are added to conflicting volumes. There are a number of detailed footnotes, but none of them apply to the sample calculation. Conflicting volumes for minor street through and left-turn movements are shown in two stages, accommodating major streets with a wide median. Since that does not exist in our sample calculation, the conflicting volume for these cases is the sum of the two stages shown in Fig. 4.3.

A more subtle, but important change is that for minor street TH and LT movements, the left turns from the major street are now multiplied by 2, doubling their impact on conflicting volumes from previous edition. This accounts for the fact that such vehicles may proceed into the intersection while "waiting" for a gap—thus increasing their impact in blocking minor street TH and LT movements.

Note that the problem statement indicates that the pedestrian flow in *each* crosswalk is 20 peds/h. Thus, the conflicting volumes shown in Table A.2 must all be increased to include the impact of pedestrians, and the increase in the impact of major street LTs. The latter is accomplished by adding major street LTs again wherever they are included in the previous computation.

Movements 1, 4, 9, and 12 each involve pedestrian conflicts in *one* crosswalk. Thus, previous conflicting volumes must be increased by 20 peds/h in each case. Movements 8, 11, 7, and 10 each involve pedestrian conflicts in *two* crosswalks. Thus, previous conflicting volumes must be increased by $20 + 20 = 40$ peds/h in each case. LT's from the minor street are also added for minor street TH and LT movements. The resulting conflicting volumes are:

$$\begin{aligned}
v_{c1} &= 411 + 20 &&= 431 \text{ conflicts/h} \\
v_{c4} &= 439 + 20 &&= 459 \text{ conflicts/h} \\
v_{c9} &= 220 + 20 &&= 240 \text{ conflicts/h} \\
v_{c12} &= 206 + 20 &&= 226 \text{ conflicts/h} \\
v_{c8} &= 890 + 40 + 21 + 29 &&= 980 \text{ conflicts/h} \\
v_{c11} &= 887 + 40 + 21 + 29 &&= 977 \text{ conflicts/h} \\
v_{c7} &= 1020 + 40 + 21 + 29 &&= 1110 \text{ conflicts /h} \\
v_{c10} &= 1019 + 40 + 21 + 29 &&= 1109 \text{ conflicts /h}
\end{aligned}$$

Finding Critical Gaps and Follow-Up Times

The process for determining critical gaps and follow-up times is more complex in the 1997 update and 2000 HCM. They are determined using Eqs. 4.5 and 4.6. The equations allow for base critical gaps and base follow-up times to be adjusted to account for such conditions as heavy vehicle presence, grades, two-stage movements, and certain movements at T-intersections. Base values are drawn from Table 4.8, while adjustment factors are drawn from Table 4.9.

The equations used are:

$$t_c = t_{cb} + t_{cHV} P_{HV} + t_{cG} G - t_{cT} - t_{3LT}$$
$$t_f = t_{fb} + t_{fHV} P_{HV}$$

For this sample calculation, the only adjustment that applies is for heavy vehicles. The intersection is at level grade, there are no two-stage movements, and it is not a T-intersection. Thus, the values of t_{cG}, t_{cT}, t_{3LT} are all "0" for all movements. Table A.7 shows the values of base critical gap, base follow-up time, and heavy vehicle adjustments selected from Tables 4.8 and 4.9.

Using these values in Eqs. 4.5 and 4.6:

$$\begin{aligned}
t_{c1,4} &= 4.1 + (0.10^{*}2) = 4.3 \text{ s/veh} \\
t_{f1,4} &= 2.2 + (0.10^{*}1) = 2.3 \text{ s/veh} \\
t_{c9,12} &= 6.9 + (0.10^{*}2) = 7.1 \text{ s/veh} \\
t_{f9,12} &= 3.3 + (0.10^{*}1) = 3.4 \text{ s/veh} \\
t_{c8,11} &= 6.5 + (0.10^{*}2) = 6.7 \text{ s/veh}
\end{aligned}$$

Table A.7 Base values and adjustments—1997 update/2000 HCM

Movement	t_{cb} (s/veh)	t_{fb} (s/veh)	t_{cHV} (s/veh)	t_{fHV} (s/veh)
LT Major (1, 4)	4.1	2.2	2	1
RT Minor (9, 12)	6.9	3.3	2	1
TH Minor (8, 11)	6.5	4.0	2	1
LT Minor (7, 10)	7.5	3.5	2	1

$$t_{f8,11} = 4.0 + (0.10^{*}1) = 4.1 \text{ s/veh}$$
$$t_{c7,10} = 7.5 + (0.10^{*}2) = 7.7 \text{ s/veh}$$
$$t_{f7,10} = 3.5 + (0.10^{*}1) = 3.6 \text{ s/veh}$$

Finding Potential Capacity

Potential capacity in the 1997 update and 2000 HCM is computed using Eq. 4.8. The computation is shown for Movement 1, and the results are tabulated in Table A.7.

$$c_{px} = v_{cx}\left[\frac{e^{-(v_{cx}t_{cx}/3600)}}{1 - e^{-(v_{cx}t_{fx}/3600)}}\right]$$

$$c_{p1} = 431\left[\frac{e^{-(431^{*}4.3/3600)}}{1 - e^{-(431^{*}2.3/3600)}}\right] = 1070\,\text{pc/h}$$

There is one major difference in Table A.8 from previous solutions: the potential capacity of the right turn movements from the minor street is significantly less than in Circular 212, the 1985 HCM, or the 94 Update. In all previous solutions, the potential capacity of LTs from the major street is *less* than the potential capacity of RTs from the minor street. In 1997, this is reversed. The primary reason for this is a significant increase in the base critical gap for right turns from the minor street from previous editions; the relative increase compared to critical gaps for major street left turns is also significantly higher. This reduces the potential capacity of right turns from the minor street.

Finding Impedance and Movement Capacity

Impedance became a more complicated issue in the 1997 update. Impedance due to pedestrian movements was added. The implication of this is that *no* movements are now unimpeded, as even LTs from the major street and RTs from the minor street are impeded by pedestrian movements. All impedance effects are summarized in Table 4.17.

In terms of pedestrian impedance, LT movements from the major street are impeded by *one* pedestrian movement each. All other movements are impeded by

Table A.8 Computed values of potential capacity—1997 update/2000 HCM

Movement	v_{cx} (veh/h)	t_{cx} (s/veh)	t_{fx} (s/veh)	c_{px} (pc/h)
1	431	4.3	2.3	1070
4	459	4.3	2.3	1044
9	240	7.1	3.4	737
12	226	7.1	3.4	752
8	980	6.7	4.1	235
11	977	6.7	4.1	236
7	1110	7.8	3.6	154
10	1109	7.8	3.6	154

two pedestrian movements each. Pedestrian movements are numbered Mvts 13, 14, 15, and 16. In our case, the numbering is irrelevant, as all crosswalks have the same demand: 20 peds/h.

The pedestrian impedance factor is given in Eq. 4.13:

$$P_y = 1 - \left[\frac{v_j(w/S_p)}{3600}\right]$$

where "w" is the lane width for the impeded movement, and S_p is the walking speed of pedestrians. Because no lane widths were specified in the problem statement, we will assume that standard 12-ft apply to all lanes. A walking speed of 4 ft/s is the default value suggested in the 2000 HCM, and will be used. Because the pedestrian flow rate in all crosswalks is the same, the impedance factor for all pedestrian movements (13-16) will be the same:

$$P_{13-16} = 1 - \left[\frac{20^*(12/4)}{3600}\right] = 0.98$$

The LT movements from the major street are impeded only by one pedestrian movement (such left turns only pass through one crosswalk). Thus:

$$c_{m1} = c_{p1}\ P_{13-16} = 1070^*0.98 = 1049 \text{ pc/h}$$
$$c_{m4} = c_{p4}\ P_{13-16} = 1044^*0.98 = 1023 \text{ pc/h}$$

RT movements from the minor street are impeded by two pedestrian movements (such turns pass through two crosswalks). Thus:

$$c_{m9} = c_{p9}\ P_{13-16}\ P_{13-16} = 737^*0.98^*0.98 = 708 \text{ pc/h}$$
$$c_{m12} = c_{p12}\ P_{13-16}\ P_{13-16} = 752^*0.98^*0.98 = 722 \text{ pc/h}$$

TH movements from the minor street are now impeded by both major street LT movements, and by pedestrians in two crosswalks. Equation 4.10 is used to compute the impedance factors for the major street left turns (Mvts 1 and 4):

$$P_1 = 1 - (v_1/c_{m1}) = 1 - (21/1049) = 0.98$$
$$P_4 = 1 - (v_4/c_{m4}) = 1 - (29/1023) = 0.97$$

Then:

$$c_{m8} = c_{p8}\ P_1\ P_4 P_{13-16}\ P_{13-16} = 235^*0.98^*0.97^*0.98^*0.98 = 215 \text{ pc/h}$$
$$c_{m11} = c_{p11} P_1\ P_4 P_{13-16}\ P_{13-16} = 236^*0.98^*0.97^*0.98^*0.98 = 215 \text{ pc/h}$$

LTs from the minor street are impeded by both major street left turns, the opposing minor street through movement and right-turn movement, and, in 1997, two pedestrian crosswalk flows. This would then result in the following:

$$c_{m7} = c_{p7} P_1 \; P_4 \; P_{11} \; P_{12} P_{13-16} P_{13-16}$$
$$c_{m10} = c_{p10} P_1 \; P_4 \; P_8 \; P_9 P_{13-16} P_{13-16}$$

Impedance factors for Mvts 9, 12, 8, and 11 are computed using Eq. 4.10:

$$P_9 = 1 - (v_9/c_{m9}) = 1 - (22/708) = 0.97$$
$$P_{12} = 1 - (v_{12}/c_{m12}) = 1 - (20/722) = 0.97$$
$$P_8 = 1 - (v_8/c_{m8}) = 1 - (123/215) = 0.43$$
$$P_{11} = 1 - (v_{11}/c_{m11}) = 1 - (117/215) = 0.46$$

However, just as was the case in the 1994 update, impedance factors P_1, P_4, and P_{11} overlap when estimating the movement capacity of Mvt. 7. Similarly, impedance factors P_1, P_4, and P_8 overlap when estimating the movement capacity of Mvt. 10. Thus, these values must be adjusted to account for this. The adjustments are made to the product of the overlapping impedance factors, or:

$$P_7'' = P_1 \; P_4 \; P_{11} = 0.98^*0.97^*0.46 = 0.44$$
$$P_{11}'' = P_1 \; P_4 \; P_8 = 0.98^*0.97^*0.43 = 0.41$$

Then:

$$P' = 0.65\, P'' - \left(\frac{P''}{P'' + 3}\right) + 0.6\sqrt{P''}$$
$$P_7' = 0.65(0.44) - \left(\frac{0.44}{0.44 + 3}\right) + 0.6\sqrt{0.44} = 0.286 - 0.128 + 0.415 = 0.57$$
$$P_{10}' = 0.65(0.41) - \left(\frac{0.41}{0.41 + 3}\right) + 0.6\sqrt{0.41} = 0.267 - 0.120 + 0.384 = 0.53$$

Then:

$$c_{m7} = c_{p7} P_7' P_{12} P_{13-16} P_{13-16} = 154^*0.57^*0.97^*0.98^*0.98 = 82 \text{ pc/h}$$
$$c_{m10} = c_{p10} P_{10}' \; P_9 P_{13-16} P_{13-16} = 154^*0.53^*0.97^*0.98^*0.98 = 76 \text{ pc/h}$$

Movement capacities for the right turns from the minor street are significantly lower than previous solution, but this is primarily due to lower potential capacities, which are in turn related to significantly higher critical gaps specified for these movements. Movement capacities for left turns from the minor street are also significantly lower. This is due to lower potential capacities and the increased impact of impedances on these movements.

Finding Shared-Lane Capacity
As in all previous solutions, Mvts 7, 8, and 9 share one STOP-controlled lane (NB) and Mvts. 10, 11, and 12 share one STOP-controlled lane (SB). The shared-lane capacities are:

$$c_{SHNB} = \frac{25 + 123 + 22}{\left(\frac{25}{82}\right) + \left(\frac{123}{215}\right) + \left(\frac{22}{708}\right)} = \frac{170}{0.304 + 0.572 + 0.031} = \frac{170}{0.907} = 187 \text{ pc/h}$$

$$c_{SHSB} = \frac{20 + 117 + 12}{\left(\frac{20}{722}\right) + \left(\frac{117}{215}\right) + \left(\frac{12}{76}\right)} = \frac{149}{0.028 + 0.544 + 0.158} = \frac{149}{0.730} = 204 \text{ pc/h}$$

The NB and SB capacities are significantly lower than previous solutions, reflecting the lower values of movement capacities computed earlier.

Finding Delay and Level of Service
The service measure for LOS was changed to control delay in 1994, and it remains the criteria in 1997/2000. There was a slight change in the delay prediction algorithm: 5 s was added to reflect observations that indicated the 1994 equation was systematically under-predicting delay. Equation 4.21 is used in the 1997 edition:

$$D_x = \frac{3600}{c_{mx}} + 900T\left[\frac{v_x}{c_{mx}} - 1 + \sqrt{\left(\frac{v_x}{c_{mx}} - 1\right)^2 + \frac{\left(\frac{3600}{c_{mx}}\right)\left(\frac{v_x}{c_{mx}}\right)}{450T}}\right] + 5$$

$$D_1 = \frac{3600}{1049} + 900{*}0.25\left[\frac{21}{1049} - 1 + \sqrt{\left(\frac{21}{1049} - 1\right)^2 + \frac{\left(\frac{3600}{1049}\right)\left(\frac{21}{1049}\right)}{450{*}0.25}}\right] + 5$$

$$D_1 = 3.432 + 225\left[-0.980 + \sqrt{0.960 + \frac{3.432{*}0.020}{112.5}}\right] + 5 = 8.5 \text{ s/veh}$$

$$D_4 = \frac{3600}{1023} + 900{*}0.25\left[\frac{29}{1023} - 1 + \sqrt{\left(\frac{29}{1023} - 1\right)^2 + \frac{\left(\frac{3600}{1023}\right)\left(\frac{29}{1023}\right)}{450{*}0.25}}\right] + 5$$

$$D_4 = 3.519 + 225\left[-0.972 + \sqrt{0.945 + \frac{3.519{*}0.028}{112.5}}\right] + 5 = 8.7 \text{ s/veh}$$

$$D_{NB} = \frac{3600}{187} + 900{*}0.25\left[\frac{170}{187} - 1 + \sqrt{\left(\frac{170}{187} - 1\right)^2 + \frac{\left(\frac{3600}{187}\right)\left(\frac{170}{187}\right)}{450{*}0.25}}\right] + 5$$

$$D_{NB} = 19.251 + 225\left[-0.091 + \sqrt{0.008 + \frac{19.251{*}0.909}{112.5}}\right] + 5 = 94.6 \text{ s/veh}$$

$$D_{SB} = \frac{3600}{204} + 900{*}0.25\left[\frac{149}{204} - 1 + \sqrt{\left(\frac{149}{204} - 1\right)^2 + \frac{\left(\frac{3600}{204}\right)\left(\frac{149}{204}\right)}{450{*}0.25}}\right] + 5$$

Table A.9 Delay and LOS—1997 update and 2000 HCM

Movement	Delay (s/veh)	LOS
1	8.5	A
4	8.7	A
NB	94.6	F
SB	59.3	F

$$D_{SB} = 17.647 + 225\left[-0.270 + \sqrt{0.073 + \frac{17.647*0.730}{112.5}}\right] + 5 = 59.3\ \text{s/veh}$$

Level of service criteria for the 1997 update and 2000 HCM are shown in Table 4.15. They represent some minor changes from the criteria used in 1994. Delays and resulting levels of service are shown in Table A.9.

All delays are substantially increased from previous solutions. For Mvts 1 and 4, however, there is no change in the projected LOS. Both will operate well. For the NB and SB STOP-controlled approaches, however, delays rise to the point where LOS F exists on both the NB and SB approaches. While the LOS for these approaches was not good in previous solutions, they were in the stable range of operations. In 1997/2000, the projection is that both have failed, although the SB approach is close to the LOS E/F boundary (LOS E/F boundary is at 50 s/veh of control delay).

The 2010 and 2016 Highway Capacity Manuals

There is one major change in 2010 that permeates the entire solution: There is no conversion of *any* volumes to pc/h. All computations, including potential, movement, and shared-lane capacities are carried out directly in veh/h. There are no major changes in the 2016 HCM, with the exception that application of impedances due to pedestrians is optional. For the sample problem, they will still be applied.

Finding Conflicting Volumes

The form in which conflicting volumes are presented in the 2010/2016 HCMs has changed somewhat. Firstly, U-turns from the major street are specifically recognized as a separate movement. In addition, some of the material presented in footnotes in previous editions are directly incorporated into equations.

For the sample calculation, there are no U-turns, conflicting volumes are not affected by this change. Thus, conflicting volumes remain the same as in the 1997/2000 solution:

$$\begin{aligned} v_{c1} &= 431\ \text{conflicts/h} & v_{c4} &= 459\ \text{conflicts/h} \\ v_{c9} &= 240\ \text{conflicts/h} & v_{c12} &= 226\ \text{conflicts/h} \\ v_{c8} &= 980\ \text{conflicts/h} & v_{c11} &= 977\ \text{conflicts/h} \\ v_{c7} &= 1110\ \text{conflicts/h} & v_{c10} &= 1109\ \text{conflicts/h} \end{aligned}$$

Finding Critical Gaps (Headways) and Follow-Up Time
While there were some minor changes in base critical gaps and follow-up times (primarily for 2-stage movements), none affect the sample calculation. The values are the same as for the 1997/2000 solution, summarized in Table A.10.

Finding Potential Capacity
The algorithm for determining potential capacity did not change in the 2010/2016 HCMs. Thus, using the same algorithm and the same input values of conflicting volume, critical gap (headway) and follow-up time, the potential capacities remain as computed in the 1997/2000 solution—with one major proviso: the results are now in units of *vehicles per hour*, not passenger cars per hour.

$$\begin{aligned} c_{p1} &= 1070\,\text{veh/h} \quad c_{p4} = 1044\,\text{veh/h} \\ c_{p9} &= 737\,\text{veh/h} \quad c_{p12} = 752\,\text{veh/h} \\ c_{p8} &= 235\,\text{veh/h} \quad c_{p11} = 236\,\text{veh/h} \\ c_{p7} &= 154\,\text{veh/h} \quad c_{p10} = 154\,\text{veh/h} \end{aligned}$$

Finding Impedance and Movement Capacity
There is one small change in the impedance factors (now called adjustment factors) in the 2010/2016 HCMs. The impedance factor caused by major street left turns is adjusted to account for the short left-turn lane provided for them in this sample calculation. Equation 4.17 is used for this adjustment.

All vehicle impedances, however, also change slightly, as the units shift to veh/h. Demand flow rates in veh/h must be used instead of pc/h. This results in slight changes in many impedance factors used in the solution.

Again, because pedestrian movements cause impedance, *all* Rank 2–4 movements have impedance. LTs from the major street (Mvts 1, 4) and RTs from the minor street (Mvts 9, 12) are only affected by pedestrian impedance. The pedestrian impedance factor for all crosswalks (P_{13-16}) remains the same as in the 1997/2000 solution: 0.98. Movements 1 and 4 are impeded by one pedestrian crosswalk flow, Movements 9 and 12 are impeded by two pedestrian crosswalk flows. Their movement capacities remain the same as in the 1997/2000 solution, except for the change in units to veh/h:

$$\begin{aligned} c_{m1} &= 1049\,\text{veh/h} \quad c_{m4} = 1023\,\text{veh/h} \\ c_{m9} &= 708\,\text{veh/h} \quad c_{m12} = 722\,\text{veh/h} \end{aligned}$$

Table A.10 Critical gaps (Headways) and follow-up times—2010/2016 HCMs

Movements	t_c (s/veh)	t_f (s/veh)
LT from Major Street	4.3	2.3
RT from Minor Street	7.1	3.4
TH from Minor Street	6.7	4.1
LT from Minor Street	3.6	3.6

Through movements from the minor street (Mvts 8, 11) are impeded by two pedestrian crosswalk flows, and by both major street left turns (Mvts 1, 4). The base impedance factors for Mvts 1 and 4 change slightly due to the difference in units:

$$P_1 = 1 - (v_1/c_{m1}) = 1 - (18/1049) = 0.98$$
$$P_4 = 1 - (v_4/c_{m4}) = 1 - (25/1023) = 0.98$$

However, because these left turns operate out of a short exclusive lane, they must be adjusted, as indicated by Eq. 4.17:

$$P^*_{1/4} = 1 - \left(1 - P_{1/4}\right)\left[(n_L - 1)\sqrt{1 + \frac{X_M^{(n_L-1)}}{1 - X_M}}\right]$$

$$X_M = \left(\frac{v_{mTH}}{s_{mTH}}\right) + \left(\frac{v_{mRT}}{s_{mRT}}\right)$$

In this equation, n_L is the number of vehicles that can be stored in the exclusive LT lane on the major street. For the sample calculation, $n_L = 2$. Saturation flow rates (s_i) use default values: 1800 veh/hg/ln for TH movements and 1500 veh/hg/ln for RT movements. Note that the saturation default is in units of vehicles per hour; thus, flow rates (v) must also be expressed in units of vehicles per hour.

Then:

$$P_1^* = 1 - (1 - 0.98)\left[(2 - 1)\sqrt{1 + \frac{0.221^{(2-1)}}{1 - 0.221}}\right]$$

$$P_1^* = 1 - 0.02^*1.31 = 0.97$$

$$X_1 = \left(\frac{v_2}{s_{mTH}}\right) + \left(\frac{v_3}{s_{mRT}}\right) = \left(\frac{380}{1800}\right) + \left(\frac{15}{1500}\right) = 0.221$$

$$X_4 = \left(\frac{v_5}{s_{mTH}}\right) + \left(\frac{v_6}{s_{mRT}}\right) = \left(\frac{350}{1800}\right) + \left(\frac{18}{1500}\right) = 0.206$$

$$P_4^* = 1 - (1 - 0.98)\left[(2 - 1)\sqrt{1 + \frac{0.206^{(2-1)}}{1 - 0.206}}\right]$$

$$P_4^* = 1 - 0.03^*1.22 = 0.96$$

These changes do impact the movement capacities for Mvts 8 and 11, which are now computed as:

$$c_{m8} = c_{p8}\, P_1^*\, P_4^*\, P_{13-16}\, P_{13-16} = 235^*0.97^*0.96^*0.98^*0.98 = 210 \text{ veh/h}$$
$$c_{m11} = c_{p11}\, P_1^*\, P_4^*\, P_{13-16}\, P_{13-16} = 236^*0.97^*0.96^*0.98^*0.98 = 211 \text{ veh/h}$$

While the change (from the 1997/2000 solution) is slight, it produces a similar change in the impedance factors for Mvts 8 and 11, which must be used in determining the movement capacities of Mvts 7 and 10:

$$P_8 = 1 - (v_8/c_{m8}) = 1 - (105/210) = 0.50$$
$$P_{11} = 1 - (v_{11}/c_{m11}) = 1 - (100/211) = 0.53$$

LTs from the minor street are impeded by both major street left turns, the opposing minor street through movement and right-turn movement, and two pedestrian crosswalk flows. This would then result in the following:

$$c_{m7} = c_{p7} P_1^* \; P_4^* \; P_{11} \; P_{12} P_{13-16} P_{13-16}$$
$$c_{m10} = c_{p10} P_1^* \; P_4^* \; P_8 \; P_9 P_{13-16} P_{13-16}$$

Impedance factors for RT movements 9 and 12 are:

$$P_9 = 1 - (v_9/c_{m9}) = 1 - (19/708) = 0.97$$
$$P_{12} = 1 - (v_{12}/c_{m12}) = 1 - (17/722) = 0.98$$

However, just as was the case in the 1994 update and 1997/2000 solutions, impedance factors for Mvts 1, 4, and 11 overlap when estimating the movement capacity of Mvt. 7. Similarly, impedance factors for Mvts 1, 4, and 8 overlap when estimating the movement capacity of Mvt. 10. Thus, these values must be adjusted to account for this. The adjustments are made to the product of the overlapping impedance factors, or:

$$P_7'' = P_1^* \; P_4^* \; P_{11} = 0.97{}^*0.96{}^*0.53 = 0.49$$
$$P_{11}'' = P_1^* \; P_4^* \; P_8 = 0.97{}^*0.96{}^*0.50 = 0.47$$

Then:

$$P' = 0.65\, P'' - \left(\frac{P''}{P''+3}\right) + 0.6\sqrt{P''}$$

$$P_7' = 0.65(0.44) - \left(\frac{0.44}{0.44+3}\right) + 0.6\sqrt{0.44} = 0.286 - 0.128 + 0.415 = 0.57$$

$$P_{10}' = 0.65(0.41) - \left(\frac{0.41}{0.41+3}\right) + 0.6\sqrt{0.41} = 0.267 - 0.120 + 0.384 = 0.53$$

Then:

$$c_{m7} = c_{p7} P_7' P_{12} P_{13-16} P_{13-16} = 154{}^*0.60{}^*0.98{}^*0.98{}^*0.98 = 87 \text{ pc/h}$$
$$c_{m10} = c_{p10} P_{10}' P_9 P_{13-16} P_{13-16} = 154{}^*0.58{}^*0.97{}^*0.98{}^*0.98 = 83 \text{ pc/h}$$

Finding Shared-Lane Capacities

As in previous solutions, Mvts 7, 8 and 9 share one NB lane, while Mvts 10, 11, and 12 share one SB lane. Shared-lane capacities for these are computed as:

$$c_{SHNB} = \frac{22+105+19}{\left(\frac{22}{87}\right)+\left(\frac{105}{210}\right)+\left(\frac{19}{708}\right)} = \frac{146}{0.252+0.500+0.027} = \frac{146}{0.779} = 187 \text{ pc/h}$$

$$c_{SHSB} = \frac{17+100+10}{\left(\frac{17}{722}\right)+\left(\frac{100}{211}\right)+\left(\frac{10}{83}\right)} = \frac{127}{0.024+0.474+0.120} = \frac{127}{0.618} = 206 \text{ pc/h}$$

Finding Delay and Level of Service

Delay is estimated using the Eq. 4.21, which is unchanged from the 1997/2000 version. Many of the input values, however, have changed, particularly due to the reversion to veh/h units.

$$D_x = \frac{3600}{c_{mx}} + 900T\left[\frac{v_x}{c_{mx}} - 1 + \sqrt{\left(\frac{v_x}{c_{mx}} - 1\right)^2 + \frac{\left(\frac{3600}{c_{mx}}\right)\left(\frac{v_x}{c_{mx}}\right)}{450T}}\right] + 5$$

$$D_1 = \frac{3600}{1049} + 900{*}0.25\left[\frac{18}{1049} - 1 + \sqrt{\left(\frac{18}{1049} - 1\right)^2 + \frac{\left(\frac{3600}{1049}\right)\left(\frac{18}{1049}\right)}{450{*}0.25}}\right] + 5$$

$$D_1 = 3.432 + 225\left[-0.983 + \sqrt{0.966 + \frac{3.432{*}0.017}{112.5}}\right] + 5 = 8.4 \text{ s/veh}$$

$$D_4 = \frac{3600}{1023} + 900{*}0.25\left[\frac{25}{1023} - 1 + \sqrt{\left(\frac{25}{1023} - 1\right)^2 + \frac{\left(\frac{3600}{1023}\right)\left(\frac{25}{1023}\right)}{450{*}0.25}}\right] + 5$$

$$D_4 = 3.519 + 225\left[-0.976 + \sqrt{0.953 + \frac{3.519{*}0.024}{112.5}}\right] + 5 = 8.7 \text{ s/veh}$$

$$D_{NB} = \frac{3600}{187} + 900{*}0.25\left[\frac{146}{187} - 1 + \sqrt{\left(\frac{146}{187} - 1\right)^2 + \frac{\left(\frac{3600}{187}\right)\left(\frac{146}{187}\right)}{450{*}0.25}}\right] + 5$$

$$D_{NB} = 19.251 + 225\left[-0.219 + \sqrt{0.048 + \frac{29.251{*}0.781}{112.5}}\right] + 5 = 80.5 \text{ s/veh}$$

$$D_{SB} = \frac{3600}{206} + 900{*}0.25\left[\frac{127}{206} - 1 + \sqrt{\left(\frac{127}{206} - 1\right)^2 + \frac{\left(\frac{3600}{206}\right)\left(\frac{127}{206}\right)}{450{*}0.25}}\right] + 5$$

$$D_{SB} = 17.476 + 225\left[-0.383 + \sqrt{0.147 + \frac{17.476{*}0.617}{112.5}}\right] + 5 = 47.2 \text{ s/veh}$$

Table A.11 Delay and LOS for sample calculation A.1 2010/2016 HCMs

Movement	Delay (s/veh)	LOS
1	8.5	A
4	8.7	A
NB	80.5	F
SB	47.2	E

Delays and levels of service for these movements are shown in Table A.11.

The delay and LOS have not changed for Mvts 1 and 4, which remain quite good. Delays for the NB and SB approaches are somewhat better than the 1997/2000 solution suggests, which occurs primarily because of the change in base units from pc/h to veh/h. The LOS on the SB approach now appears to be E, which is stable, but still not good.

Comparison and Comments

Table A.12 compares the primary results of the analyses from the various editions of the HCM. Capacity and LOS for each of the affected lanes are shown.

The general result is quite consistent for all of the solutions. The major street left turns work well, while the NB and SB approaches experience some difficulties. The extent of the difficulties vary, with the most recent methodologies revealing the worst operations. The NB approach is the worst throughout, and operates at LOS F for the 1997/2000 and 2010/2016 methodologies. The SB approach also fails using the 1997/2000 methodology, but just squeeks by with LOS E for the 2010/2016 methodology. The bottom line, however, is consistent throughout: this is a marginal operation for TWSC approaches.

Improving the Intersection

The results of the Sample Calculation reveal that the two STOP-controlled approaches do not operate well. With the 2010/2016 methodology, the NB approach fails, while the SB approach is close to failure.

When STOP-controlled intersections do not work well, the most obvious question is whether signalization should be considered. It is doubtful that any of the signal warrants would be triggered by the demands shown or the delays indicated by the analysis. Warrants such as the "Crash Warrant" also need to be considered, but would require detailed accident data for the location. A field study of delay could also be done to evaluate total delay during the peak hour. In this case, however, it is doubtful that signalization would be the solution selected.

Table A.12 Comparison of solutions for sample calculation

Mvt	Circular 212		1985 HCM		1994 HCM		1997/2000 HCM		2010/2016 HCM	
	c (pc/h)	LOS	C (pc/h)	LOS	C (pc/h)	LOS	c (pc/h)	LOS	c (veh/h)	LOS
1	690	A	680	A	1031	A	1049	A	1049	A
4	670	A	660	A	996	A	1023	A	1024	A
NB	234	E	232	E	254	E	187	F	187	F
SB	254	D	252	D	318	D	204	F	206	E

That leaves a variety of physical changes to the design of the intersection. The simplest approach would be addition of a lane or lanes on the STOP-controlled approaches. If space were available, three lanes might be considered on each STOP-controlled approach, creating exclusive lanes for the three movements. If only two lanes could be provided on each approach, then the lane usage could be LT – TH/RT or LT/TH – RT. Because the left turn is the most difficult movement, it is likely that providing an exclusive LT lane would bring more improvement than providing an exclusive RT lane.

Each of the options noted above can be analyzed using the TWSC methodologies. We will examine each using the 2010/2016 methodology (it would require too many pages to display solutions with previous methodologies as well).

As a general note: The proposed design changes do not affect the operation of Mvts 1 and 4. They also do not involve changes in computations up to and including the determination of movement capacities. Shared-lane capacities, delays, and LOS would change, however.

Option 1: Three Lanes on Each STOP-Controlled Approach

In this case, each STOP-controlled approach would have a separate lane for each movement: LT, TH, and RT. The movement capacities remain as computed for Sample Calculation 1, but there are *no* shared-lane capacities, as minor street movements do not share lanes. Delay computations, using Eq. 4.21 are separately computed for each lane. A weighted average can then be taken for each approach. These computations are summarized in Table A.13.

On average, it appears that some improvement has occurred. The NB and SB approaches, taken as a whole, operate stably with significantly lower delay that when operating out of a shared lane. The two left-turn movements (7 and 10), however, still fail, and have unacceptably high delays.

Option 2: Two Lanes—Each Approach has an Exclusive LT Lane and a shared RT/TH Lane

This approach recognizes that there might not be enough space to provide exclusive lanes for all movements on the STOP-controlled approaches. In this case, the LT movements are left in an exclusive lane, while the TH and RT movements are combined into a shared lane. The shared-lane capacity of each of these lanes must

Table A.13 Delays and LOS for option 1—2010/2016 HCM

Mvt	v_x (veh/h)	c_{mx} (veh/h)	D (s/veh)	LOS	Av D (s/veh)	LOS
7	22	87	59.8	F	37.8	E
8	105	210	38.1	E		
9	19	708	10.2	B		
10	10	83	54.2	F	34.0	D
11	100	211	36.6	E		
12	17	722	10.1	B		

be determined:

$$c_{SH8/9} = \frac{105+19}{\left(\frac{105}{210}\right)+\left(\frac{19}{708}\right)} = \frac{124}{0.500+0.027} = 235 \text{ veh/h}$$

$$c_{SH11/12} = \frac{100+17}{\left(\frac{100}{211}\right)+\left(\frac{17}{722}\right)} = \frac{117}{0.474+0.024} = 235 \text{ veh/h}$$

Delays are again computed, this time for the LT lanes and the shared TH/RT lanes on the STOP-controlled approaches. The results are summarized in Table A.14.

In the proposed configuration, the operation of the STOP-controlled approaches is LOS E. The left turns, which continue to operate in separate lanes, continue to fail—their operation was not affected by combining through and right turns in one shared lane. This configuration is slightly worse (in terms of average delay) than the situation in Option 1 in which each movement had its own lane, but is still a considerable improvement over the case in which all movements shared a single lane.

Option 3: Two Lanes—Each Approach has a Shared LT/TH Lane and a RT Lane

In this scenario, the shared lane on each STOP-controlled approach is for LT and TH vehicles. Their shared-lane capacities are computed as:

$$c_{SH8/9} = \frac{105+22}{\left(\frac{105}{210}\right)+\left(\frac{22}{87}\right)} = \frac{127}{0.500+0.253} = 169 \text{ veh/h}$$

$$c_{SH11/12} = \frac{100+10}{\left(\frac{100}{211}\right)+\left(\frac{10}{83}\right)} = \frac{110}{0.474+0.120} = 185 \text{ veh/h}$$

Once again, delays for each lane are computed, and averages for the NB and SB approaches found. Results are summarized in Table A.15.

Combining the LT and TH movements into a shared lane on each approach is not at all helpful. The NB approach continues to fail, and the SB approach is still LOS E, approaching LOS F.

In summary, the intersection can be improved by providing exclusive lanes for each movement on each STOP-controlled approach, but left turns will still experience unacceptable delay. If only one lane can be added on each approach, an exclusive

Table A.14 Delays and LOS for option 2—2010/2016 HCMs

Mvt	v_x (veh/h)	c_{mx} (veh/h)	D (s/veh)	LOS	Av D (s/veh)	LOS
7	22	87	59.8	F	39.8	E
8, 9	124	210	36.2	E		
10	10	83	54.2	F	36.2	E
11, 12	117	211	34.6	D		

Table A.15 Delays and LOS for option 3—2010/2016 HCMs

Mvt	v_x (veh/h)	c_{mx} (veh/h)	D (s/veh)	LOS	Av D (s/veh)	LOS
7, 8	127	169	72.0	F	64.0	F
9	19	708	10.2	B		
10, 11	110	185	49.5	E	44.3	E
12	17	722	10.1	B		

LT lane would be provided, and RTs and TH movements would share the second lane. The LTs would still experience unacceptable delay, but the approach as a whole would see better operations than if the existing situation of one shared lane on each approach continues.

Another potential solution to this case would be the construction of a roundabout to replace the intersection. This would be much more costly than adding approach lanes. This alternative will be illustrated in a sample calculation for roundabouts (Appendix C).

Appendix B
Sample Calculations—AWSC Intersections

Because the methodology for AWSC intersections in more recent manuals is very detailed and iterative, hand computations are best illustrated using a very simple problem involving two one-way streets. This will limit the number of potential scenarios of arriving vehicles that can exist, and will simplify iterations so that they can be adequately presented and explained.

A Simple Problem: Two One-Way Streets with AWSC

Figure B.1 illustrates a simple AWSC intersection with two one-lane approaches. Demands are shown as flow rates for the peak 15 min of the analysis hour. No pedestrian demands are shown, as these are not accounted for in any of the analysis methodologies.

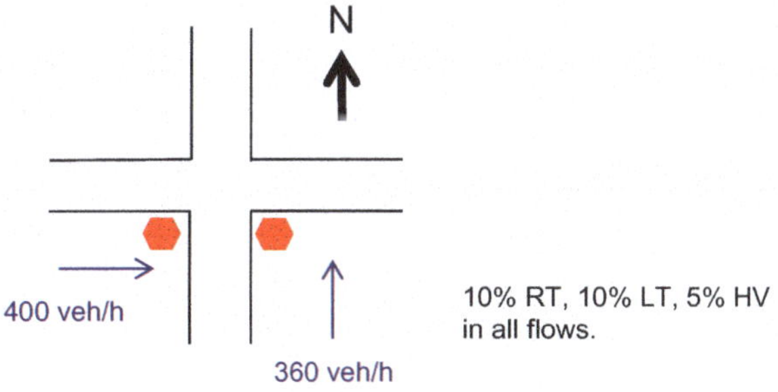

Fig. B.1 Intersection for sample calculation

E. S. Prassas and R. P. Roess, *The Highway Capacity Manual: A Conceptual and Research History Volume 2*, Springer Tracts on Transportation and Traffic 12,
https://doi.org/10.1007/978-3-030-34480-1

The 1985 Highway Capacity Manual

The 1985 HCM offers a very general means of approximately evaluating the performance of an AWSC intersection. It is based solely on the number of lanes on each approach, and the balance of demand on the two intersecting streets.

For our sample calculation, there are only two approaches, one on each one-way street. The NB approach carries $360/(360 + 400) = 0.47$ or 47% of the demand. The demand split for the intersection is, therefore, 53/47.

From Table 5.2, the capacity (in terms of total intersection flow) for such an intersection is 1800 veh/h (using the 55/45 split). From Table 5.3, the approximate LOS C service flow rate is 1140 veh/h (again using the 55/45 split). Since the given demand flow rate is only $400 + 360 = 760$ veh/h, we can expect the operation to be *better* than LOS C. No more detailed conclusion can be made.

The 1994 Update to the 1985 HCM

The 1994 update based capacity and level of service on Eqs. 5.1 and 5.2. Equation 5.2 estimates the delay for each AWSC approach. LOS is then based upon the estimated delay:

$$\begin{aligned} c_s &= 1000\ P_s + 700\ P_o + 200\ N_s - 100\ N_o - 300\ P_{LT_o} \\ &\quad + 200\ P_{RT\,o} - 300\ P_{LT_c} + 300\ P_{RT_c} \\ D_x &= e^{3.8(v_s/c_s)} \end{aligned}$$

There are only two approaches in the sample calculation: NB and EB. Table B.1 gives the values of the variables in the capacity equation for each of these approaches, along with the definition of each variable.

Table B.1 Critical variables—1994 update

Variable	Definition	Value NB	Value EB
P_s	Proportion of demand in subject approach	$360/760 = 0.474$	$400/760 = 0.526$
P_o	Proportion of demand on opposing approach	0	0
N_s	Number of lanes on subject approach	1	1
N_o	Number of lanes on oppossing approach	0	0
P_{LTo}	Proportion of LTs on opposing approach	0	0
P_{RTo}	Proportion of RTs on opposing approach	0	0
P_{LTc}	Proportion of LTs on conflicting approach	0.10	0.10
P_{RTc}	Proportion of RTs on conflicting approach	0.10	0.10

Then:

$$c_{sNB} = (1000^*0.474) + (700^*0.526) + (200^*1) - (100^*0) - (300^*0) + (200^*0) - (300^*0.10) + (300^*0.10) = 1042\text{ veh/h}$$

$$c_{sEB} = (1000^*0.526) + (700^*0.47) + (200^*1) - (100^*0) - (300^*0) + (200^*0) - (300^*0.10) + (300^*0.10) = 1058\text{ veh/h}$$

Using these computed values of capacity, the delay to vehicles on each of the approaches may now be estimated:

$$D_{NB} = e^{3.8(360/1042)} = 3.7\text{ s/veh}$$

$$D_{EB} = e^{3.8(400/1058)} = 4.2\text{ s/veh}$$

The level of service is obtained from Table 5.5. As these delays are lower than 5 s/vehicle, the LOS for both approaches is A.

The 1997 Update, and the 2000–2016 HCMs

The methodologies of the 1997 update, the 2000 HCM, 2010 HCM, and the 2016 HCM are virtually identical, except for some minor editorial changes.

The first step in the process is to identify the "Geometry Group," as defined in Table 5.7. Unfortunately, the table *does not list* a simple 2-approach intersection. We will treat our intersection as Geometry Group 1, in which two of the approaches (SB and WB) have "0" demand.

The iterative approach begins by assuming that all departure headways are 3.2 s/veh. With only two active approaches, we need to start by computing the degree of utilization (X) for each approach. Remember that we will hypothetically be including four approaches, two of which have no demand. Equation 5.5 is used to do this:

$$X_j = \frac{v_j h_{dj}}{3600}$$

$$X_{NB} = \frac{360^*3.2}{3600} = 0.320$$

$$X_{SB} = \frac{0^*3.2}{3600} = 0.000$$

$$X_{EB} = \frac{400^*3.2}{3600} = 0.356$$

$$X_{WB} = \frac{0^*3.2}{3600} = 0.000$$

Table B.2 Scenarios for sample calculation—1997 to 2016

Subject approach	NB	SB	EB	WB
NB	X	0	1	0
	X	0	0	0
EB	1	0	X	0
	0	0	X	0

For *each* approach under analysis (NB and EB in our case), there are only two scenarios that can occur. In both cases, the probability that there is no SB or WB vehicle is always 1.0, as these approaches do not exist. Thus, for the NB approach, either a vehicle is present on the EB approach, or it is not. For the EB approach, either a vehicle is present on the NB approach, or it is not. Table B.2 shows the possible scenarios, based upon a vehicle being present (1) or not (0).

The probability that a vehicle is present in a given approach lane is X. The probability that a vehicle is *not* present in a given approach lane is 1 − X. Thus, the probability that each of these scenarios exists is computed:

$$p_{NB}(1) = 1.0^{*}0.356^{*}1.0 = 0.356$$
$$p_{NB}(2) = 1.0^{*}(1 - 0.356)^{*}1.0 = 0.644$$
$$p_{EB}(1) = 1.0^{*}0.320^{*}1.0 = 0.320$$
$$p_{EB}(2) = 1.0 * (1 - 0.320)^{*}1.0 = 0.680$$

The next step in the process is to estimate the departure headway for *each* scenario. This is a two-step process, with a base headway selected from Table 5.9, and an adjustment computing using Eq. 5.4 and values from Table 5.10.

From Table 5.7, *both* NB and EB scenarios 2 are a degree-of-conflict (DOC) 1, and have 0 vehicles on opposing and conflicting approaches. *Both* NB and EB scenarios 2 have degree-of-conflict 2, and have 1 vehicle on opposing and conflicting approaches. Thus, the base departure headway for Scenarios 1 is 4.7 s/veh, while the base departure headway for Scenarios 2 is 3.9 s/veh. In both cases, the Geometry Group designation is 1 as stated previously.

From Table 5.8, the adjustment to headway for right turns (h_{adjRT}) is −0.6, the adjustment to headway for left turns (h_{adjLT}) is 0.2, and the adjustment to headway for heavy vehicles (h_{adjHV}) is 1.7. Then, using Eq. 5.4:

$$h_{adj} = h_{LTadj}P_{LT} + h_{RTradj}P_{RT} + h_{HVadj}P_{HV}$$
$$h_{adj} = (0.2^{*}0.10) + (-0.6^{*}0.10) + (1.7^{*}0.10) = 0.13$$

As both NB and EB movements have the same percent RT, LT, and HV, this adjustment is applied to *all* base departure headways. Thus:

$$h_{NB1} = h_{EB1} = 4.7 + 0.13 = 4.83\,\text{s/veh}$$

$$h_{NB2} = h_{EB2} = 3.9 + 0.13 = 4.03\,\mathrm{s/veh}$$

The headways must now be associated with the probability of the scenario actually existing. These were computed earlier. They must, however, be adjusted using Eq. 5.7. To do this, scenarios have to be "grouped" by degree of conflict (DOC). For each approach (NB, EB), Scenario 1 is DOC 2, while Scenario 2 is DOC 1. Thus:

$$P_{NB}(1) = 0.644$$
$$P_{NB}(2) = 0.356$$
$$P_{EB}(1) = 0.680$$
$$P_{EB}(2) = 0.320$$

By definition, because there are no other scenarios possible, the probability that DOC 3, 4, 5, or 6 exists for either approach is 0.

Equations 5.7 indicate that there would be no adjustment to DOC 1 cases, as the adjustment involves only DOCs higher than 2, all of which are "0" for the sample calculation. There is, however, an adjustment for DOC 2 cases, which are involved here:

$$Adj\,P(2-4) = \frac{\alpha[P(C_3) + 2P(C_4) + 3P(C_5) - P(C_2)]}{3}$$

$$Adj\,P_{NB}(2) = \frac{0.01[0.0 + (2^*0.0) + (3^*0.0) - 0.356]}{3} = -0.012$$

$$Adj\,P_{EB}(2) = \frac{0.01[0.0 + (2^*0.0) + (3^*0.0) - 0.320]}{3} = -0.010$$

Then:

$$P_{NB}(1) = 0.644$$
$$P_{NB}(2) = 0.356 - 0.012 = 0.344$$
$$P_{EB}(1) = 0.680$$
$$P_{EB}(2) = 0.320 - 0.010 = 0.310$$

The estimated departure headways are now computed using Eq. 5.3:

$$h_{NB} = (0.644^*4.83) + (0.344^*4.03) = 3.11 + 1.39 = 4.50\,\mathrm{s/veh}$$
$$h_{EB} = (0.680^*4.83) + (0.310^*4.03) = 3.28 + 1.25 = 4.53\,\mathrm{s/veh}$$

Unfortunately, this is not the end. These results must be compared with the starting assumption that h = 3.2 s/veh for all scenarios. The assumption and the result are clearly significantly different (diff > 0.01), so iterations must continue until closure. The process begins again with the resulting departure headways used as a starting point.

It is not possible to efficiently show all of the iterations needed to reach equilibrium. However, it is possible to close in this case with 3 iteration cycles, each of which repeats the computations shown, starting with the previous result for departure headway. The final results are that $h_{NB} = h_{EB} = 4.389$ s/veh.

Finding capacity of each approach is a multiply iterative nightmare for hand computation. Literally, the entire process for departure headway is repeated, each time incrementally increasing the demand on the subject approach (while holding demand on other approaches fixed at their original value). When a value of X = 1.00 is reached, the iteration is complete. Since the computational process involves iterations within each iteration, they are not shown here. However, each iteration within each iteration proceeds using the exact same steps as the first iteration shown herein.

The computations involved approximately 20 iterations at two levels (assisted with a spreadsheet), and resulted in the following capacities:

$$c_{NB} = 806 \text{ veh/h}$$
$$c_{EB} = 815 \text{ veh/h}$$

Level of service is determined on the basis of delay, where delay is estimated using Eq. 5.11:

$$D_i = t_{si} + 900T\left[(X_i - 1) + \sqrt{(X_i - 1)^2 + \frac{h_{di}X_i}{450T}}\right] + 5$$

where $t_{si} = t_{di} - t_m$, where t_m is the move-up time, which has a default value of 2.0 s/veh. Degree of utilization is estimated as v_i/c_i. Therefore:

$$t_{sNB} = 4.389 - 2.0 = 2.389 \text{ s/veh}$$
$$t_{sEB} = 4.389 - 2.0 = 2.389 \text{ s/veh}$$
$$t_{dNB} = t_{dSB} = 4.389 \text{ s/veh}$$
$$X_{NB} = 360/806 = 0.447$$
$$X_{EB} = 400/815 = 0.491$$
$$T = 0.25\,h$$

and

$$D_{NB} = 2.389 + 900{*}0.25\left[(0.447 - 1) + \sqrt{(0.447 - 1)^2 + \frac{4.389{*}0.447}{450{*}0.25}}\right] + 5$$

$$D_{NB} = 2.389 + 225\left[-0.553 + \sqrt{0.306 + 0.017}\right] + 5 = 10.8 \text{ s/veh}$$

$$D_{EB} = 2.389 + 900{*}0.25 + \left[(0.491 - 1) + \sqrt{(0.491 - 1)^2 + \frac{4.389{*}0.491}{450{*}0.25}}\right] + 5$$

Table B.3 Comparing results for sample calculation

Item	1985 HCM	1994 update	1997–2016 HCM
c_{NB}	NA	1042 veh/h	806 veh/h
c_{EB}	NA	1058 veh/h	815 veh/h
c_{TOT}	1800	2100 veh/h	1621 veh/h
D_{NB}	NA	3.7 s/veh	10.8 s/veh
D_{EB}	NA	4.2 s/veh	11.4 s/veh
LOS_{NB}	Better than C	A	B
LOS_{EB}	Better than C	A	B

$$D_{EB} = 2.389 + 225\left[-0.509 + \sqrt{0.259 + 0.019}\right] + 5 = 11.4 \text{ s/veh}$$

From Table 5.12, both approaches operate at LOS B.

Comparisons and Discussion

Table B.3 compares the results of the three analyses in terms of the capacity of each approach, the total capacity, delay and LOS for each approach.

Despite significant differences in the three methodologies, the results are not all that different. The range in capacities is fairly broad (from 1621 to 2100 or 479 veh/h), but they are "in the same ballpark" as it were. Delays are significantly higher in the 1997/2016 method than in the 1994 method, but the LOS projected is still very good—and it is also consistent with the expected operations from the 1985 HCM.

In the final analysis, this is a case that works fairly well as an AWSC intersection. Note, however, that pedestrians are not accounted for in any way. Many AWSC intersections are part of school crossing programs, where pedestrian interference would be quite high during certain hours of the day.

The complexity of the 1997–2016 methodology is difficult to present clearly, given its pattern of many nested iterations that are necessary to come to closure. For simple problems, a spreadsheet can be relatively easily created, but for more complex cases, use of software (primarily the HCS Package) is required.

Appendix C
Sample Calculations—Roundabouts

Roundabouts are specifically treated only in the 1997/2000 and 2010/2016 HCMs. The 2010 and 2016 manuals are identical except for editorial improvements made in the 2016 presentation. Previous editions were silent on the subject. It is difficult to compare results between the two because the 1997/2000 version was limited to one-lane roundabouts with one-lane approaches, and yielded only a range of capacities that might be expected. No levels of service were defined. The 2010/2016 method is more complete, and can handle a mix of options with one- or two-lane roundabouts and one- or two-lane approaches. Level of Service measures were added as well.

A Simple One-Lane Roundabout

Figure C.1 illustrates a simple one-lane roundabout with one-lane approaches. All demands are shown as *flow rates* for the worst 15 min of the analysis hour, and are in units of vehicles per hour.

A. **The 1997/2000 Highway Capacity Manual**

The 1997/2000 HCM only permits examination of the range of capacities that are likely to be served by the roundabout shown. The capacity of each one-lane approach is estimated. The key computation is the determination of the circulating flow rate (v_c) for each approach.

The pattern of conflicting flows is shown in Table 6.1 (Chap. 6). In each case, the conflicting flow consists of the through and left-turn flows from the approach to the left (of the subject approach), and the left-turn movement from the opposing approach. Note that the 2000 HCM *does not* deal directly with U-turn movements, which is not an issue here, as there are no such turns shown. The conflicting flows for the four approaches are:

E. S. Prassas and R. P. Roess, *The Highway Capacity Manual: A Conceptual and Research History Volume 2*, Springer Tracts on Transportation and Traffic 12,
https://doi.org/10.1007/978-3-030-34480-1

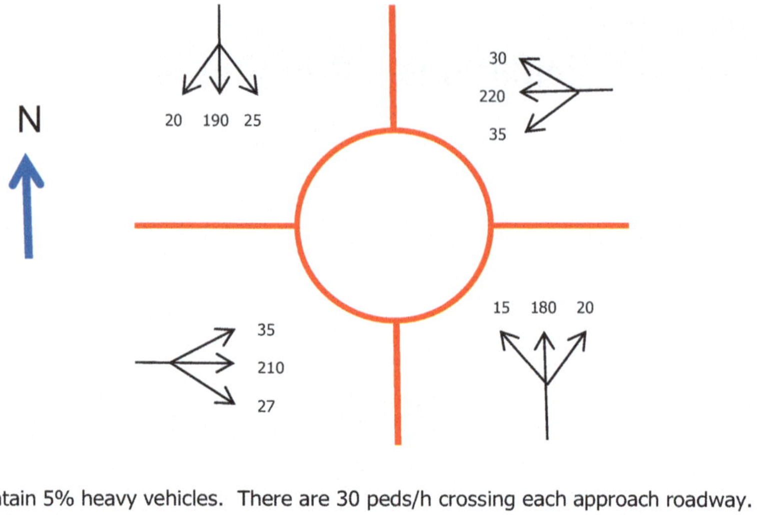

Fig. C.1 One-Lane roundabout for sample calculation

$$v_{cNB} = v_{THEB} + v_{LEB} + v_{LSB} = 210 + 35 + 25 = 270 \text{ veh/h}$$
$$v_{cWB} = v_{THNB} + v_{LNB} + v_{LEB} = 180 + 15 + 35 = 230 \text{ veh/h}$$
$$v_{cSB} = v_{THWB} + v_{LWB} + v_{LNB} = 220 + 35 + 15 = 270 \text{ veh/h}$$
$$v_{cEB} = v_{THSB} + v_{LSB} + v_{LWB} = 190 + 25 + 35 = 250 \text{ veh/h}$$

The capacity for each approach is given by Eq. 6.1:

$$c_a = \frac{v_c e^{-v_c t_c/3600}}{1 - e^{-v_c t_f/3600}}$$

Unfortunately, the 1997/2000 HCM only provides a range of values for the critical gap (t_c) and the follow-up time (t_f). The critical gap ranges between 4.1 and 4.6 s/veh, while the follow-up time ranges between 2.6 and 3.1 s/veh. The highest capacity results from using the low values of t_c and t_f, while the lowest capacity results from using the high values of t_c and t_f. Thus, each of the four approaches has two capacity values (high and low). The computations for the NB approach are shown below. The results for each approach are shown in Table C.1.

Table C.1 Results for One-Lane sample calculation—1997/2000 HCM

Approach	c (high) (veh/h)	c (low) (veh/h)	v (veh/h)
NB	1121	922	215
WB	1156	954	285
SB	1121	922	235
EB	1138	938	272

$$c_{NB(high)} = \frac{270\,e^{-270*4.1/3600}}{1-e^{-270*2.6/3600}} = 1112 \text{ veh/h}$$
$$c_{NB(low)} = \frac{270\,e^{-270*4.6/3600}}{1-e^{-270*4.6/3600}} = 922 \text{ veh/h}$$

As shown in Table C.1, there is no reason to think that this roundabout will not operate very well under the existing demands. The actual approach flow rates are well below the estimates of capacity. Unfortunately, no more specific predictions concerning operations can be made using the 1997/2000 methodology.

B. **The 2010/2016 Highway Capacity Manuals**

For a one-lane roundabout with one-lane approaches, the basic structure of the 2010/2016 HCM methodology is quite similar to the 2000 version. There are two major differences: computations are carried out in pc/h, and pedestrian impacts are included in the methodology.

Circulating flow rates are the same as in the 1997/2000 solution, except that they must be adjusted to reflect pc/h. Demand flows must also be adjusted to reflect pc/h. As all flows in the sample calculation have 5% heavy vehicles, we can simply convert the v_c values and demand flow rate directly. For the 2010/2016 HCMs, the equivalent for heavy vehicles (E_{HV}) is 2. The heavy vehicle adjustment factor is, therefore:

$$f_{HV} - \frac{1}{1 + P_{HV}(E_{HV} - 1)} = \frac{1}{1 + 0.05(2 - 1)} = 0.952$$

All demand and circulating flow rates are converted by dividing the values in veh/h by this adjustment factor. These computations are shown in Table C.2, with results rounded to the nearest whole pc.

Unlike the 1997/2000 method, the 2010/2016 method provides a single estimate of capacity for each approach. For various configurations, equations for this estimate are found in Table 6.2. For one-lane roundabouts with one-lane approaches, the equation for estimating capacity (in pc/h) is given by:

$$c_{epce} = 1130\,e^{(-1*10^{-3})v_{cpce}}$$
$$c_{NB} = 1130\,e^{(-0.001*284)} = 851 \text{ pc/h}$$
$$c_{EB} = 1130\,e^{(-0.001*26)} = 869 \text{ pc/h}$$
$$c_{SB} = 1130\,e^{(-0.001*284)} = 851 \text{ pc/h}$$
$$c_{WB} = 1130\,e^{(-0.001*273)} = 860 \text{ pc/h}$$

Table C.2 Conversion of flow rates to pc/h—One-Lane sample calculation 2010/2016 HCMs

Approach	v (veh/h)	v_c (veh/h)	v (pc/h)	v_c (pc/h)
NB	215	270	215/0.952 = 226	270/0.952 = 284
EB	272	250	272/0.952 = 286	250/0.952 = 263
SB	235	270	235/0.952 = 247	270/0.952 = 284
WB	285	230	285/0.952 = 299	230/0.952 = 273

These capacities, however, must be adjusted to account for pedestrian interference. These adjustments are computed as shown in Table 6.3. The sample calculation is a case of a one-lane roundabout with circulating flows less than 881 veh/h and pedestrian flows less than 101 peds/h. Then:

$$f_p = 1 - 0.000137 n_{ped} = 1 - (0.000137{*}30) = 0.996$$

Then:

$$\begin{aligned} c_{NBa} &= 851{*}0.996 = 848\ \text{pc/h} \\ c_{EBa} &= 869{*}0.996 = 866\ \text{pc/h} \\ c_{SBa} &= 851{*}0.996 = 848\ \text{pc/h} \\ c_{WBa} &= 860{*}0.996 = 857\ \text{pc/h} \end{aligned}$$

The 2010/2016 HCMs provide for an estimation of delay on each approach, using Eq. 6.6:

$$D_i = \frac{3600}{c_i} + 900T\left[X_i - 1 + \sqrt{(X_i - 1)^2 + \frac{\left(\frac{3600}{c_i}\right)X_i}{450T}}\right] + 5{*}\min[X_i, 1]$$

This requires that the values of X_i (the v/c ratio for each approach) be computed. Capacities in pc/h are shown above, and demand flows in pc/h were computed in Table C.2. Then:

$$\begin{aligned} X_{NB} &= 226/848 = 0.267 \\ X_{EB} &= 286/866 = 0.330 \\ X_{SB} &= 247/848 = 0.291 \\ X_{WB} &= 299/857 = 0.349 \end{aligned}$$

In all cases, the analysis period is 15 min or 0.25 h. Then:

$$D_{NB} = \frac{3600}{848} + 900{*}0.25\left[0.267 - 1 + \sqrt{(0.267 - 1)^2 + \frac{\left(\frac{3600}{848}\right)0.267}{450{*}0.25}}\right] + 5{*}0.267$$

$$D_{NB} = 4.245 + 225{*}\left[-0.733 + \sqrt{0.537 + \frac{4.254{*}0.267}{112.5}}\right] + 1.335 = 7.1\ \text{s/veh}$$

The results for all approaches are shown in Table C.3. Levels of service are defined in Table 6.4.

With under 10 s/veh of delay on all approaches, the roundabout will operate very well under the existing demands. This is not inconsistent with the view provided by the 1997/2000 HCM, but provides more specificity in terms of operating parameters.

Table C.3 Results for One-Lane sample calculation C.1—2010/2016 HCMs

Approach	X_i	D_i (s/veh)	LOS
NB	0.267	7.12	A
EB	0.330	7.84	A
SB	0.291	7.43	A
WB	0.349	8.18	A

A Roundabout Solution for the TWSC Sample Calculation of Appendix A

The sample calculation in Appendix A involved a TWSC intersection with a 4-lane major roadway. Virtually all of the HCMs indicated that the STOP-controlled approaches would not operate well, and (in some cases) failed (LOS F). Several approaches to improving the intersection as a TWSC location did not produce significant gains, and still left a poorly-operating intersection.

Another alternative (short of signalization, which did not seem warranted) would be to install a roundabout. Because the main roadway has 4 lanes, it would have to be a 2-lane roundabout, where two of the approaches have one lane, and two of the approaches have two lanes. Because a 2-lane roundabout is involved, only the 2010/2016 methodology can be applied to the analysis of such an intersection.

Figure C.2 shows a schematic of the roundabout that would result. Of course, this solution requires that sufficient land be available to allow construction of a roundabout.

The demand volumes shown in Figure C.2 are in veh/h over the full hour. These must be converted to flow rates in pc/h:

$$v_i = \frac{V_i}{PHF^* f_{HV}}$$

As shown in Fig. C.2, the PHF is 0.90, and there are 10% heavy vehicles in all movements. Using the standard heavy vehicle equivalent (E_{HV}) of 2 recommended in the 2010/2016 HCM (for roundabouts).

$$f_{HV} = \frac{1}{1 + P_T(E_T - 1)} = \frac{1}{1 + 0.10(2 - 1)} = 0.909$$

Table C.4 shows the conversion of demand volumes in veh/h to flow rates in pc/h.

The total demand flow on each approach is obtained by adding the left, through, and right turn flows on each approach:

$$v_{NB} = 27 + 128 + 23 = 178 \text{ pc/h}$$
$$v_{EB} = 22 + 464 + 18 = 504 \text{ pc/h}$$
$$v_{SB} = 12 + 122 + 21 = 155 \text{ pc/h}$$
$$v_{WB} = 31 + 428 + 24 = 483 \text{ pc/h}$$

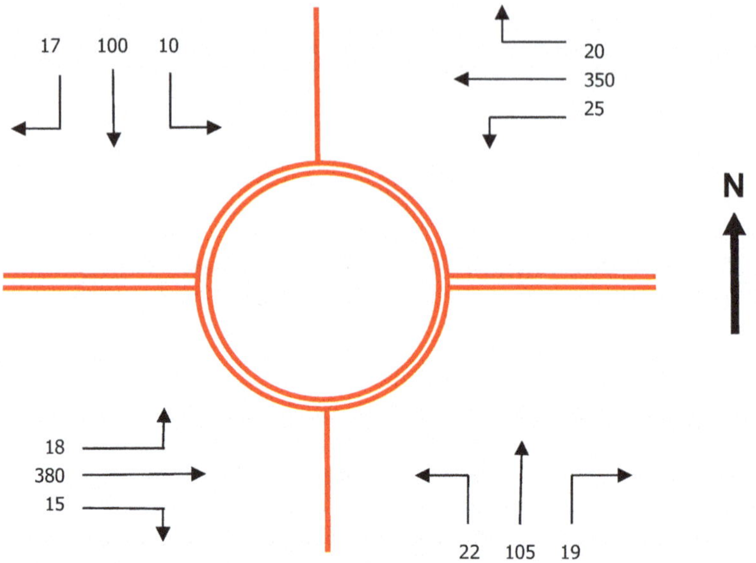

10% SUT in all mvts; PHF = 0.90, Random flow on major street; 20 peds/h, all xwalks, level terrain; major street speed = 30 mi/h.

Fig. C.2 A proposed roundabout solution for a TWSC intersection

Table C.4 Conversion of demand volumes to flow rates—sample calculation C.2

Movement	v (veh/h)	PHF	f_{HV}	v (pc/h)
NBL	22	0.90	0.909	27
NBT	105	0.90	0.909	128
NBR	19	0.90	0.909	23
EBL	18	0.90	0.909	22
EBT	380	0.90	0.909	464
EBR	15	0.90	0.909	18
SBL	10	0.90	0.909	12
SBT	100	0.90	0.909	122
SBR	17	0.90	0.909	21
WBL	25	0.90	0.909	31
WBT	350	0.90	0.909	428
WBR	20	0.90	0.909	24

The next step is the computation of circulating flows that will conflict with entering vehicles on each approach. With no U-turns present, circulating volume for each approach is comprised of through and left-turn flows from the adjacent approach to the left, and left-turn flows from the opposing approach. Then:

$$v_{cNB} = v_{EBL} + v_{EBT} + v_{SBL} = 22 + 464 + 12 = 498\,\text{pc/h}$$
$$v_{cEB} = v_{SBL} + v_{SBT} + v_{WBL} = 12 + 122 + 31 = 165\,\text{pc/h}$$
$$v_{cSB} = v_{WBL} + v_{WBT} + v_{NBL} = 31 + 428 + 27 = 486\,\text{pc/h}$$
$$v_{cWB} = v_{NBL} + v_{NBT} + v_{EBL} = 27 + 128 + 27 = 182\,\text{pc/h}$$

The capacity of each approach lane may now be computed. Note that the EB and WB approaches have *two* lanes each. Thus, two capacities are computed. The total capacity of the approach is the sum of the two lane capacities computed. Equations for capacity are found in Table 6.2.

The NB and SB approaches have one lane each. The roundabout itself has two lanes. The equation for capacity in such situations (from Table 6.2) is:

$$c_{epce} = 1130e^{(-0.07^{*}10^{-3})v_{cpce}}$$
$$c_{NB} = 1130e^{-0.0007^{*}498} = 797\,\text{pc/h}$$
$$c_{SB} = 1130e^{-0.0007^{*}486} = 804\,\text{pc/h}$$

The EB and WB approaches have *two* lanes, and they enter a two-lane roundabout. The equations governing the capacity of the right and left lanes of these approaches (from Table 6.2) are:

$$c_{eRpce} = 1130\,e^{(-0.7^{*}10^{-3})v_{cpce}}$$
$$c_{eLpce} = 1130\,e^{(-0.75^{*}10^{-3})v_{cpce}}$$
$$c_{EBR} = 1130\,e^{-0.0007^{*}165} = 1007\,\text{pc/h}$$
$$c_{EBL} = 1130\,e^{-0.00075^{*}165} = 1007\,\text{pc/h}$$
$$c_{EB} = 1007 + 998 = 2005\,\text{pc/h}$$
$$c_{WBR} = 1130\,e^{-0.0007^{*}182} = 995\,\text{pc/h}$$
$$c_{WBL} = 1130\,e^{-0.00075^{*}182} = 986\,\text{pc/h}$$
$$c_{WB} = 995 + 986 = 1981\,\text{pc/h}$$

Two steps remain. Capacities must be adjusted for the presence of pedestrians. Then, X_i (or v/c ratios) for each lane must be computed. In the latter case, the 2010/2016 HCMs give some general guidelines as to how to divide the demand on two-lane approaches into the two lanes. Given the demand splits between turning and through movements, nothing suggests that demand will not divide itself roughly in the proportion of the capacity of the two lanes. Thus, a single value of X_i based on total demand and total capacity will be used for the EB and WB approaches.

The problem statement shows that there are 20 pedestrians crossing each approach. Table 6.3 provides equations for the adjustment factor that should be applied. For the

NB and SB approaches, the equation for one-lane approaches with pedestrian flows under 101 peds/h and circulating flows under 881 pc/h is used:

$$f_p = 1 - 0.000137 n_{ped}$$
$$f_p = 1 - (0.000137^*20) = 0.997$$

For the EB and WB approaches, the equation for two-lane approaches with pedestrian flows less than 1000 peds/h is used:

$$f_{ped} = \min\left[1 - \tfrac{n_{ped}}{100}\left(1 - \tfrac{1260.6 - 0.329 v_{cpce} - 38.1}{1380 - 0.50 v_{cpce}}\right), 1\right]$$
$$f_{pedEB} = 1 - \tfrac{20}{100}\left(1 - \tfrac{1260.6-(0.329^*165)-38.1}{1380-(0.50^*165)}\right) = 0.980$$
$$f_{pedWB} = 1 - \tfrac{20}{100}\left(1 - \tfrac{1260.6-(0.329^*182)-38.1}{1380-(0.50^*182)}\right) = 0.980$$

Then:

$$c_{NBa} = 797^*0.997 = 795 \text{ pc/h}$$
$$c_{EBa} = 2005^*0.980 = 1965 \text{ pc/h}$$
$$c_{SBa} = 804^*0.997 = 802 \text{ pc/h}$$
$$c_{WBa} = 1981^*0.98 = 1941 \text{ pc/h}$$

The final step in the process is the estimation of delay on each approach using Eq. 6.6:

$$D_i = \frac{3600}{c_i} + 900T\left[X_i - 1 + \sqrt{(X_i - 1)^2 + \frac{\left(\frac{3600}{c_i}\right)X_i}{450T}}\right] + 5^*\min[X_i, 1]$$

This equation has been illustrated in previous sample calculations. The results for each approach are shown in Table C.5.

The proposed roundabout works quite well, with all approaches at LOS A., experiencing low delays. This compares to the TWSC approach (in Sample Calculation A.1, where the minor approaches (NB and SB) experience very high delays of LOS E or F.

Table C.5 Results for roundabout solution to TWSC sample calculation C.2

Movement	c_i (pc/h)	v_i (pc/h)	X_i	D_i (s/veh)	LOS_i
NB	795	178	0.224	6.95	A
EB	1965	504	0.256	3.74	A
SB	802	155	0.193	6.52	A
WB	1941	483	0.249	3.71	A

Table C.6 Comparative delays for sample calculations C.1 and C.2

Movement	Flow (pc/h)*	Delay (TWSC) (s/veh)	Delay (Roundabout) (s/veh)
WBL	31	8.5	3.71
WBTR	452	0.0	3.71
EBL	22	8.7	3.74
EBTR	482	0.0	3.74
NB	178	80.5	6.95
SB	155	47.2	6.52

*Based upon TWSC Sample Calculation conversion from veh/h to pc/h

Is the roundabout clearly a better solution? Maybe. Before making a final judgment, the issue of total delay should be addressed. In the TWSC option, there is *no* delay to through and right-turn vehicles on the EB and WB approaches. They obviously fare better under the TWSC option. The NB and SB approaches clearly have a much better deal using the roundabout.

If we use the output of the 2010/2016 HCM analysis for the TWSC intersection, we can compare the expected delays resulting from a TWSC intersection and a roundabout. This is shown in Table C.6.

A weighted average delay can be computed for each case by adding the products of flow * delay and dividing by the total flow rate, which totals 1320 pc/h in this case. To conserve space, this arithmetic computation is not shown here. The results, however, are interesting: The TWSC approach yields an average delay for all vehicles of 16.7 s/veh; the roundabout approach reduces this to 4.49 s/veh.

Therefore, the roundabout approach clearly works better in this case, as the delay added to EB and WB through and right-turning vehicles is rather small, while the reduction to NB and SB vehicles is significant. The remaining issues, of course, are the cost to construct the roundabout, and the availability of the ROW to do so.

Appendix D
Sample Calculations—Signalized Intersections

Historically speaking, the conceptual development of analysis methodologies for signalized intersections has changed radically between 1950 and 2016. Since Circular 212 in 1980, the fundamental approach has been critical lane or critical movement analysis. However, both the 1950 and 1965 HCMs had well-developed approaches to the analysis of signalized intersections, and the implementation of critical movement analysis has also changed radically in the years since 1980.

Because of the great complexity of more recent methodologies, only relatively simple problems can be presented and illustrated in a comprehensible manner. Software is required in most cases to make computations, some of which are now iterative in nature.

We have selected a relatively simple sample calculation to illustrate, as best as possible, the various methodologies applied to signalized intersection analysis over the years.

Sample Calculation: A Simple Signalized Intersection

The intersection shown in Fig. D.1 is to be analyzed to determine the likely quality of operation that will result from the demands indicated.

The intersection shown consists of a two-way N-S artery with two lanes in each direction, and a one-way (EB) E-W street with three lanes. There is no parking on any of the approaches (or departure roadways). The signal is a simple two-phase signal, with Phase A handling the E-W artery and Phase B handling the EB one-way street. There are no protected turns, but only one left turn (the SB left turn) faces an opposing flow due to the one-way street.

The signal timing is shown in Fig. D.1. Each of the two phases has a green (G), a yellow (y), and an all-red (ar) interval.

E. S. Prassas and R. P. Roess, *The Highway Capacity Manual: A Conceptual and Research History Volume 2*, Springer Tracts on Transportation and Traffic 12,
https://doi.org/10.1007/978-3-030-34480-1

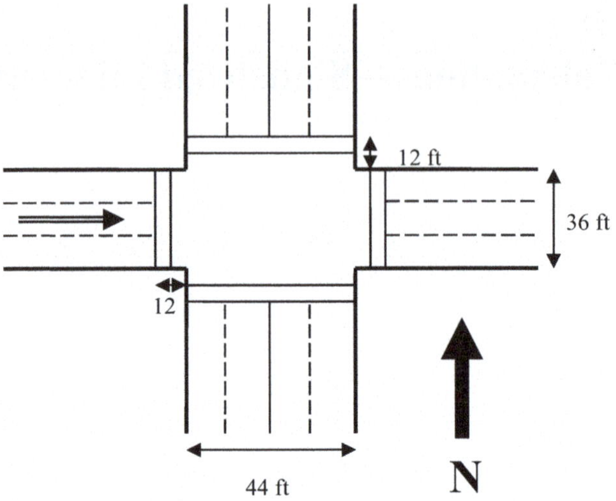

Traffic Volumes (veh/h)

Approach	Left	Through	Right
EB	25	1,200	20
NB	N/A	700	25
SB	10	800	N/A

PHF = 0.95; 10% Trucks; 0% RVs; Low Pedestrian Activity; Average Approach Speed – All Approaches = 35 mi/h; Crosswalk Width = 10 ft.; Pedestrian Walking Speed = 4.0 ft/s; Grade (G) = 0.0%; Start-Up Lost Time (l_1) = 2.0 s; Vehicle Use of Yellow +All Red (e) = 2.0 s.

Signal Timing

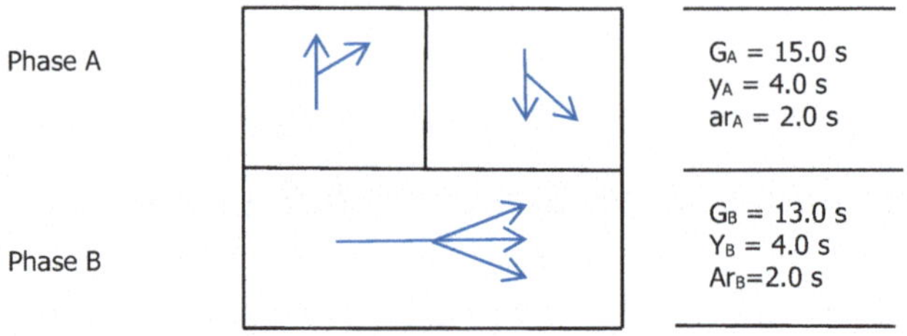

Fig. D.1 Intersection for sample calculation

The 1950 Highway Capacity Manual

The 1950 HCM takes a broad-brush approach to estimating *possible capacity* and *practical capacity* of each intersection approach. It begins by estimating a saturation flow rate for the entire approach based upon Fig. 7.1 (from Chap. 7).

Figure 7.1 has two parts: (a) for two-way streets, and (b) for one-way streets. In both cases, the figure is based upon total curb-to-curb street width (ft), parking conditions (permitted or prohibited), and the location of the intersection (downtown, intermediate, outlying). For our intersection, Fig. 7.1a is used for NB and SB approaches, while Fig. 7.1b is used for the EB approach. The results are illustrated in Fig. D.2. We will assume an intermediate location. The given street width for NB

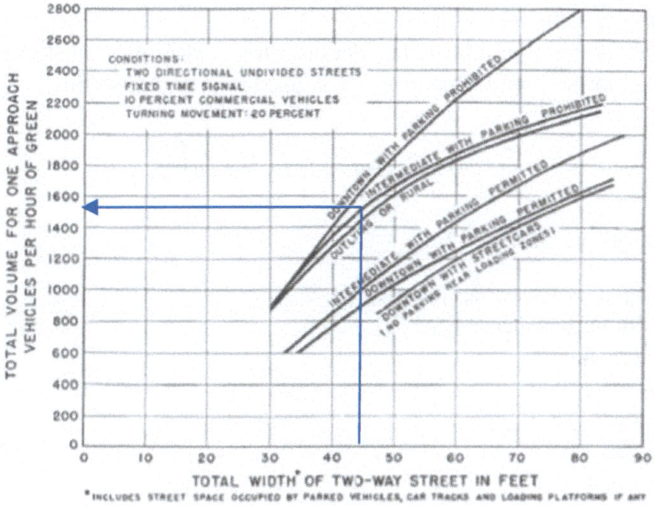

(a) NB and SB Approaches using Figure 7.1a

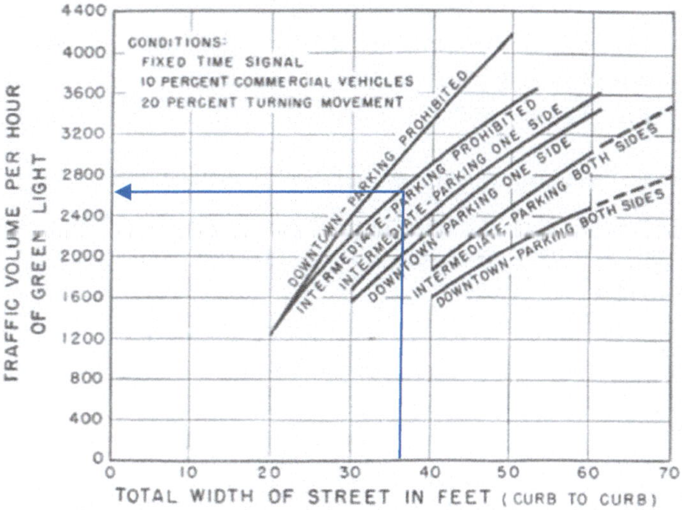

(b) EB Approach using Figure 7.1b

Fig. D.2 Determination of total saturation flow—1950 HCM and SB approaches is 44 ft, while the street width for the EB approach is given as 36 ft.

From Fig. D.1, the capacity of *each* approach (NB and SB) is 1525 veh/hg. This is based on the condition of a 44-ft street width that is assumed to be evenly split between the two directions, and the standard conditions of the figure, which includes 10% RT, 10%LT, and 10%HV. Similarly, the capacity of the EB approach is 2650 veh/hg, with the same base conditions.

These values must be adjusted to account for conditions that do not match the base included in the figure. Equation 7.1 is used to accomplish this:

$$c = s^*_{TB} f^*_c f^*_{HV} f^*_{RT} f^*_{LT} f^*_{LB} f^*_P \left(\frac{G}{C}\right)$$

Adjustments are provided for:

- Type of capacity (possible, practical)—f_c
- Heavy vehicle presence—f_{HV}
- Right turns—f_{RT}
- Left turns—f_{LT}
- Local buses—f_{LB}
- Parking—f_P
- Signal timing—G/C.

The actual proportion of right- and left-turning vehicles will be needed to obtain appropriate adjustment factors for turns. These are computed from the given input volumes:

$$\begin{aligned} \text{EB}: P_{RT} &= 20/1245 = 0.016;\ P_{LT} = 25/1245 = 0.020 \\ \text{NB}: P_{RT} &= 25/725 = 0.034;\ P_{LT} = 0.000 \\ \text{SB}: P_{RT} &= 0.000;\ P_{LT} = 10/810 = 0.012 \end{aligned}$$

For our case, the adjustments for *all* approaches include:

$f_c = 1.10$ for possible capacity; 0.90 for practical capacity (Eq. 7.2)
$f_{HV} = 1 - (P_{HV} - 0.10) = 1 - (0.10 - 0.10) = 1.00$ (Eq. 7.3)
$f_{LB} = 1.000$ (Table 7.1)
$f_P = 1.000$ (Table 7.1).

Adjustment factors for right turns, left turns, and the G/C ratio vary for each approach. Equation 7.4 estimates f_{RT}:

$$f_{RT} = 1.00 - \left(\frac{P_{RT} - 0.10}{2}\right)$$

$$f_{RTNB} = 1.00 - \left(\frac{0.034 - 0.10}{2}\right) = 1.033$$

$$f_{RTSB} = 1.00 - \left(\frac{0.000 - 0.10}{2}\right) = 1.050$$

$$f_{RTEB} = 1.00 - \left(\frac{0.016 - 0.10}{2}\right) = 1.042$$

Adjustment factors for left turns are estimated using Eq. 7.5:

$$\begin{aligned}
f_{LT} &= 1.00 - (P_{LT} - 0.10)\\
f_{LTNB} &= 1 - (0.000 - 0.10) = 1.100\\
f_{LTSB} &= 1 - (0.012 - 0.10) = 1.088\\
f_{LTEB} &= 1 - (0.020 - 0.10) = 1.080
\end{aligned}$$

G/C ratios are based upon the signal timing shown in Fig. D.1. Note that the 1950 HCM does not deal with the concept of *effective green time*, so that actual green times are used directly. The signal as shown has a 40-s cycle length.

$$\begin{aligned}
G/C_{NB,SB} &= 15/40 = 0.375\\
G/C_{EB} &= 13/40 = 0.325
\end{aligned}$$

With all of the input values estimated, Equation 7.1 can be implemented to compute both *possible capacity* and *practical capacity*. These computations are illustrated in Table D.1.

In terms of later editions of the manual, the term "possible capacity" is fundamentally the same as "capacity under prevailing conditions." "Practical capacity" is approximately a LOS C service volume, although there is no clear definition of what that would mean operationally. Table D.2 compares possible and practical capacities for each approach with the total demand volume, with all terms in vehicles per hour.

From Table D.2, it is clear that this intersection would be expected to fail, i.e., experience LOS F (even though that is not yet defined in the 1950 HCM). The actual demand volumes on every approach are over the possible capacities, which is indicative of a failure of all approaches.

The 1965 Highway Capacity Manual

The methodological approach of the 1965 HCM is quite similar to that of the 1950 manual, except that base conditions contain 5% trucks, 10% RT, and 10% LT vehicles, and the concept of LOS has been introduced. LOS is based upon the parameter "load factor," defined as the proportion of fully-loaded cycles within the peak hour (or analysis hour). The definition of "fully-loaded" was somewhat hard to establish, and led to many misunderstandings with the methodology.

Table D.1 Capacities for sample calculation—1950 HCM

App*	s_{TB}	f_c	f_{HV}	f_{RT}	f_{LT}	f_{LB}	f_P	G/C	c**
NB (po)	1525	1.10	1.000	1.033	1.100	1.000	1.000	0.375	715
NB (pr)	1525	0.90	1.000	1.033	1.100	1.000	1.000	0.375	585
SB (po)	1525	1.10	1.000	1.050	1.080	1.000	1.000	0.375	713
SB (pr)	1525	0.90	1.000	1.050	1.080	1.000	1.000	0.375	584
EB (po)	2650	1.10	1.000	1.042	1.088	1.000	1.000	0.325	1074
EB (pr)	2650	0.90	1.000	1.042	1.088	1.000	1.000	0.325	879

*po = possible capacity, pr = practical capacity; ** vehicles per hour

Table D.2 Comparison of capacities and demand volumes—1950 HCM

Approach	Possible capacity (veh/h)	Practical capacity (veh/h)	Demand volume (veh/h)
NB	715	585	725
SB	713	584	810
EB	1074	879	1245

The method estimated the service volume (SV) that could be accommodated at each level of service. Demand volumes were then compared to these criteria to establish the expected LOS.

Figure 7.2 is used to establish unadjusted base service volumes. Note that for two-way streets, the unit of width is no longer curb-to-curb distance, but curb-to-division line. Figure D.3 illustrates these determinations.

Note that in Fig. D.3, levels of service are correlated to the load factors in the figure: LOS E = 1.00, LOS D = 0.70, LOS C = 0.30, LOS B = 0.10, LOS A = 0.0. Thus, the unadjusted base service volumes for the three approaches are as shown in Table D.3.

Equation 7.6 is then used to adjust these base values:

$$SV_i = s^*_{TBi} f^*_{POP} f^*_L f^*_{HV} f^*_{LB} f^*_{RT} f^*_{LT} \left(\frac{G}{C}\right)$$

Adjustments for population (f_{POP}) and location within the city (f_L) are found on Fig. 7.2, and can be seen in Fig. D.3. If the location is taken as "fringe area," which corresponds to the "intermediate" of the 1950 HCM, and the population is stated as 80,000, with a PHF = 0.95, then these adjustment factors are:

NB and SB Approaches: $f_{POP=0.99}$(interpolating); $f_L = 1.25$
EB Approach: $f_{POP} = 1.03$(interpolating); $f_L = 1.10$

The heavy vehicle adjustment factor is obtained from Table 7.3. For all approaches, with 10% trucks, the adjustment is $f_{HV} = 0.95$. There are no local buses, so the adjustment is $f_{LB} = 1.00$ for all approaches.

Turn adjustment factors are obtained from Tables 7.4 and 7.5. Right turns and left turns from one-way streets are obtained from Table 7.4, while left turns from two-way streets are obtained from Table 7.5. Turn proportions are the same as for the 1950 solution:

$$\begin{aligned} NB &: P_{LT} = 0.000, P_{RT} = 0.034 \\ SB &: P_{LT} = 0.012, P_{RT} = 0.000 \\ EB &: P_{LT} = 0.020, P_{RT} = 0.016 \end{aligned}$$

Using Tables 7.4 and 7.5, the following adjustment factors result:

Fig. D.3 Unadjusted service volumes—1965 HCM

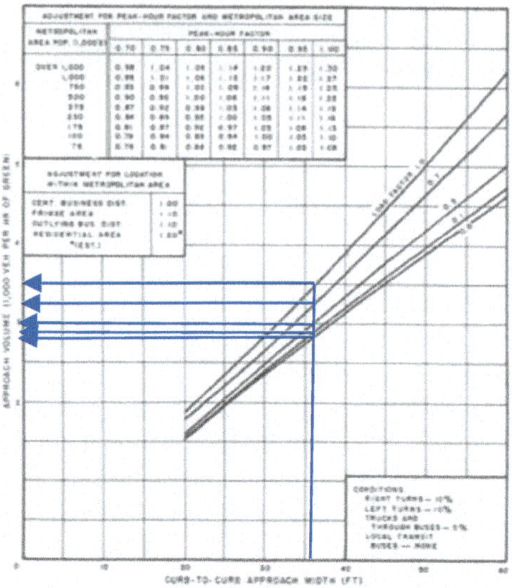

(a) Service Volumes for EB Approach

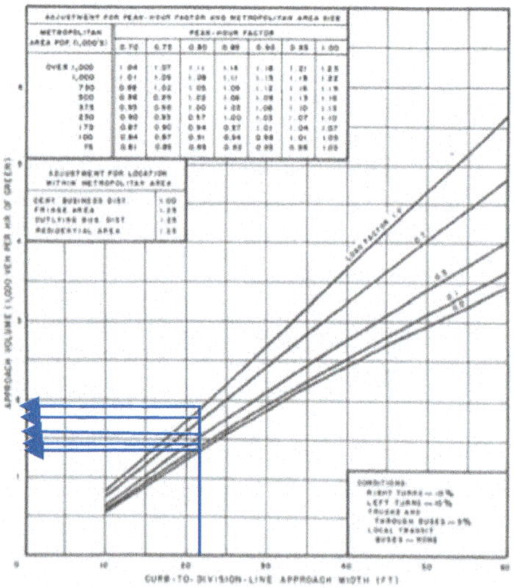

(b) Service Volumes for NB and SB Approaches

NB : $f_{LT} = 1.10$, $f_{RT} = 1.033$ (interpolating)
SB : $f_{LT} = 1.078$(interpolating), $f_{RT} = 1.050$
EB : $f_{LT} = 1.02$, $fRT = 1.020$

As in the 1950 solution, G/C = 15/40 = 0.375 for NB and SB approaches, and 13/40 = 0.325 for the EB approach.

Table D.3 Unadjusted service volumes for three approaches—1965 HCM

Level of service	NB and SB approaches (veh/hg)	EB approach (veh/hg)
A	1300	2750
B	1400	2850
C	1550	3000
D	1750	3250
E	1950	3500

Equation 7.6 essentially takes the base saturation flows of Table D.3 and multiplies them by the product of all the adjustment factors. The product of adjustment factors (F) is:

$$F_{NB} = 0.99^{*}1.25^{*}0.95^{*}1^{*}1.10^{*}1.033^{*}0.375 = 0.501$$
$$F_{SB} = 0.99^{*}1.25^{*}0.95^{*}1^{*}1.078^{*}1.050^{*}0.375 = 0.499$$
$$F_{EB} = 1.03^{*}1.10^{*}0.95^{*}1^{*}1.02^{*}1.02^{*}0.325 = 0.364$$

Multiplying the base saturation volumes of Table D.3 by these products produces the values shown in Table D.4.

The service volumes of Table D.4 must be compared to the actual demand volumes on the various approaches to determine the expected LOS:

$$V_{NB} = 725\ \text{veh/h} \quad \text{LOS} = \text{C}$$
$$V_{SB} = 810\ \text{veh/h} \quad \text{LOS} = \text{D}$$
$$V_{EB} = 1245\ \text{veh/h} \quad \text{LOS} = \text{E}$$

The results are quite different from those of the 1950 HCM. While the 1950 HCM predicted a failure for all three approaches, the 1965 HCM indicates that all will work, although the EB approach will be at LOS E, which is poor.

It must be remembered, however, that the 1965 HCM is still a relatively rough instrument. While it provides an adjustment that takes PHF into account, it really doesn't deal with peak 15-min flow rates. Further, it still uses actual green time and

Table D.4 Adjusted service volumes—1965 HCM

Level of service	NB approach (veh/h)	SB approach (veh/h)	EB approach (veh/h)
A	651	649	1001
B	701	699	1037
C	777	773	1092
D	877	873	1183
E	977	973	1274

Note All values rounded to the nearest integer

does not account for the differences between actual and effective green time, which becomes important in later editions of the manual.

Nevertheless, it is seen that the 1965 HCM predicts higher capacities than the 1950 manual, and provides the first measure of level of service—load factors. Unfortunately, load factor does not provide a great deal of information on operations, and was never a well-understood parameter. Its measurement in the field for calibration and verification purposes was quite difficult, as it hinged on the issue of a "fully-loaded green phase," which involved considerable judgment to be exercised.

Circular 212

Circular 212 introduced the profession to *critical movement analysis* as a general approach to signalized intersection analysis. A planning level and design/operations level procedure was introduced. This solution uses the design/operations procedure. The methodology has two unique features: (a) all demands are converted to peak 15-min flow rates in *passenger car equivalents*, and (b) effective green time is used.

The latter is interesting in that effective green time (g) is taken to be the sum of actual green (G), yellow (y), and all-red (ar) times—which suggests that there is no lost time in the signal cycle.

The solution begins by converting every demand movement to pc/h and peak flow rates using Eq. 7.8:

$$v_{pci} = \frac{(V_{PCi} + 2V_{Ti} + 5V_{LBi})^* E_M^* W^* U}{PHF}$$

In our problem, the PHF is given as 0.95, with 10% trucks and 90% passenger cars. If we apply the truck percentages to the demand volumes, we will get fractional vehicles. While this will not conform exactly to reality, it will, nevertheless, provide a more consistent view of converted flow rates.

Passenger car equivalents for trucks (E_T) and local buses (E_{LB}) are taken from Table 7.8, but are already built into Eq. 7.8, as they are constants for all conditions.

Movement equivalents (E_M) are obtained from Table 7.9. There are no exclusive turn phases or lanes in our problem. The opposing volumes are 800 veh/h for the NB approach, 700 veh/h for the SB approach, and 0 for the EB approach. "Low pedestrian activity" suggests approximately 50 peds/h in each crosswalk. Thus:

$$E_{LTNB/SB} = 4.00;\ E_{LTEB} = 1.00$$
$$E_{RTNB/SB} = 1.00;\ E_{RTEB} = 1.00$$
$$E_{TH} = 1.00\ \text{(all approaches)}$$

Lane width and lane utilization adjustment factors are drawn from Table 7.10. Lane width on NB and SB approaches is 44/4 = 11 ft. Lane width is 36/3 = 12 ft on the EB approach. Then:

$$W = 1.00(\text{all approaches})$$
$$U_{NB/SB} = 1.05;\ U_{EB} = 1.10$$

Conversion of all demand volumes is now completed using Eq. 7.8.

$$v_{NBTH} = \frac{(630^*1 + 70^*2 + 0^*5)^*1^*1^*1.05}{0.95} = 851 \text{ pc/h}$$
$$v_{NBRT} = \frac{(18^*1 + 2^*2 + 0^*5)^*1^*1^*1.05}{0.95} = 24 \text{ pc/h}$$
$$v_{SBLT} = \frac{(9^*1 + 1^*2 + 0.5)^*4^*1^*1.05}{0.95} = 12 \text{ pc/h}$$
$$v_{SBTH} = \frac{(720^*1 + 80^*2 + 0^*5)^*1^*1^*1.05}{0.95} = 973 \text{ pc/h}$$
$$v_{EBLT} = \frac{(22.5^*1 + 2.5^*2 + 0^*5)^*1^*1^*1.10}{0.95} = 32 \text{ pc/h}$$
$$v_{EBTH} = \frac{(1080^*1 + 120^*2 + 0^*5)^*1^*1^*1.10}{0.95} = 1528 \text{ pc/h}$$
$$v_{EBRT} = \frac{(18^*1 + 2^*2 + 0^*5)^*1^*1^*1.10}{0.95} = 25 \text{ pc/h}$$

Converted demand flow rates are summarized in Table D.5.

Lane flow rates must be estimated from the converted demand flow rates of Table D.5. With no exclusive turning lanes involved, the total approach flow rate is divided by the number of lanes to get the lane flow rate. It is assumed that equivalent vehicles distribute themselves uniformly across all available lanes. Remember that a lane utilization factor was applied, which accounts for any unequal use of lanes that may exist. Then:

$$v_{LNB} = (851 + 24)/2 = 438\,\text{pc/h}$$
$$v_{LSB} = (12 + 973)/2 = 493\,\text{pc/h}$$
$$v_{LFB} = (32 + 1538 + 25)/3 = 532\,\text{pc/h}$$

Figure 7.4 is used to determine the critical variable: The sum of critical lane flow rates. In our case, each street has one phase and no exclusive turn lanes. For Phase A of the signal (NB and SB), the critical lane flow rate is the maximum of the lane flow rates in the NB and SB approaches. In our case, this is the SB approach with a lane flow rate of 493 pc/h. The EB approach is critical by definition, and has a lane

Table D.5 Converted lane flow rates—Circular 212

Approach	LT (pc/h)	TH (pc/h)	RT (pc/h)
NB	0	851	24
SB	12	973	0
EB	32	1528	25

flow rate of 532 pc/h. The sum of critical lane flow rates is, therefore, 493 + 532 = 1025 pc/h.

The sum of critical lane flow rates determines the level of service using the criteria of Table 7.6. For a two-phase signal, our intersection operation is expected to be LOS B, which produces an approximate delay in the range of 16.1–22.0 s/veh (Table 7.7).

This result is markedly better than the 1950 and 1965 solutions, and projects an intersection working quite well. It must be noted, however, that the Circular 212 methodology produces an overall intersection LOS, and does not look at the details of signal timing or its impact on individual approaches. This is the major weakness of the procedure.

Signal timing is not used in the analysis. This is because fundamental assumptions on effective green time, base saturation flow rates, and an assumed reasonable apportionment of green time are "baked into" the methodology.

The 1985 Highway Capacity Manual

The 1985 HCM provided the fundamental structure of signalized intersection analysis that is still used today—with many detailed revisions. It was organized into five computational modules, and is based upon critical lane or critical movement analysis. As computations are completed in modules, they are presented in modular form.

Input Module

The input module is merely the organization of input information needed to complete computations. The 1985 HCM required a great deal more detailed input information than previous editions. For the sample calculation at hand, the input information remains as shown in Fig. D.1. The 1985 HCM, however, introduced a completely new variable into the process—*Arrival Type*. Arrival type depends on the coordination of signals with adjacent signals. As the problem statement does not include this, we will assume for this (and all subsequent solutions) that the Arrival Type is 3—essentially, the base case of random arrivals at the subject intersection.

The input module also includes a computation of the minimum green time required by pedestrians to safely cross the street. It is used as a check against the actual signal timing to insure that pedestrians will not be at risk or more of a problem than the methodology suggestions. Minimum green times for pedestrians are computed using Eq. 8.2:

$$G_p = 7.0 + \frac{L_p}{4} - Y$$

where L_p is the length of the crosswalk (from the curb to the *midpoint* of the farthest travel lane), Y is the sum of yellow (y) and all-red (ar) intervals, and 4 ft/s is the assumed walking speed of pedestrians.

Table D.6 Demand flow rates for sample calculation—1985 HCM

Traffic Flow Rates (veh/h)			
Approach	Left	Through	Right
EB	25/0.95 = 26	1200/0.95 = 1263	20/0.95 = 21
NB	N/A	700/0.95 = 737	25/0.95 = 26
SB	10/0.95 = 11	800/0.95 = 842	N/A

Note All values rounded to the nearest integer

In the sample calculation, pedestrians cross the EB approach during Phase A. The crossing length would be 2 * 12 + 6 = 30 ft. Pedestrians cross the NB and SB approaches in Phase B, with a crossing length of 3 * 11 + 5.5 = 38.5 ft. Then:

$$G_{pA} = 7.0 + \frac{30}{4} - (4 + 2) = 8.5\,\text{s}$$

$$G_{pB} = 7.0 + \frac{38.5}{4} - (4 + 2) = 10.6\,\text{s}$$

In the sample calculation, the actual green times (15 and 13 s respectively) both exceed the minimums required for safety. No special pedestrian problems would be expected.

Volume Adjustment Module

In the volume adjustment module, four things are done: demand volumes (in veh/h) are converted to demand flow rates (in veh/h) using the PHF, analysis "lane groups" are established, the potential presence of "defacto left-turn lanes" is investigated, and a lane utilization factor is applied to demand flow rates.

Flow rates are computed by dividing demand volumes by the PHF, as shown in Table D.6.

As there are no exclusive turning lanes, all lanes on each approach comprise a single "lane group" for analysis.

There is only one opposed left turn in the sample calculation: the SB left turn. The possibility exists that a significant enough number of left turns in this case could essentially create a "defacto" left-turn lane, with no other vehicles using it. In such a case, a LT lane would be established for the SB approach as a separate lane group for analysis.

Equations 8.5 and 8.6 are used to assess this possibility. The equivalent left-turn flow rate (on the SB approach) is equal to:

$$v_{LE} = v_L(1800/1400 - v_o) = 11\left(\frac{1800}{1400 - 737}\right) = 30\,\text{pc/h}$$

A defacto LT lane will be said to exist if the equivalent left-turn flow rate is greater than the average flow in other lanes of the lane group. For our sample calculation:

Table D.7 Adjusted demand flow rates for sample calculation—1985 HCM

Traffic flow rates (veh/h)			
Approach	Left	Through	Right
EB	26*1.10 = 29	1263*1.10 = 1389	21*1.10 = 23
NB	N/A	737*1.05=774	26*1.05 = 24
SB	11*1.05=12	842*1.05 = 884	N/A

Note All values rounded to the nearest integer

$$v_{LE} = 30 \geq \frac{v - v_L}{N - 1} = \frac{853 - 11}{2} = 421 \text{ pc/h}$$

As the left-turn equivalents *do not* exceed the recommended limit, no defacto left-turn lane will exist. Therefore, the use of a single lane group for each approach is appropriate.

Lane utilization factors (U) are found in Table 8.5. They are the same as in Circular 212—1.05 for 2-lane lane groups (NB, SB) and 1.10 for 3-lane lane groups (EB). The demand flow rates in Table D.6 are multiplied by these factors, resulting in the adjusted demand flow rates shown in Table D.7.

These are the demand flow rates that will be used in the remainder of this solution.

Saturation Flow Module

In the saturation flow module, a base saturation flow rate is adjusted to account for a variety of factors. Generally described by Eq. 8.8, the equation for the 1985 HCM is:

$$s = s_o N f_w f_{HV} f_g f_p f_{bb} f_a f_{RT} f_{LT}$$

The base saturation flow rate for the 1985 HCM is 1800 pc/hg/ln.

Most of these adjustment factors are obtained from Table 8.6. For the sample calculation:

$$\mathrm{f_{wNB/SB}} = 0.97\,(11 - \text{ft lanes})\quad \mathrm{f_{wEB}} = 1.00\,(12 - \text{ft lanes})$$
$$\mathrm{f_{HV}} = 0.95\,(\text{all approaches, 10\% trucks})$$
$$\mathrm{f_g} = 1.00\,(\text{all approaches, level terrain})$$
$$\mathrm{f_p} = 1.00\,(\text{all approaches, no parking})$$
$$\mathrm{f_{bb}} = 1.00\,(\text{all approaches, no local buses})$$
$$\mathrm{f_a} = 1.00\,(\text{all approaches, all other areas})$$

The right-turn adjustment factor is obtained from Table 8.7, and is given by equation an equation for a shared RT lane with permitted phasing.

$$f_{RT} = 1.0 - P_{RT}[0.15 + (peds/2100)]$$
$$f_{RTNB} = 1.0 - 0.034\,[0.15 + (50/2100)] = 0.994$$
$$f_{RTSB} = 1.0 - 0.000\,[0.15 + (50/2100)] = 1.000$$
$$f_{RTEB} = 1.0 - 0.016\,[0.15 + (50/2100)] = 0.997$$

The left-turn adjustment factor is quite complicated in the 1985 HCM, particularly for permitted left turns. In our case, only the SB left turn is truly "permitted" as no opposing flows exist for the NB and EB approaches. The latter two would be considered to have signal protection as a result of the intersection geometry. Table 8.10 is used to find the left-turn adjustment factor for the NB and EB approaches, under the category of protected turns in a shared lane. Then:

$$f_{LT} = 1/(1.0 + 0.05 P_{LT})$$
$$f_{LTNB} = 1/(1.0 + 0.05{*}0.0) = 1.000$$
$$f_{LTEB} = 1/(1.0 + 0.05{*}0.02) = 0.999$$

An alternative interpretation of the EB LT, made from a one-way street, is that it should be treated as if it were a right turn, with the principle consideration being pedestrian interference. In that case, the equation for right turn factors would be applied, and:

$$f_{LTEB} = 1.0 - 0.02\,[0.15 - (50/2100)] = 0.996$$

While there is not much difference between the two potential LT factors for the EB approach, we will use the more conservative value of 0.996 for the remainder of the sample calculation.

This leaves the final adjustment factor to be determined: The left-turn factor for the permitted SB left turn, which has an opposing flow. The procedure is based upon identifying three different portions of the green phase, and determining when left turns can be made, and at what rates. The complex process begins with the estimation of the saturation flow rate in the opposing lane group, using Eq. 8.24:

$$s_{op} = \frac{1800\,N_o}{1 + P_{LTo}(400 + v_M)/(1400 - v_M)}$$
$$s_{op} = \frac{1800{*}2}{1 + 0.0{*}(400 + 884)/(1400 - 884)} = 3600\ \text{pc/h}$$

Next, the ratio of opposing flow rate to opposing saturation flow rate is computed using Eq. 8.25:

$$Y_o = v_o/s_{op} = 774/3600 = 0.215$$

The unsaturated portion of the green time is given by Eq. 8.26. This equation, however, does not use actual green times. It uses effective green times. Effective green time is basically the actual green time minus the start-up lost time, plus the yellow and all red time used by vehicles to enter the intersection, or:

$$g = G - \ell_1 + e$$

As indicated in the problem statement, both ℓ_1 and e are 2.0 s, so the effective green time and the actual green time are numerically equivalent. Then, using Eq. 8.26:

$$g_u = \frac{g - CY_o}{1 - Y_o} = \frac{15 - 40*0.215}{1 - 0.215} = 8.15\,s$$

Equation 8.27 is now used to determine the proportion of left-turning vehicles in the left lane of the SB approach:

$$P_L = P_{LT}\left[1 + \frac{(N-1)\,g}{f_s g_u + 4.5}\right]$$

The value of P_{LT} computed as 12/(12 + 884) = 0.13. Then:

$$f_s = \frac{875 - 0.625v_o}{1000} = \frac{875 - 0.625*774}{1000} = 0.391$$

$$P_{L,SB} = 0.013*\left[1 + \frac{(2-1)*15}{0.39*8.15 + 4.5}\right] = 0.039$$

The two other portions of the green phase may now be computed using Eq. 8.28:

$$g_q = g - g_u = 15.0 - 8.15 = 6.85$$

$$g_f = \frac{2P_T{}^*}{P_L}\left(1 - P_T^{0.5g_q}\right) = \frac{2*0.961}{0.039}\left(1 - 0.961^{0.8*6.85}\right) = 6.27\,s$$

where g_q is the portion of the green phase blocked by the clearance of an opposing queue, and g_f is the time until the arrival of the first left-turning vehicle in the shared left-turn/through lane. Note that P_T is proportion of through vehicles in the shared lane, or $1 - P_L$.

Equation 8.30 is used to determine the LT adjustment factor *as applied to the left lane of the shared approach*. It has three components: (a) before the first left-turning vehicles arrive, there is no impact on flow, i.e., $f_{LT} = 1.00$, (b) during g_u, left turns move through an unsaturated opposing flow at some reduced rate, and (c) "sneakers" wait in the intersection until the opposing queue clears, and then proceed through during the yellow and all-red intervals at the end of the phase.

The LT adjustment factor for the left lane is then:

$$f_m = \frac{g_f}{g} + \frac{g_u}{g}\left[\frac{1}{1 + P_L(E_L - 1)}\right] + \frac{2}{g}(1 + P_L)$$

The value of E_L is normally estimated using Eq. 8.29. Then:

$$E_L = \frac{1800}{1400 - v_o} = \frac{1800}{1400 - 774} = 2.74$$

$$f_m = \frac{6.27}{15} + \frac{8.15}{15}\left[\frac{1}{1 + 0.039^*(2.74 - 1)}\right] + \frac{2}{g}(1 + 0.039)$$

$$= 1.065 \leq 1.00 = 1.00$$

Since the value is greater than 1.0, f_m is set to 1.0. To be applied to the SB approach as a whole, this factor for the *left lane* has to be averaged with other lanes on the approach that have a LT adjustment of 1, using Eq. 8.31:

$$f_{LT,SB} = \frac{f_m + (N - 1)}{N} = 1.0$$

Note that there are only 12 left-turn vehicles per hour out of the 896 vph on the approach so it makes sense that they would have a negligible impact.

We now have all of the adjustment factors needed to estimate the saturation flow rate in each lane group (or approach in this case), using Eq. 8.8 presented earlier:

$$s = s_o N f_w f_{HV} f_g f_{pp} f_{bb} f_a f_{RT} f_{LT}$$

$$s_{NB} = 1800^*2^*0.97^*0.95^*1^*1^*1^*1^*0.994^*1 = 3297 \text{ veh/hg}$$

$$s_{SB} = 1800^*2^*0.97^*0.95^*1^*1^*1^*1^*1^*1 = 3317 \text{ veh/hg}$$

$$s_{EB} = 1800^*3^*1^*0.95^*1^*1^*1^*1^*0.997^*0.996 = 5094 \text{ veh/hg}$$

Capacity Analysis Module

In the capacity analysis module, for each approach, the capacity and v/c ratio (X) are computed. Also computed is the critical v/c ratio for the intersection (X_c), which is based upon identifying the critical lane groups.

Capacity is a simple computation based upon saturation flow rates, using Eq. 8.45:

$$c = s(g/c)$$

$$c_{NB} = 3297(15/40) = 1236 \text{ veh/h}$$

$$c_{SB} = 3317(15/40) = 1244 \text{ veh/h}$$

$$c_{EB} = 5094(13/40) = 1656 \text{ veh/h}$$

Volume-to-capacity ratios are simply computed by dividing the demand flow rate for each approach (total) by the capacity of the approach:

$$X_i = v_i/c_i$$
$$X_{NB} = 798/1236 = 0.646$$
$$X_{SB} = 798/1244 = 0.720$$
$$X_{EB} = 1441/1656 = 0.870$$

The critical v/c ratio is computed using Eq. 8.47, which is based upon the sum of the critical v/s ratios (flow ratios). In our case, there is a two-phase signal, meaning that there will be two critical approaches (or lane groups). Phase B of the signal is clearly controlled by the EB approach, with a v/s ratio of 1441/5094 = 0.283. Phase A is controlled *either* by the NB *or* the SB approach, depending upon which one has the larger v/s ratio. The v/s ratio for the NB approach is 798/3297 = 0.242; the v/s ratio for the SB approach is 896/3317 = 0.270. The SB approach clearly controls Phase A. The sum of critical v/s ratios is, therefore, 0.283 + 0.270 = 0.553. Then, using Eq. 8.47:

$$X_c = \sum (v/s)_c \left[\frac{C}{C-L}\right]$$

where L is the lost time per cycle. Because actual green and effective green are numerically the same in this case, the lost time is the sum of yellow and all-red intervals, or 4 + 2 + 4 + 2 = 12 s/cycle. Then:

$$X_c = 0.551 \left[\frac{40}{40-12}\right] = 0.787$$

At this point, we know that the capacity of all three approaches is below capacity, so that unstable queues will not be developing. The critical X ratio is also good, and simply indicates that there is enough total capacity in the critical lane groups (or approaches) to handle the total demand in them.

Level of Service Module

Level of service in the 1985 HCM is based upon stopped-time delay per vehicle, and is separately reported for each lane group. Aggregations for each approach (where there are more than 1 lane group per approach) and for the intersection as a whole can also be estimated.

Stopped-time delay is estimated using Eq. 8.51:

$$d = d_i\, PF$$

$$d_i = 0.38\, C \left[\frac{(1-g/C)^2}{1-(g/C)\, X_i}\right] + 173\, X_i^2 \left[(X_i - 1) + \sqrt{(X_i - 1)^2 + \left(\frac{16\, X_i}{c_i}\right)}\right]$$

Table D.8 Stopped time delay and LOS for sample calculation—1985 HCM

Approach	C (s)	g/C	X	c (veh/h)	d (s/veh)	LOS
NB	40	0.375	0.646	1236	8.67	B
SB	40	0.375	0.720	1244	9.57	B
EB	40	0.325	0.870	1656	13.47	B

The progression factor (PF) adjusts for the impact of progressed signal systems on delay. It is obtained from Table 8.15. For a pretimed signal with Arrival Type 3, the factor is 1.00, so that for our case $d = d_i$.

For the sample calculation, $C = 40$ s, $g/C_{NB/SB} = 0.375$, $g/C_{EB} = 0.325$, and values of X_i and c_i were computed in the *Capacity Analysis Module*. The NB computation is shown here, and the results for all approaches are summarized in Table D.8.

$$d_{NB} = 0.38^{*}40^{*}\left[\frac{(1-0.375)^2}{1-0.375^{*}0.646}\right] + 173^{*}0.646^2 \overline{+\sqrt{(0.646-1)^2+\left[\frac{16^{*}0.646}{1236}\right]}} = 8.67\ \text{s/veh}$$

Levels of service are obtained from the criteria of Table 8.14. All approaches operate at LOS B, which covers delays from 5.1 to 15.0 s/veh. The predicted LOS is the same as was indicated in the Circular 212 analysis (LOS B), but better than predicted by the 1965 and 1950 HCMs. Surprisingly, the delay values are in the same range as predicted by the Circular 212 methodology, but the numbers don't compare directly. The Circular 212 estimate is general and approximate, and does not specify the definition of delay. The 1985 HCM deals with stopped-time delay, a parameter that would be revised in later editions.

The 1994 Update to the 1985 HCM

The 1994 update retained the basic modular structure as the 1985 HCM. There were a number of incremental changes made to the procedure. The solution here focuses on those aspects where changed from the 1985 HCM.

Input Module

There were no major changes to the input module that affect the sample calculation. Specific identification of lost time per phase was added. As discussed previously, in our case, lost time is the sum of yellow and all-red intervals, so both Phase A and Phase B involve $4 + 2 = 6$ s of lost time in each phase.

Volume Adjustment Module

There were some changes introduced in the volume adjustment module, but none of them affect the sample calculation. The check for a defacto left-turn lane was removed to the saturation flow module, but does not affect our calculations. The results shown in Table D.7 remain unchanged.

Saturation Flow Rate Module

There were several changes in this module that affect the sample calculation. The most important is that the base saturation flow rate was increased from 1800 pc/hg/ln to 1900 pc/hg/ln. This, of course, changes all of the computations for saturation flow rates. Heavy vehicle adjustment factors changed, as the base truck equivalent was increased from 1.5 pc/truck to 2.0 pc/truck.

The 1994 update replaced the tables for adjustment factors with equations. Except for the heavy vehicle adjustment and the left-turn adjustment for permitted left turns, the results remained the same as in 1985. Thus, the following adjustment factors remain:

$$
\begin{aligned}
&f_{wNB/SB} = 0.97\,(11-\text{ft lanes}) \quad f_{wEB} = 1.00\,(12-\text{ft lanes})\\
&f_g = 1.00\,(\text{all approaches, level terrain})\\
&f_p = 1.00\,(\text{all approaches, no parking})\\
&f_{bb} = 1.00\,(\text{all approaches, no local buses})\\
&f_a = 1.00\,(\text{all approaches, all other areas})\\
&f_{RTNB} = 0.994 \quad f_{RTSB} = 1.000 \quad f_{RTEB} = 0.997\\
&f_{LTNB} = 1.000 \quad f_{LTEB} = 0.996
\end{aligned}
$$

The heavy vehicle adjustment for all approaches is found using the equation from Table 8.8, with $E_T = 2.0$. Then:

$$f_{HV} = \frac{100}{100 + P_T\,(E_T - 1)} = \frac{1}{1 + 10\,(2-1)} = 0.909$$

The most significant change, however, involves the left-turn adjustment for permitted phases, which exists in the SB approach of the sample calculation. The process begins by estimating the portions of the green phase (as in 1985). Initial equations, however, are based upon field studies, rather than purely analytic constructs.

The time until the first left-turning vehicle arrives (g_f) is estimated using Eq. 8.32:

$$g_f = G\,e^{-0.882\,LTC^{0.717}} - t_L$$

In this equation, LTC is the number of left turns per cycle, computed as:

$$LTC_{SB} = \frac{v_{LTSB}\,C}{3600} = \frac{12{*}40}{3600} = 0.133\,\text{LT/cycle}$$

Time lost per phase was noted previously, $t_L = 6.0$ s. Then:

$$g_f = 15\,e^{-0.882{*}0.133^{0.717}} - 6 = 6.18\,\text{s}$$

The time during which left turns are blocked by the clearance of the opposing queue (g_q) is estimated using Eq. 8.34:

$$g_q = \frac{v_{OLC}\, qr_o}{0.5 - \left[v_{OLC}\left(\frac{1-qr_o}{g_o}\right)\right]}$$

The variable v_{OLC} is the opposing flow rate per lane per cycle. It is computed as v_o C/3600 N_o. For the SB approach, this is (774 * 40)/(3600 * 2) = 4.3 veh/cycle/ln. The variable qr_o is the opposing queue ratio, which is found as $1 - R_{po}$ (g_o/C). The opposing green time, g_o, is 15 s. The platoon ratio (R_{po}) is the ratio of the percent of vehicles arriving on green and the percent of the cycle that is green. For Arrival Type 3 (random arrivals), this ratio is 1.00. Then:

$$qr_o = 1 - 1^*\left(\frac{15}{40}\right) = 0.625$$

$$g_q = \frac{4.3^*0.625}{0.5 - \left[4.3^*\left(\frac{1-0.625}{15}\right)\right]} = 6.85\ \text{s}$$

Equation 8.35 is used to determine the value of g_u. There are two cases presented, depending upon whether g_f is less or more than g_q. In our case, g_f is 6.18 s and g_q is 6.85 s. Thus, $g_f < g_q$, and:

$$g_u = g - g_q = 15 - 6.85 = 8.15\ \text{s}$$

which is also the same result as obtained in the 1985 HCM solution.

With the three portions of the green phase (g_u, g_f, and g_q) determined, the LT adjustment factor for the left lane (f_m) and the lane group (f_{LT}) can now be completed. Both steps of the process, however, were revised in the 1994 update. The LT adjustment factor for the left lane is determined using Eq. 8.36. Compared to the 1985 equation, the term for "sneakers" has been eliminated, and the left-turn equivalent values (E_{LT}) were revised:

$$f_m = \frac{g_f}{g} + \frac{g_u}{g}\left[\frac{1}{1 + P_L\,(E_{L1} - 1)}\right]$$

The value of E_{L1} is found by interpolating the values from Table 8.11 as

$$E_{L1} = 3.6 + 2.4^*\left(\frac{774 - 600}{800 - 600}\right) = 5.688$$

The value of f_m then becomes:

$$f_m = \frac{6.18}{15} + \frac{8.15}{15}\left[\frac{1}{1 + .039^*(5.688 - 1)}\right] = 0.871$$

Equation 8.37 is used to compute the adjustment factor for the entire lane group (or approach in our case). It differs from the 1985 solution in that it is assumed that lanes other than the left lane are also affected by left turns. The assumed factor applying to other lanes is 0.91 in 1994, compared to 1.00 in 1985:

$$f_{LT,SB} = \frac{f_m + 0.91^*(N-1)}{N} = \frac{0.871 + 0.91^*(2-1)}{2} = 0.891$$

At this point all of the adjustment factors have been determined, and the saturation flow rates for each of the approaches can be computed. The base equation does not change, although some of the input values are different from the 1985 solution, as discussed:

$$\begin{aligned} s &= s_o N f_w f_{HV} f_g f_p f_{bb} f_a f_{RT} f_{LT} \\ S_{NB} &= 1900^*2^*0.97^*0.909^*1^*1^*1^*1^*0.994^*1 = 3330 \text{ veh/hg} \\ S_{SB} &= 1900^*2^*0.97^*0.909^*1^*1^*1^*1^*1^*0.891 = 2985 \text{ veh/hg} \\ S_{EB} &= 1900^*3^*1^*0.909^*1^*1^*1^*1^*0.997^*0.996 = 5145 \text{ veh/hg} \end{aligned}$$

There are no major changes in these values from 1985.

The capacity of each approach may now be computed:

$$\begin{aligned} c &= s(g/C) \\ c_{NB} &= 3330^*0.375 = 1249 \text{ veh/h} \\ c_{SB} &= 2985^*0.375 = 1119 \text{ veh/h} \\ c_{EB} &= 5145^*0.325 = 1672 \text{ veh/h} \end{aligned}$$

which yields the following volume-to-capacity ratios:

$$\begin{aligned} X_i &= v_i/c_i \\ X_{NB} &= 798/1249 = 0.639 \\ X_{SB} &= 896/1119 = 0.801 \\ X_{NB} &= 1441/1672 = 0.862 \end{aligned}$$

The critical v/c ratio (X_c) is:

$$X_c = \sum (v/s)_c \left[\frac{C}{C-L}\right]$$

As in the 1985 solution, the critical lane groups are the SB and EB approaches. The sum of critical v/s ratios is, therefore, (896/2985) + (1441/5145) = 0.580. Then:

$$X_c = 0.580\left[\frac{40}{40-12}\right] = 0.829$$

Table D.9 Stopped time delay and LOS for sample calculation—1994 update

Approach	C (s)	g/C	X	c (veh/h)	d (s/veh)	LOS
NB	40	0.375	0.639	1249	8.60	B
SB	40	0.375	0.801	1120	11.47	B
EB	40	0.325	0.862	1672	13.72	B

Level of Service Module

The delay formula in the 1994 update was slightly changed from the 1985 version. The progression adjustment factor (PF) became the delay adjustment factor (DF), and was applied only to the first term of the delay equation. Other minor changes were also made. Equations 6.53 through 6.55 are used:

$$d_1 = 0.38\,C\left[\frac{(1-g/C)^2}{1-[g/C][\min(X,1)]}\right]$$

$$d_2 = 173\,X^2\left[(X-1)+\sqrt{(X-1)^2\frac{mX}{c}}\right]$$

$$d = d_1 DF + d_2$$

Values of DF and m are selected from Table 8.17. For our case (Arrival Type 3), DF = 1.0 and m = 16. For the sample calculation, therefore, the equation is the same as in 1985. Some of the input variables (X, c) have changed in the 1994 solution for the southbound approach. The computations are not shown to save space, but the results are summarized in Table D.9. Levels of service criteria are the same as in 1985, shown in Table 8.14.

None of the approaches change a great deal from the 1985 results. The delays are slightly lower on the northbound and eastbound approaches, and the LOS remains B for both. The SB approach delay is slightly higher, but also remains at LOS B.

The 1997 Update to the 1985 HCM

Input Module

There were no fundamental changes in the input module in 1997. A minor change made clearance lost time a variable dependent upon the yellow and all-red times, and vehicle use of yellow and all-red time (e). Previously, start-up and clearance lost times had constant default values. In 1997, the parameter "e" was given a constant default value (2 s), which meant that clearance lost time varied with yellow and all-time times:

$$\ell_2 = y + ar - e$$

Table D.10 Demand flow rates for sample calculation—1997 update

Approach	Left (veh/h)	Through (veh/h)	Right (veh/h)
EB	25/0.95 = 26	1200/0.95 = 1263	20/0.95 = 21
NB	N/A	700/0.95 = 737	25/0.95 = 26
SB	10/0.95 = 11	800/0.95 = 842	N/A

Note All values rounded to the nearest integer

This, however, does not change any of the values used in the sample calculation.
Volume Adjustment Module
One major change occurred in this module in 1997. The application of a lane utilization factor (U) to demand volumes was removed in favor of an additional adjustment factor on saturation flow rate (f_{LU}). This meant that the demand volumes were numerically adjusted only for the PHF. Thus, the values used in subsequent computations changed. The values of demand flow rate (V/PHF) were previously computed and are shown again in Table D.10.

Saturation Flow Rate Module
There were only a few changes in the 1997 update in this module. The first was the addition of an adjustment factor for lane utilization (which had been removed from the Volume Adjustment Module. It is obtained from Table 8.9, and is dependent upon the number of lanes in the lane group (or approach in our case). Thus:

$$f_{LU,NB/SB} = 0.95 \qquad f_{LU,EB} = 0.91$$

There were several changes in the determination of the variables P_L and E_{L1}. E_{L1} is obtained by interpolating from Table 8.13.

$$E_{L1} = 2.5 + (3.1 - 2.5)^{*}\left(\frac{774-600}{800-600}\right) = 3.022$$

$$P_L = P_{LT}\left[1 + \left(\frac{(N-1)g}{g_f + \frac{g_u}{E_{L1}} + 4.24}\right)\right] = 0.012^{*}\left[1 + \left(\frac{(2-1)^{*}15}{6.18 + \frac{8.15}{3.022} + 4.24}\right)\right] = 0.026$$

$$f_m = \frac{6.18}{15} + \frac{8.15}{15}\left[\frac{1}{1+.026^{*}(3.022-1)}\right] = 0.928$$

$$f_{LT,SB} = \frac{f_m + 0.91^{*}(N-1)}{N} = \frac{0.928 + 0.91^{*}(2-1)}{2} = 0.919$$

All other adjustment factors remain the same as in the 1994 update:

$$f_{wNB/SB} = 0.97\,(11 - \text{ft lanes}) \quad f_{wEB} = 1.00\,(12\text{ - ft lanes})$$
$$f_{HV} = 0.909\,(\text{all approaches, 10\% heavy vehicles})$$
$$f_g = 1.00\,(\text{all approaches, level terrain})$$
$$f_p = 1.00\,(\text{all approaches, no parking})$$
$$f_{bb} = 1.00\,(\text{all approaches, no local buses})$$
$$f_a = 1.00\,(\text{all approaches, all other areas})$$
$$f_{RTNB} = 0.994 \quad f_{RTSB} = 1.000 \quad f_{RTEB} = 0.997$$
$$f_{LTNB} = 1.000 \quad f_{LTSB} = 0.919 \quad f_{LTEB} = 0.996$$

Then, adding the lane utilization factor to the saturation flow rate equation:

$$s = s_o N f_w f_{HV} f_g f_p f_{bb} f_a f_{RT} f_{LT} f_{LU}$$
$$s_{NB} = 1900{*}2{*}0.97{*}0.909{*}1{*}1{*}1{*}1{*}0.994{*}1{*}0.95 = 3164\text{ veh/hg}$$
$$s_{SB} = 1900{*}2{*}0.97{*}0.909{*}1{*}1{*}1{*}1{*}1{*}0.919{*}0.95 = 2925\text{ veh/hg}$$
$$s_{EB} = 1900{*}3{*}1{*}0.909{*}1{*}1{*}1{*}1{*}1{*}0.997{*}0.996{*}0.91 = 4682\text{ veh/hg}v$$

Capacity Analysis Module
The capacity analysis module was unchanged in the 1997 update. Because of the changes in the Volume Adjustment and Saturation Flow Rate Modules, some of the input values do change.

The capacity of each approach is:

$$c = s(g/C)$$
$$c_{NB} = 3164{*}0.375 = 1187\frac{\text{veh}}{\text{h}}$$
$$c_{SB} = 2925{*}0.375 = 1097\frac{\text{veh}}{\text{h}}$$
$$c_{EB} = 4682{*}0.325 = 1522\frac{\text{veh}}{\text{h}}$$

The v/c ratios for each approach are:

$$X_i = v_i/c_i$$
$$X_{NB} = (737 + 26)/1187 = 0.643$$
$$X_{SB} = (842 + 11)/1097 = 0.778$$
$$X_{EB} = (26 + 1263 + 21)/1522 = 0.861$$

and the critical v/c ratio for the intersection is:

$$X_c = \sum (v/s)_c \left[\frac{C}{C - L}\right]$$

As in previous cases (1985, 1994), the critical lane groups are the SB and EB approaches. The sum of critical v/s ratios is, therefore, (853/2925) + (1310/4682) = 0.571. Then:

$$X_c = 0.571^* \left(\frac{40}{40 - 12} \right) = 0.816$$

These results mirror closely the 1994 solution. It is expected that the 1994 and 1997 solutions for the sample calculation would be similar. The only change affecting the sample calculation was the transfer of the lane utilization adjustment from the Volume Adjustment Module to the Saturation Flow Rate Module, which is basically an arithmetic, not a substantive, change.

Level of Service Module

The level of service module underwent significant changes in 1997. By far the most significant was the move from *average individual stopped-time delay* as a service measure to *control delay*. Control delay is basically time-in-queue plus time loss due to deceleration and acceleration.

Variables were added that make delay sensitive to the type of controller used and the effects of upstream feeder signals in coordinated systems. As will be seen, these do not affect the sample calculation, which involves a pre-timed controller at an isolated location (Arrival Type 3). A term was added to include delay due a pre-existing queue at the subject signal at the start of the analysis period. As the sample calculation is considering a single 15-min flow period, this change also does not affect the sample calculation.

The 1997 delay equations are:

$$d = d_1\, PF + d_2 + d_3$$

$$d_1 = \frac{0.50\, C(1 - g/C)^2}{1 - [g/C][\min(X, 1]}$$

$$d_2 = 900\, T\left[(X - 1) + \sqrt{(X - 1)^2 + \left[\frac{8kIX}{cT}\right]}\right]$$

$$d_3 = delay\, due\, to\, pre\!-\!existing\, queue.$$

For the sample calculation, the *PF* continues to be 1.00 (Arrival Type 3). There is no pre-existing queue, so $d_3 = 0.0$ s/veh. *T* is the length of the analysis period in hours, or 0.25 h. *k* is an adjustment based upon controller type and other conditions. It is obtained from Table 8.17, and is 0.50 for pre-timed controllers. *I* is estimated using Eq. 8.60, but is 1.00 for all isolated signals.

Using the equations for d_I and d_2, the resulting delays are shown in Table D.11. Levels of service are obtained from Table 8.20.

All of the delays increase over the 1994 solution, due to the change from stopped-time to control delay. The LOS criteria, however, have been adjusted to reflect this, and the approach LOS designations, therefore, are unchanged from 1994.

Table D.11 Control delay and LOS for sample calculation—1997 update

Approach	C (s)	g/C	X*	c (veh/h)	d (s/veh)	LOS
NB	40	0.375	0.643	1187	12.98	B
SB	40	0.375	0.778	1097	16.47	B
EB	40	0.325	0.861	1521	19.27	B

*min[X,1]

The 2000 Highway Capacity Manual

Input Module
There were no changes in the Input Module in the 2000 HCM.

Volume Adjustment Module
There were no changes in the Volume Adjustment Module in the 2000 HCM. The input demand flows remain as shown in Table D.10.

Saturation Flow Rate Module
The 2000 HCM added two new adjustment factors to the saturation flow rate equation. They accounted for the specific impact of bicycle and pedestrian interference on right-turning and left-turning vehicles. This was a dramatic step that added some complexity to the determination of saturation flow rates.

The impact of bicycles had never been incorporated into signalized intersection analysis. The effect of pedestrians on right-turning vehicles had been implicitly embedded in the right-turn adjustment factor (f_{RT}) in various ways over the years, however, their impact on left-turning vehicles was never included.

Because pedestrian impacts had been embedded in f_{RT}, the estimation of this factor (which remains in effect) had to be altered to avoid over-counting the impact. Note that because the LT adjustment for the EB approach was based upon its consideration as a right turn, that too is changed.

All other adjustment factors, including the LT adjustment, were unchanged from 1997 for the sample calculation:

$$f_{wNB/SB} = 0.97\,(11 - \text{ft lanes}) \quad f_{wEB} = 1.00\,(12\text{ - ft lanes})$$
$$f_{HV} = 0.909\,(\text{all approaches, }10\%\text{ heavy vehicles})$$
$$f_g = 1.00\,(\text{all approaches, level terrain})$$
$$f_p = 1.00\,(\text{all approaches, no parking})$$
$$f_{bb} = 1.00\,(\text{all approaches, no local buses})$$
$$f_a = 1.00\,(\text{all approaches, all other areas})$$
$$f_{LTNB} = 1.000 \quad f_{LTSB} = 0.919$$

The new values for f_{RT} (and f_{LTEB}) are computed using Eq. 8.23:

$$f_{RT} = 1.0 - 0.15P_{RT}$$

$$f_{RTNB} = 1.0 - (0.15^*0.034) = 0.996$$
$$f_{RTSB} = 1.0 - (0.15^*0.000) = 1.000$$
$$f_{RTEB} = 1.0 - (0.15^*0.016) = 0.998$$
$$f_{LTEB} = 1.0 - (0.15^*0.020) = 0.997$$

The determination of the new adjustment factors (f_{Rpb} and f_{Lpb}) involves a number of computational steps that can get quite complicated. In our case, there are no bicycles, so a number of steps are eliminated.

The computation of f_{Rpb} begins with an estimation of pedestrian flow rates in all crosswalks, using Eq. 8.14.

$$v_{pedg} = v_{ped}\left(C/g_{ped}\right)$$

The g_{ped} is the sum of the pedestrian green plus flashing DON'T WALK (pedestrian clearance) intervals. For the sample calculation, there are no pedestrian signals. To be conservative, we will assume that the sum of the WALK and flashing DON'T WALK intervals is equal to the vehicular green time (g). In effect, there will be no pedestrians in the crosswalk during yellow and all-red intervals. Remember that the sample calculation involves low pedestrian flow, which defaults to 50 peds/h in each crosswalk. Then:

$$v_{pedgNB/SB} = 50(40/15) = 133 \text{ peds/hg}$$
$$v_{pedgEB} = 50(40/13) = 154 \text{ peds/hg}$$

Next, the "occupancy" of the conflict area (turning path conflicting with crosswalk) is computed using Eq. 8.16:

$$OCC_{pedg} = \frac{v_{pedg}}{2000}$$
$$OCC_{pedgNB/SB} = \frac{133}{2000} = 0.067$$
$$OCC_{pedgEB} = \frac{154}{2000} = 0.077$$

Where there is both bike *and* pedestrian interference, the variable OCC_r must be estimated. In our case, only pedestrians exist, and $OCC_r = OCC_{pedg}$ for all approaches.

Next, an adjustment factor that applies during the permitted portion of the phase is estimated using Eq. 8.20. Note that because of the sample calculation geometry, $N_{rec} > N_{turn}$. Then:

$$A_{pbT} = 1 - 0.6\,OCC_r$$
$$A_{pbTNB/SB} = 1 - (0.6^*0.067) = 0.960$$

$$A_{pbT\,EB} = 1 - (0.6^{*}0.077) = 0.954$$

The right-turn interference adjustment factor is then estimated using Eq. 8.12:

$$f_{Rpb} = 1 - P_{RT}^{*}\left(1 - A_{pbT}\right)(1 - P_{RTA})$$

where P_{RTA} is the proportion of right turns made on a protected phase. For the sample problem, this is 0.0. Then:

$$\begin{aligned} f_{Rpb,NB} &= 1 - 0.034^{*}(1 - 0.960)(1 - 0) = 0.999 \\ f_{Rpb,SB} &= 1 - 0.000^{*}(1 - 0.960)(1 - 0) = 1.000 \\ f_{Rpb,EB} &= 1 - 0.016^{*}(1 - 0.954)(1 - 0) = 0.999 \end{aligned}$$

The left-turn interference factor is found in a similar way, but has a significant complication when the left turns are served by permitted phasing and the opposing queue clearance time (g_q).

In the sample calculation, only one opposed left turn exits (SB approach). The opposing queue clearance time has not changed from the 1997 computation, $g_q = 6.85\ s$. The relevant occupancy must take into account that while the opposing queue is clearing, there is no chance for conflict, and is estimated using Eq. 8.22:

$$\begin{aligned} OCC_r &= OCC_{pedg}\left[1 - 0.50\left(g_q/g_p\right)\right]\ \left[e^{-(5/3600)\,v_o}\right] \quad for\ g_q < g_p \\ OCC_{r,SB} &= 0.077^{*}\left[1 - 0.50(6.85/15)\left[e^{-\left(\frac{5}{3600}\right)^{*}763}\right]\right] = 0.021 \\ A_{pbT} &= 1 - 0.60CC_r = 1 - 0.6^{*}0.021 = 0.988 \\ f_{Lpb} &= 1 - P_{LT}^{*}\left(1 - A_{pbT}\right)(1 - P_{LTA}) \\ f_{Lpb,SB} &= 1 - 0.012^{*}(1 - 0.988)(1 - 0) = 0.999 \end{aligned}$$

The other two approaches have unopposed left turns, the methodology for right-turn interference is used. The NB approach has no left turns, and therefore, $f_{Lpb,NB} = 1.000$. The EB approach has some left turns, but the value of A_{pbT} is the same as for right turns because it is a one-way approach. Thus:

$$\begin{aligned} f_{Lpb} &= 1 - P_{LT}^{*}\left(1 - A_{pbT}\right)(1 - P_{LTA}) \\ f_{Lpb,EB} &= 1 - 0.020^{*}(1 - 0.954)(1 - 0) = 0.999 \end{aligned}$$

Ultimately, because of low pedestrian flows, the impact of pedestrian interference on operations is minimal.

All of the adjustment factors have now been developed, and the saturation flow rate for each approach may be estimated using the saturation flow rate equation, which now includes two new adjustments:

$$s = s_o N f_w f_g f_g f_p f_a f_{RT} f_{LT} f_{LU} f_{Rpb} f_{Lpb}$$
$$s_{NB} = 1900^*2^*0.97^*0.909^*1^*1^*1^*1^*0.994^*1^*0.95^*0.999^*1 = 3161$$
$$s_{SB} = 1900^*2^*0.97^*0.909^*1^*1^*1^*1^*0.919^*0.95^*1^*0.999 = 2,922$$
$$s_{EB} = 1900^*3^*1^*0.909^*1^*1^*1^*0.998^*0.996^*0.91^*0.999^*0.999 = 4,677$$

In the final analysis, these values are almost the same as in the 1997 solution, as the impact of pedestrian interference on operations is expectedly minimal.

Capacity Analysis Module
There were no changes to this module in the 2000 HCM, but there were some minor changes in the input values. Capacities for the three approaches are:

$$c = s(g/C)$$
$$c_{NB} = 3161^*0.375 = 1185 \text{ veh/h}$$
$$c_{SB} = 2922^*0.375 = 1096 \text{ veh/h}$$
$$c_{EB} = 4677^*0.325 = 1520 \text{ veh/h}$$

Volume-to-capacity ratios for the three approaches are:

$$X_i = v_i/c_i$$
$$X_{NB} = 763/1185 = 0.644$$
$$X_{SB} = 853/1096 = 0.778$$
$$X_{EB} = 1310/1520 = 0.862$$

and the critical v/c ratio for the intersection is:

$$X_c = \sum (v/s)_c \left[\frac{C}{C-L}\right]$$

As in previous cases (1985, 1994, 1997), the critical lane groups are the SB and EB approaches. The sum of critical v/s ratios is, therefore, (853/2922) + (1310/4677) = 0.572. Then:

$$X_c = 0.572\left(\frac{40}{40-12}\right) = 0.817$$

Once again, the results do not change much from the 1994 or 1997 solutions.

Level of Service Module
There were no changes to delay calculations in the 2000 HCM. Thus, results also do not vary much from the 1997 results. Delays were recalculated using the minor changes in input variables, and are shown in Table D.12.

Table D.12 Control delay and LOS for sample calculation—2000 HCM

Approach	C (s)	g/C	X	c (veh/h)	d (s/veh)	LOS
NB	40	0.375	0.644	1185	13.0	B
SB	40	0.375	0.778	1096	16.5	B
EB	40	0.325	0.862	1520	19.3	B

The 2010 Highway Capacity Manual

Given that worksheets were removed from the 2010 HCM, there are no clear module definitions. However, for a pretimed example, there is less need for iterations and we can describe the problem in terms similar to the 2000 HCM.

Inputs
There were no changes in the Inputs in the 2010 HCM that affect the current problem. However, in the 2000 HCM, it was possible to enter PHF by movement. This was removed in the 2010 HCM.

Volume Adjustment and Lane Groups
There were no changes to the Volume Adjustment made in the 2010 HCM. The input demand flows remain as shown in Table D.10.

The lane groups, however, have changed in the 2010. The eastbound approach will be analyzed as three lane groups: left/through lane, through-only lane, and through/right lane. The northbound and southbound approaches will be analyzed as two lane groups each: a through-only lane and a through/turn lane. Northbound has a through/right lane and southbound has a left/though lane.

Each of the approach legs have more than one lane group serving the same movement, and the through vehicles can choose either a shared lane or a through only lane. Dividing the through volume among the various lane groups is accomplished by equalizing the v/s ratios of the lane groups. This step is iterative and within the iteration, saturation flow rate will need to be calculated. Thus assigning volume to the lane groups is done within the next step of finding saturation flow rate.

Saturation Flow Rate
The 2010 HCM changed the lane width adjustment factor. The new adjustment has no effect unless the lane width is less than 10 ft or greater than 12.9 ft. Since the north/south lanes in this problem are 11-ft wide, beginning in 2010, there is no negative effect due to lane width, and the adjustment factor is effectively 1.00.

Table D.13 shows the basic factors and the adjusted saturation flow rate for each lane group. Turning factors and the ped/bike adjustment factors are not included yet. For the shared turning/through lanes, the proportion of turns in the lane is needed to calculate the right- and left-turn adjustment factors. That will only be known after the vehicles are assigned to a lane group, which is done using the iterative procedure to equalize v/s ratios. These calculations are shown in Tables D.14, D.15, and D.16. The calculations for the ped/bike factors are shown in the same tables.

Table D.13 Adjustments to saturation flow rate: basic factors

Factor	EB lane groups			NB lane groups		SB lane groups	
	LT	T	TR	T	TR	LT	T
s	1900	1900	1900	1900	1900	1900	1900
f_w	1.000	1.000	1.000	1.000	1.000	1.000	1.000
f_{HV}	0.909	0.909	0.909	0.909	0.909	0.909	0.909
f_g	1.000	1.000	1.000	1.000	1.000	1.000	1.000
f_{bb}	1.000	1.000	1.000	1.000	1.000	1.000	1.000
f_a	1.000	1.000	1.000	1.000	1.000	1.000	1.000
f_{LU}	1.000	1.000	1.000	1.000	1.000	1.000	1.000
Lane group saturation flow rate		1727		1727			1727

Each column of Tables D.14, D.15, and D.16 represents one iteration. Because the turning volume is very low, it takes only a few iterations to reach equilibrium. Tables D.14 and D.15 are for the northbound and southbound approaches, respectively. Both are two-lane approaches with a single through-only lane and a shared through-turn lane. Northbound has a shared through/right lane and southbound has a shared left/through lane. Table D.16 shows the eastbound iterations. The eastbound approach is a one-way street with three lanes: a shared left/through lane, a through-only lane, and a shared through/right lane.

In the first iteration, the total approach volume is divided equally among all the lanes, with zero turners in the shared lanes. As discussed in Chap. 9, these numbers, which are very counter-intuitive, will change quickly with the iterations. Notice that the value for g_f on the southbound approach is 4 s higher than it was in the earlier HCM versions. The equation is the same, but beginning in 2010, the lost time was accounted for where it actually occurs in the phase and thus only the 2 s startup lost time is subtracted and not the full 6 s total lost time for the phase. The final volumes in each lane group are shown in Table D.17.

Capacity, Delay and Level of Service

Because this is a fairly simple intersection, only one lane group needs to use the incremental queue accumulation methodology (IQA) to find capacity and delay. This is the southbound shared left/through lane which is served by permitted only phasing. Figure D.4 shows the QAP of the southbound shared left-turn lane-group phasing. There are three polygons that make up the QAP. The first polygon is the triangle when the queue grows during the red phase and the saturation flow rate is zero. The second polygon is the time until the first left turner arrives. The saturation flow rate during this time is the saturation flow rate for through-only vehicles, which is 1752 veh/h/g (Table D.15). We know from Table D.15 that the opposing queue

Table D.14 Calculations for equalizing v/s ratios of lanes on NB Approach (permitted right turns)

	Iteration 1	Iteration 2	Iteration 3
Volume in Shared Lane, Vsr	382	369	378
Right-turn volume in Shared Lane, Vsr,rt	0	26	26
Volume in Exclusive Through Lane, $V_T = V_{TH} - (V_{sr} - Vsr,rt)$	355	395	385
Proportion of Rights in Shared Lane, P_R	0	0.071	0.069
Cycle Length, C	40	40	40
Green phase for pedestrians, g_{ped}	15	15	15
Demand flow rate for pedestrians, V_{ped}	50	50	50
Flow rate for pedestrians during green phase, $V_{pedg} = V_{ped} * (C/g_{ped})$	133	133	133
Avg ped occupancy in conflict zone, $OCC_{pedg} = V_{pedg}/2000$	0.067	0.067	0.067
Relevant Conflict Zone Occupancy, $OCC_r = OCC_{pedg}$	0.067	0.067	0.067
Unblocked portion of phase, $A_{pbT} = 1 - 0.6OCC_r$	0.960	0.960	0.960
Right-turn ped/bike interference factor, $f_{Rpb} = A_{pbT}$	0.960	0.960	0.960
Demand Volume on Approach per lane, v_{app} (Upstream of any bays)	382	382	382
Probability of a lane change, $P_{lc} = 1 - ((2v_{app}/(3600/3.7)) - 1)^2$	0.953	0.953	0.953
Effective permitted green time, g_p	15	15	15
Adjusted saturation flow rate of through-only lane, s_{th}, from Table D.12	1752	1752	1752
Modified right-turn equivalency value, $E_{R,m} = ((E_R/f_{Rpb}) - 1)P_{lc} + 1; E_R = 1.18$	1.219	1.219	1.219
Adjusted Saturation flow rate of shared lane, $s_{sr} = s_{TH} * (1/(1 + P_R(E_{R,m} - 1))$	1752	1719	1720
Flow Ratio, $y = (V_{sr} + V_T)/(s_{sr} + s_{TH})$	0.210	0.220	0.220
New $Vsr = y * s_{sr}$	369	378	378

[1] Through demand flow rate, $V_{TH} = 737$ vph; right-turn demand flow rate, $V_R = 26$ vph

clears before the first left turn vehicle in the subject direction arrives. Thus there is no delay caused by $g_{q,opp}$. After the first left-turner arrives at the stopline, the saturation flow rate is the saturation flow rate of a shared lane, calculated below. To calculate capacity of the lane group, the capacity of each of the three polygons that make up the QAP plus sneakers is computed and summed for a total capacity of the lane group.

Calculate capacity

During the red time, the saturation flow rate and thus the capacity is zero. During g_f the saturation flow rate is the adjusted through-lane saturation flow rate, s_{TH}, since

Table D.15 Calculations for equalizing v/s ratios of lanes on SB Approach (permitted left turns)

	Iteration 1	Iteration 2	Iteration 3
Volume in Shared Lane, *Vsl*	426	441	445
Left-turn volume in Shared Lane, *Vsl,lt*	0	11	11
Volume in Exclusive Through Lane, $V_T = V_{TH} - (V_{sl} - V_{sl,lt})$	416	412	408
Proportion of Rights in Shared Lane, P_L	0	0.0250	0.0247
Cycle Length, C	40	40	40
Green phase for pedestrians, g_{ped}	15	15	15
Demand flow rate for pedestrians, V_{ped}	50	50	50
Flow rate for pedestrians during green phase, $V_{pedg} = V_{ped} * (C/g_{ped})$	133	133	133
Opposing queue clearance time, $g_{q,opp}$	6.96	6.96	6.96
Avg ped occupancy in conflict zone, $OCC_{pedg} = V_{pedg}/2000$	0.067	0.067	0.067
Ped occupancy after opposing queue clears, $OCC_{pedu} = OCC_{pedg}(1 - (0.5g_{q,opp}/g_{ped})$	0.051	0.051	0.051
Relevant Conflict Zone Occupancy, $OCC_R = OCC_{pedu}e^{-5.00v_{opp}/3600}$	0.018	0.018	0.018
Unblocked portion of phase, $f_{Lpb} = A_{pbT} = 1 - 0.6OCC_r$	0.989	0.989	0.989
Demand Volume on Approach per lane, v_{app} (Upstream of any bays)	426	426	426
Probability of a lane change, $P_{lc} = 1 -((2v_{app}/(3600/3.7)) - 1)$^2	0.985	0.985	0.985
Effective permitted green time, g_p	15	15	15
Permitted saturation flow rate, s_p	715	715	715
Modified right-turn equivalency value, $E_{L1,m} = ((E_{L1}/f_{Lpb}) - 1)P_{lc} + 1; E_{L1} = s_o/s_p; s_o = 1900$	2.690	2.690	2.690
First left arrives at stop line, $g_f = G_p e^{-0.882LTC^{0.717}} - l_1$	10.41	10.41	10.41
Unsaturated green time, $g_u = \min(g - g_f, g - g_q)$	4.587	5.91	5.91
Expected number of sneakers, $n_s^* = \frac{P_L}{1-P_L}(1 - P_L^{n_s})$	0	0.0256	0.0253
Adjusted saturation flow rate of through-only lane, s_{th}, from Table D.12	1752	1752	1752
Adjusted Saturation flow rate of shared lane, $s_{sl} = \frac{s_{TH}}{g_p} * \left(g_f + \frac{g_u}{1+P_L^*(E_{L1,m}-1)} + \frac{3600*n_s^*}{s_{TH}}\right)$	1752	1715	1715
Adjusted through sat flow, $s_T = s_{TH} * 0.91$	1594	1594	1594
Flow Ratio, $y = (V_{sl} + V_T)/(s_{sl} + s_T)$	0.252	0.256	0.256
New *Vsl* $= y * s_{sl}$	441	445	445

*Through demand flow rate, $V_{TH} = 842$ vph; left-turn demand flow rate, $V_{Lt} = 11$ vph

Table D.16 Calculations for equalizing v/s ratios of lanes on EB Approach (one-way approach)

	Iteration 1	Iteration 2	Iteration 3
Volume in Shared Lane, Vsl	437	421	437
Left-turn volume in Shared Lane, Vsl,lt	0	26	26
Volume in Shared Lane, Vsr	437	421	434
Right-turn volume in Shared Lane, Vsr,rt	0	21	21
Volume in Exclusive Through Lane, $V_T = V_{TH} - (V_{sr} - Vsr,rt)$	389	468	440
Proportion of Lefts in Shared Lane, P_L	0	0.062	0.059
Proportion of Rights in Shared Lane, P_R	0	0.050	0.048
Cycle Length, C	40	40	40
Green phase for pedestrians, g_{ped}	13	13	13
Demand flow rate for pedestrians, V_{ped}	50	50	50
Flow rate for pedestrians during green phase, $V_{pedg} = V_{ped} * (C/g_{ped})$	154	154	154
Avg ped occupancy in conflict zone, $OCCr = OCC_{pedg} = V_{pedg}/2000$	0.077	0.077	0.077
Unblocked portion of phase, $A_{pbT} = 1 - 0.6OCC_r$	0.954	0.954	0.954
Right and Left turn ped/bike interference factor, $f_{Rpb} = f_{Lpb} = A_{pbT}$	0.954	0.954	0.954
Demand Volume on Approach per lane, v_{app} (Upstream of any bays)	437	437	437
Probability of a lane change, $P_{lc} = 1 - ((2v_{app}/(3600/3.7)) - 1)\^2$	0.990	0.990	0.990
Modified Left-turn Equivalency, $E_{L1,m} = ((E_{L1}/f_{Lpb}) - 1) * P_{lc} + 1; E_{L1} = 1.05$	1.10	1.10	1.10
Time till first left turn arrives, g_f (by definition g_f =0 for one-way streets or split phasing)	0	0	0
Effective permitted green time, g_p	13	13	13
Unsaturated green, g_u (because no opposing volume)	13	13	13
Expected number of sneakers per cycle, n_s^*, sneakers = 0 on one-way streets	0	0	0
Adjusted saturation flow rate of through-only lane, s_{th}, from Table D.12	1752	1752	1752
Adjusted Saturation flow rate of shared left/through lane, $s_{sl} = (s_{TH}/g_p) * (g_f + (\min(g_p - g_f), gu)/(1 + P_L(E_{L1,m} - 1)) + (3600*n_s^*/s_{TH}))$	1752	1741	1741
Modified right-turn equivalency value, $E_{R,m} = ((E_R/f_{Rpb}) - 1)P_{lc} + 1; E_R = 1.18$	1.235	1.235	1.235
Adjusted Saturation flow rate of shared through/right lane, $s_{sr} = s_{TH} * (1/(1 + P_R(E_{R,m} - 1))$	1752	1726	1727

(continued)

Table D.16 (continued)

	Iteration 1	Iteration 2	Iteration 3
Adjusted Through lane saturation flow rate, s_{TH}	1752	1752	1752
Flow Ratio, $y = (V_{lr} + V_T + V_{sr})/(s_{lr} + s_{TH} + s_{sr})$	0.240	0.251	0.251
New $Vsl = y * s_{lr}$	421	437	437
New $Vsr = y * s_{sr}$	421	433	433

*Through demand flow rate, $V_{TH} = 1263$ vph; left-turn demand flow rate, $V_L = 26$ vph; right-turn demand flow rate, $V_R = 21$ vph

Table D.17 Lane groups and final lane group volumes

	EB lane groups			NB lane groups		SB lane groups	
	LT	T	TR	T	TR	LT	T
veh/h	437	440	433	385	378	445	408

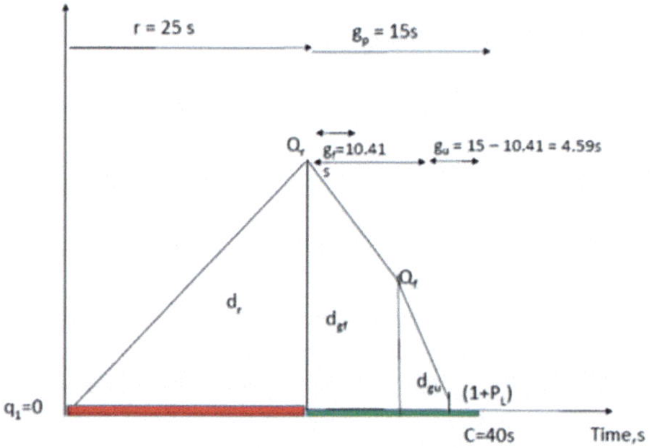

Fig. D.4 Queue accumulation polygon of the southbound left/through shared lane

only through vehicles move during this time. This value was found to be 1752 vph in Table D.16. The length of g_f was found to equal 10.4 s. Thus

$$c_{g_f} = s^*_{TH}\frac{g_f}{C} = 1752^*\frac{10.41}{40} = 456 \text{ vph}$$

Note that even though this is a two-way street, and there is an opposing queue, the opposing queue does not cause any delay because the first left turn does not arrive until after the opposing queue clears at 6.96 s, as was calculated in Table D.15. The unsaturated green time remaining after the first left-turner arrives is 15 − 10.41 = 4.59 s. During this time the saturation flow rate is found as

$$s_{g_u} = \frac{s_{TH}}{1 + P_L^*\left(\frac{E_{L1}}{f_{Lpb}} - 1\right)} = \frac{1752}{1 + 0.0247^*\left(\frac{2.69}{0.989} - 1\right)} = 1681 \text{ vph}$$

And the capacity during this time is:

$$c_{g_u} = s_{g_u}^* \frac{g_u}{C} = 1681^* \frac{4.587}{40} = 193 \text{ vph}$$

Lastly, the sneaker capacity is found as:

$$c_{sn} = \frac{3600^*(1 + P_L)}{C} = \frac{3600^*(1 + 0.0247)}{40} = 92 \text{ vph}$$

Total capacity of the shared left/through lane group on the southbound approach is the sum of these three capacities:

$$c_{SB,LT} = c_{g_f} + c_{g_u} + c_{sn} = 456 + 193 + 92 = 741 \text{ vph}$$

Volume to capacity ratio of the southbound shared left/through lane is:

$$X_{SB,LT} = \frac{v}{c} = \frac{445}{741} = 0.600$$

Calculate Uniform Delay for southbound left/through lane, d_1
Incremental delay is found for each portion of time where there is a change in either arrival rate or saturation flow rate. To find uniform delay for the SB LT lane group, the arrival rates on green and red are found as follows:

v = Lane group volume = 445 vph
P = Proportion of vehicles arriving on green
= (AT/3)*(g/C) = (3/3)*(15/40) = 0.375
v_r = Arrival rate on red = $(1 - P)v(C/r)$ = (1 – 0.375)*445*(40/25) = 445 vph
v_g = Arrival rate on green = $vP(C/g)$ = 445*0.375*(40/15) = 445 vph.

Note that since the arrival type is 3, that is arrivals are uniform with no platoons forming, the arrival rate throughout the entire cycle is 445 vph.

Following are the steps to find the incremental delays of each polygon:

1. Start at beginning of effective red, where initial queue is assumed to be zero, $q_1 = 0$.
2. Find the queue at the end of the red time, $q_2 = q_1 + (v_r - s)/3600 * \Delta t \geq 0$, where $s = 0$ because there are no departures. Δt is the effective red time for the SB phase, which is equal to the cycle length minus the effective green time, or 40 –

15 = 25 s:

$$q_2 = \mathrm{Q_R} = 0 + (445 - 0)/3600^{*}25 = 3.09 \text{ vehs}$$

3. Find the incremental delay during the effective red time:

$$d_r = \Delta t^{*}(q_1 + q_2)/2 = 25^{*}(0 + 3.09)/2 = 38.6 \text{ veh-s}$$

4. Find the queue that remains after g_f:

$$Q_{g_f} = Q_r + \left(\frac{v_g - s_{TH}}{3600}\right) * g_f = 3.09 + \left(\frac{445 - 1752}{3600}\right) * 10.41 = -0.236 \text{ veh}$$

Note that the answer is negative. This means that the queue in the subject direction clears during g_f. The next step then is to find the time that this occurs. Knowing this changes the QAP for delay, as shown in Fig. D.5, which shows that the queue ends before g_f. The next step is to find that time where the queue clears.

5. Find the time, during g_f, when the queue clears:

$$\Delta \mathrm{t}_2 = \frac{3600^{*} Q_R}{s_{TH} - v_g} = \frac{3600^{*}3.09}{1752 - 445} = 8.51 \text{ s}$$

6. Calculate delay during $\Delta \mathrm{t}_2$:

$$d_{\Delta \mathrm{t}_2} = \Delta \mathrm{t}_2^{*} \frac{Q_r + Q_{\Delta \mathrm{t}_2}}{2} = 8.61^{*} \frac{3.09 + 0}{2} = 13.15 \text{ veh-s}$$

7. Calculate total arrivals during effective green:

$$n_a = \frac{v_g}{3600}^{*} g = \frac{445}{3600}^{*} 15 = 1.85 \text{ veh}$$

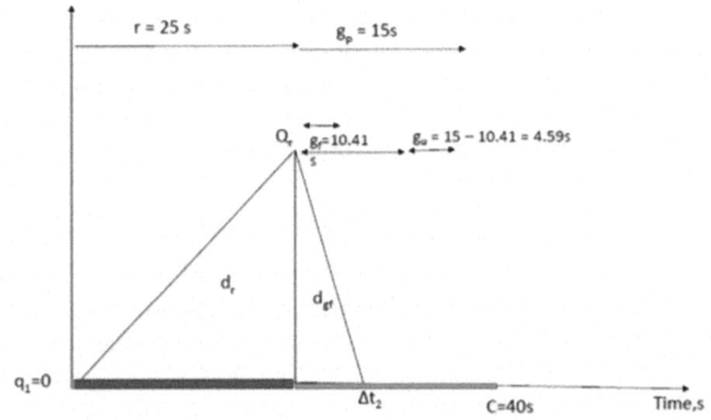

Fig. D.5 QAP for delay for southbound left/through lane

8. Calculate average uniform delay:

$$d_{1,SB,LT} = \frac{d_r + d_{\Delta t_2}}{Q_r + n_a} = \frac{38.6 + 13.15}{3.09 + 1.85} = 10.47 \text{ s/veh}$$

Note that because the QAP for delay turned out to be a triangle, the Webster equation for delay would have computed the same answer as follows:

$$d_1 = \frac{0.50C(1 - g/C)^2}{1 - g/C(\min(X, 1))} = \frac{0.50{*}40{*}(1 - 15/40)^2}{1 - (15/40)^*(1752)(15/40)} = 10.47 \text{ s/veh}$$

9. Calculate d_2:

$$d_2 = 900T^*\left[(X-1) + \sqrt{(X-1)^2 + \left(\frac{8kIX}{cT}\right)}\right]$$

$$= 900{*}0.25^*\left[(0.600-1) + \sqrt{(0.600-1)^2 + \left(\frac{8{*}0.5{*}1{*}0.600}{741{*}0.25}\right)}\right]$$

$$= 3.57 \text{ s/veh}$$

10. Calculate total control delay per vehicle:

$$d_{SB,LT} = d_1 + d_2 = 10.47 + 3.57 = 14.04 \text{s/veh}$$

Remaining Lane Groups

The equations for capacity and delay that are used for the all of the lane groups except the southbound left/through lane are those from the 2000 HCM:

$$c = s^* g/C$$

$$d_1 = \frac{0.50C(1 - g/C)^2}{1 - g/C(\min(X, 1))}$$

$$d_2 = 900T^*\left[(X-1) + \sqrt{(X-1)^2 + \left(\frac{8kIX}{cT}\right)}\right]$$

Table D.18 shows the capacity, delay, and level of service results for all lane groups and approaches.

The analysis shows that the eastbound lane groups operate at LOS C and all northbound and southbound lane groups operate a LOS B. Therefore, no problems would be expected at this intersection.

Table D.18 Capacity and level of service for certain lane groups

	Eastbound			Northbound		Southbound	
	LT	T	TR	T	TR	LT*	T
Volume, veh/h	437	440	433	385	378	445	408
Adjusted Saturation flow rate of through-only lanes, s_{TH}		1752		1752			1594
Percent turners in shared lane, $P_L or P_R$	0.059	–	0.048	–	0.069		–
Adjusted Saturation flow rate of shared lane, $s_{sr} = s_{TH} * (1/(1 + P_R((E_R/f_{Rpb}) - 1))$; $s_{sl} = s_{TH} * (1/(1 + P_L((E_L/f_{Lpb}) - 1))$; $E_R = 1.18; E_L = 1.05$	1742	–	1732	–	1724		–
Green to Cycle Length Ratio, g/C	0.325	0.325	0.325	0.375	0.375		0.375
Capacity, $c = s * g/C$	566	569	563	657	647		598
Volume to capacity ratio, $X = v/c$	0.772	0.772	0.770	0.586	0.584		0.683
Uniform delay, d_1	12.16	12.16	12.15	10.01	10.00	10.04	10.50
Incremental delay, d_2	9.83	9.78	9.78	3.80	3.84	3.57	6.22
Total control delay, d s/veh	21.99	21.94	21.93	13.81	13.84	14.04	16.70
Lane group level of service	C	C	C	B	B	B	B

The 2016 Highway Capacity Manual

As noted, there were only a few changes in the 2016 HCM, none of which would affect this sample problem. The results from the 2010 HCM would not be changed.

10 Comparing the Results

It is very difficult to compare the results of these analyses among methodologies ranging from the 1950 HCM to the 2016 HCM. The models themselves are very different, and the criteria for Level of Service are dramatically different:

- The 1950 HCM has NO level of service designations.
- The 1965 HCM defines level of service using the nebulous "Load Factor" as a measure of effectiveness.
- Circular 212 defines level of service in terms of "unused capacity."
- The 1985 HCM categorizes level of service on the basis of "stopped time delay."

- From the 1997 update forward, "control delay" has been the measure of effectiveness defining level of service, but the underlying models for its estimation have significantly changed over the years.

One thing is clear: from Circular 212, subsequent methodologies have resulted in better predicted operations, as measured by the level of service:

- The 1950 HCM estimated that all on all approaches, volume exceeded capacity. While no level of service is defined, this is clearly an indication of what is presently called LOS F—failure.
- The 1965 HCM estimated that the NB approach operated at LOS C, the SB approach at LOS D, and the EB approach at LOS E.
- From Circular 212 through the 2000 HCM, all approaches were estimated to operate at LOS B.
- The 2010 and 2016 HCMs predict all approaches operating at LOS B or C.

That is quite a range of results. It represents many things, including more sophisticated modeling of signal operations at an intersection and, perhaps, more efficient driver behavior as they become familiar with modern signal control.

Because of the desire to use a problem that could at least be solved on paper, the example used is clearly not representative of a wide range of base conditions. Perhaps different comparative results would arise from more complex problems.

Yet it must be commented: From Circular 212 through the 2000 HCM, the fundamental results do not change much. Levels of service remain B, except for the EB approach, which show LOS C using the 2010/2016 methodology. Much time, effort, and money was spent researching and developing the methodology over this period. While this problem is only illustrative, we have noted that there is no clear evidence that the accuracy of predictions has actually improved with increasing complexity in methodologies and models.

At the end of the day, there is a clear need for a significant data base of field-measured conditions across a wide range of signalized intersections be collected and used as a basis to evaluate both existing methodologies and the development of future methodologies. At the end of the day, these methods predict saturation flow rates (and therefore capacity) and delay (at least since 1985). Complexity brings acknowledged ability to consider a wide range of input information, and yields sensitivities to a variety of underlying traffic factors. Unfortunately, we really don't know whether it yields greater accuracy in our predictions, or whether the cost of data is justified by improved results.

Index

E. S. Prassas and R. P. Roess, *The Highway Capacity Manual: A Conceptual and Research History Volume 2*, Springer Tracts on Transportation and Traffic 12,
https://doi.org/10.1007/978-3-030-34480-1

The manufacturer's authorised representative in the EU is Springer Nature Customer Service Centre GmbH, Europaplatz 3, 69115 Heidelberg, Germany. If you have any concerns regarding our products, please contact ProductSafety@springernature.com

Printed and bound by CPI Group (UK) Ltd, Croydon, CR0 4YY
30/09/2026
02225647-0013